Energy Saving, Storage and Carbon Emission Mitigation Application for Buildings

Energy Saving, Storage and Carbon Emission Mitigation Application for Buildings

Editors

Zhenjun Ma
Müslüm Arıcı

MDPI

Basel • Beijing • Wuhan • Barcelona • Belgrade • Novi Sad • Cluj • Manchester

Editors

Zhenjun Ma
Sustainable Buildings
Research Centre
University of Wollongong
Wollongong
Australia

Müslüm Arıcı
Mechanical Engineering Department
Kocaeli University
Kocaeli
Turkey

Editorial Office
MDPI
St. Alban-Anlage 66
4052 Basel, Switzerland

This is a reprint of articles from the Special Issue published online in the open access journal *Buildings* (ISSN 2075-5309) (available at: www.mdpi.com/journal/buildings/special_issues/5WUE3WOZ91).

For citation purposes, cite each article independently as indicated on the article page online and as indicated below:

Lastname, A.A.; Lastname, B.B. Article Title. *Journal Name* **Year**, *Volume Number*, Page Range.

ISBN 978-3-0365-9521-4 (Hbk)
ISBN 978-3-0365-9520-7 (PDF)
doi.org/10.3390/books978-3-0365-9520-7

Contents

About the Editors

Zhenjun Ma

Dr. Zhenjun Ma is a Professor at the Sustainable Buildings Research Centre, University of Wollongong, Australia. He is a Fellow of the Australian Institute of Refrigeration, Air conditioning and Heating (AIRAH). Dr Ma currently serves as an Associate Editor of two Elsevier Journals (*Solar Energy* and *Energy for Sustainable Development*). His research is mainly focused on sustainable building energy technologies, building demand flexibility, thermal energy storage, and renewable energy in buildings. He has edited 1 book and published over 210 peer-reviewed papers in prestigious journals. He is a recipient of several prestigious awards, such as the Australian Endeavour Research Fellowship Award; the Excellence in Heating, Ventilation, Air Conditioning and Refrigeration Research Award from AIRAH; the Engineering Education Engagement Award from the Australasian Association for Engineering Education; and the Innovation Award (Energy Efficiency) from the World Society of Sustainable Energy Technologies.

Müslüm Arıcı

Dr. Müslüm Arıcı is a Professor at the Mechanical Engineering Department of Kocaeli University, Turkey. He has co-authored 220+ papers in refereed prestigious journals. He has served as an Associate Editor in Energy Sources, Part A: Recovery, Utilization, and Environmental Effects; Gazi University Journal of Engineering and Architecture; Energy, Ecology, and Environment; and the Journal of Thermal Engineering. In addition, he has acted as a Guest Editor in several Elsevier journals, such as *Journal of Cleaner Production; Energy and Buildings; Energy; Sustainable Energy Technologies and Assessments; Energy for Sustainable Development*; and *Journal of Magnetism and Magnetic Materials*. He was enlisted among the world's top 2% scientists according to Stanford University in the years 2020, 2021, and 2022. His fields of expertise include energy-efficient buildings, nanofluids, renewable energy systems, thermal energy storage, and thermal management by phase change materials.

Preface

Buildings are responsible for a significant proportion of global energy usage and generate a similar share of greenhouse gas emissions [1]. Building energy consumption will continuously increase in the coming decades [2] due to factors such as increasing floor areas, increasing population, fast urbanization, economic growth, and climate change. This increasing demand in conjunction with the likelihood of extreme weather events will pose significant challenges to existing electricity infrastructure. Therefore, a low-carbon and clean energy future could not be achieved without deep decarbonization of the building sector and providing access to affordable and clean energy to end-users. This has intensified the calls for innovative technologies and solutions for energy savings, storage, and carbon emission mitigation applications to achieve increased energy efficiency, resilience, and sustainability of the built environment.

Building decarbonization is now becoming a prerequisite for achieving the Paris Agreement objectives and many UN Sustainable Development Goals [3]. Building decarbonization can be achieved through energy efficiency, electrification, integration of distributed energy sources (DERs), and the implementation of enabling policies and regulations. Energy efficiency aims to reduce building energy demand through advanced design, optimal control, and the implementation of energy conservation technologies such as improved building envelopes, energy-efficient lighting and appliances, high-performance windows, and thermal energy storage systems. Opportunities for improved building energy efficiency are enormous, and the benefits are also significant. A case study for a two-story office building showed that 62.9% of energy savings can be achieved through the implementation of a package of cumulative energy efficiency measures [4].

DERs are small-scale energy resources that are commonly situated close to the sites of electricity usage. The main types of DERs include distribution generation such as rooftop solar panels, behind-the-meter batteries, smart charging electric vehicles, power-to-heat systems such as heat pumps and thermal storage, and controllable loads that can support demand response [5]. DERs are fast growing and are now starting to change the way buildings generate and manage electricity usage. Unlocking the potential of DERs in buildings can likely take the buildings off the grid with the assistance of flexible demand in buildings and energy-sharing mechanisms. The Electricity Network Transformation Roadmap developed by CSIRO and Energy Networks Australia estimated that DERs could contribute up to 45% of Australia's electricity generation capacity by 2050 [6]. Globally, renewable energy needs to be scaled up at least six times faster in order to meet the Paris Agreement targets [7].

Building electrification is to shift from using conventional fossil fuels to electricity to meet building energy demand. Electrification is one of the promising and effective strategies to achieve emission reductions and reduce environmental impact. As a significant proportion of energy used in buildings is for space heating and cooling, the electrification of heating and cooling will be among the key initiatives, and electric heat pumps will play a key role in building electrification. Enabling regulations are among the most fundamental policy tools to improve building energy efficiency and achieve building decarbonization. Regulations can be mandatory and voluntary. Mandatory regulations often set out the minimum energy performance and efficiency requirements for buildings to comply to, while voluntary regulations often provide best practices to follow.

To assist in achieving building decarbonization, this Special Issue aimed to collect excellent research and review articles to address some of the global imperatives and stringent environmental standards related to energy and sustainability in buildings. This Special Issue consists of eight research articles and three review articles. There are four research articles related to the use of phase change materials (PCMs) in buildings. PCMs have shown great potential in building applications for indoor temperature regulation, load shifting, and peak demand reduction. To maximize the benefit of using PCMs in buildings, appropriate optimization is needed, including the selection of optimal PCMs with optimal melting temperatures, an optimal amount of PCMs to be used, the optimal configuration of thermal energy storage systems, and the optimal location of PCMs in building structures, as well as the optimal control of PCM integrated systems. These are among the important topics to be systematically addressed. Thermal insulation is one of the energy-efficient solutions that has been frequently used in existing building retrofits. One research article investigated the potential benefits of thermal insulation in the retrofits of public buildings. CO_2 heat pumps using natural refrigerants can work more efficiently in both high temperature ranges and at low ambient conditions. One research article in this Special Issue carried out an exergy analysis of a transcritical CO_2 air-source heat pump system with a novel honeycomb gas cooler. Recycling is an essential element in circular economy. One research article examined the incorporation of recycled materials into building components to enhance building thermal insulation and carbon emissions. Conservation voltage reduction (CVR) is an efficient technique that is often used by utilities to strategically reduce demand during peak demand periods by reducing the incoming voltage at the supply side of buildings. One research article employed thermal-electric modeling to evaluate the impact of CVR on cooling systems in buildings. The results from these research articles clearly demonstrated significant opportunities for using technological advances in achieving building energy efficiency and carbon emission reductions.

Three review articles with different focuses highlighted various opportunities for increased building energy savings, storage, and carbon emissions mitigation. One article provided an overview of emerging and sustainable technologies for increased building energy efficiency. It was emphasized that collaborative efforts are needed to facilitate the integration and deployment of these technologies on a larger and wider scale. Photovoltaic thermal (PVT) systems are a nice marriage of solar photovoltaics and solar thermal collectors and have been extensively investigated. The second review article provided a comprehensive overview of the potential applications of PVT techniques in both industrial and building settings. Building energy consumption is highly impacted by climate conditions. The last review article in this Special Issue provided a literature review of climate zoning for buildings.

Many innovative solutions and technologies have been developed and tested in recent decades. Innovative solutions and technological breakthroughs are continuously needed to help create an affordable, low-emission, clean energy future and provide a highly efficient and human-centric built environment.

References:

[1] IEA (2019), Global Status Report for Buildings and Construction 2019, IEA, Paris License: CC BY 4.0.

[2] U.S. Energy Information Administration. Annual Energy Outlook 2022.

[3] Technical Brief: The Importance of Building Decarbonization and Benefits for the SDGs. Climate and SDGs Synergies Conference.

[4] NABERS. Improving energy efficiency and reducing costs for your office building – example from Victoria.

[5] IRENA, Innovation landscape brief: Market integration of distributed energy resources, International Renewable Energy Agency, Abu Dhabi, 2019.
[6] CSIRO and Energy Networks Australia 2017, Electricity Network Transformation Roadmap: Final Report.
[7] IRENA (2018), Global Energy Transformation: A roadmap to 2050, International Renewable Energy Agency, Abu Dhabi.

Zhenjun Ma and Müslüm Arıcı
Editors

buildings

MDPI

Article

Investigation of the Effect of Air Layer Thickness on the Thermal Performance of the PCM Integrated Roof

Dnyandip K. Bhamare [1], Manish K. Rathod [1,*], Jyotirmay Banerjee [1] and Müslüm Arıcı [2]

[1] Mechanical Engineering Department, Sardar Vallabhbhai National Institute of Technology, Surat 395007, Gujarat, India
[2] Mechanical Engineering Department, Engineering Faculty, Kocaeli University, 41001 Kocaeli, Turkey
* Correspondence: mkr@med.svnit.ac.in

Abstract: Recently, Phase Change Materials (PCM) have become more prevalent in improving buildings' thermal management. The relative location of the PCM layer is a valuable measure for assessing the thermal performance of building envelopes, in addition to meteorological circumstances and PCM qualities. The optimum air layers between the PCM layer and roof may significantly reduce energy consumption in buildings. In this regard, the influence of air gap layer thickness on the thermal performance of a PCM (HS 29) integrated roof of the test room is investigated experimentally. Experiments are carried out for an unconditioned test room located on the terrace of a laboratory in Surat, India, considering various air layer thickness values (0, 2, 4, and 6 cm) and a fixed PCM layer thickness. Different configurations within the research, including no- PCM and PCM with 0, 2, 4, and 6 cm air layer thickness, are investigated for the effects of diurnal change in room temperature. Results are evaluated based on the peak value, valley value temperatures of different roof layers, and an index (MKR, Measure of Key Response). It is observed that the maximum temperature difference between the PCM-integrated test room and the non-PCM test room is 4 °C to 7 °C. Results showed that, with a higher MKR index of 8.83, a PCM-integrated roof with a 2 cm air layer thickness could reduce the diurnal room temperature variations compared with the non-PCM test room. This conclusion from the current research demonstrates the significance of an air layer provided between the PCM layer and the roof of the building.

Keywords: PCM integrated roof; air layer; MKR index; peak value temperatures; solar heat gain reduction rate

Citation: Bhamare, D.K.; Rathod, M.K.; Banerjee, J.; Arıcı, M. Investigation of the Effect of Air Layer Thickness on the Thermal Performance of the PCM Integrated Roof. *Buildings* **2023**, *13*, 488. https://doi.org/10.3390/buildings13020488

Academic Editor: Ricardo M. S. F. Almeida

Received: 19 January 2023
Revised: 6 February 2023
Accepted: 8 February 2023
Published: 10 February 2023

1. Introduction

Many countries have proposed numerous measures to reduce greenhouse gas emissions as a global initiative [1]. These initiatives include developing renewable energy systems and energy conservation measures for buildings [2], which account for 30–40% of global primary energy demand [3]. Most of the energy use in buildings is due to space cooling applications. The cooling and heating energy consumption in buildings is significantly influenced by their enclosure design. The energy consumption in the buildings can be reduced by enhancing the thermal performance (TP) of building envelopes [4] and enhancing the comfort of the built environment [5]. Using envelope elements such as roofs, walls, windows, etc., with this aim, higher thermal resistance is the most common practice. Thermal resistance may be enhanced by adding thermal mass or insulation. Increasing envelope element size to increase thermal mass may not be viable for structural and economic reasons. The installation of thermal insulation in the building envelope is also limited due to the low heat storage capacity of insulating materials. However, many researchers emphasize the relevance of the air layer while investigating TP enhancement in building envelopes employing conventional thermal insulators.

Sanea et al. [6] investigated air spaces together with insulation layers for different thicknesses in buildings for the climatic conditions of the city of Riyadh and showed that the air gap has a dominant role when the insulation layer is lower than 7.5 cm. Jamal et al. [7] carried out a numerical investigation on double solid layer walls separated by an air gap layer. They noted that an optimum air gap thickness of 5 cm along with low emissivity surface material of the wall results in reduced energy consumption of the building envelope. Mahlia and Iqbal [8] studied air layer thickness and optimum insulation thickness to analyze the potential reduction in energy consumption and emissions in building walls. The results have shown that air layers of 2–6 cm in a wall and an optimum insulation thickness can reduce energy consumption and emissions by 65–77% compared to a wall without an air layer. Mavromatidis et al. [9] evaluated the effect of air layer thickness on temperature distribution in multilayer insulation walls using a numerical study. It was found that maximum thermal resistance can be achieved with a 3 cm air layer thickness. Kurt [10] studied the effect of the air layer on the insulation thickness and energy cost savings for a composite wall building in Turkey and reported that an optimum insulation thickness along with the air layer reduces the heat loss through the composite walls.

An alternative practice for improving the TP of the enclosure is to boost its heat storage capacity. In this regard, using Phase Change Materials (PCMs) with high latent heat is becoming increasingly common for augmenting the TP of building envelopes [11]. PCMs can store/retrieve a great deal of heat in the latent form in a constant or narrow temperature range during the melting/freezing process [12]. Thus, with the application of PCM in the building envelopes, maximum heat with a minimum temperature variation can be stored. PCM can be incorporated into various building envelope elements. However, for effective heat transfer and ease of installation, PCM is typically integrated into floors, walls, and roofs [13]. The TP of such a building envelope depends mainly on climatic conditions, material properties [14], and relative positioning of the layer of PCM [15]. Although climatic conditions and properties of PCM are crucial parameters, with ease in installation and prior design considerations, the relative positioning of the PCM layer is also an effective parameter for the TP assessment of building envelopes. The relative positioning includes layering (i.e., single or multi-layer) as well as the thickness and location of the PCM layer. Numerous researchers have noted the relative positioning PCMs to achieve a higher TP and reduce energy usage in building applications (Table 1). As seen from Table 1, studies related to the TP evaluation of building envelopes using the PCM layer mainly focused either on the thickness or location of the PCM layer. Few authors also studied the multiple layers of PCM in building envelopes.

Table 1. Literature summary.

Authors	PCM Used in	Climatic Conditions	Parameters Studied	Remarks
Ascione et al. [14]	wall	Five Mediterranean climatic cities	PCM layer thickness	Maximum cooling energy savings are attained for a PCM plaster of 3 cm thickness.
Li et al. [16]	Roof	Northeast and cold regions of China		Increasing PCM layer thickness decreases the peak time of the average temperature of a base layer and upper surface heat flux.
Tokuc et al. [17]	Roof	Mediterranean climate of İstanbul, Turkey		Optimum PCM layer thickness is suitable for achieving energy conservation in flat roofs of the city of Istanbul.
Saffari et al. [18]	Wall	Warm and temperate Climate of the city of Madrid		PCM layer thickness of 10 mm provides greater energy savings and a payback period.
Singh and Bhat [19]	Roof	Hot and dry climate of the city of Indore, India		Optimum PCM layer thickness of 16 mm can reduce the cooling load in the building.
Arici et al. [20]	Wall	Erzurum, Diyarbakır, and Konya cities, Turkey.		Optimum PCM layer thickness varies from 1 to 20 mm depending on the climatic zone.

Table 1. *Cont.*

Authors	PCM Used in	Climatic Conditions	Parameters Studied	Remarks
Jin et al. [21]	Wall	Dynamic wall simulator	PCM layer location	Optimal position of the PCM layer is approximately 0.2 times the total length of the wall from the internal surface of the wall. The optimal position of the PCM layer shows 41% of peak heat flux reduction.
Nidhal Ben Khedher [22]	Wall	Tropical climate of Tunisia		The PCM n-octadecane performs best for maintaining an indoor temperature close to 27 °C for the test room.
Lee et al. [23]	Wall	Warm and temperate Climate of the city of Lawrence, KS, USA		Optimal PCM location on the south wall is 2.54 cm, whereas, for a west wall, it is 1.27 cm from the location of the wallboard. Peak heat flux reduction for these optimal locations is 29.7% and 51.3% for the west and south, respectively.
Lagou et al. [24]	Wall	Southern, central, and northern European climatic conditions		PCM layer placement should be on the inside of the building for yearly energy saving.
Pasupathy and Velraj [25]	Roof	Hot and humid climate of the city of Chennai, India	Single and double-layer PCM	Double-layer PCM narrows indoor temperature swings.
Jin and Zhang [26]	Floor	Humid subtropical climate of the city of Nanjing, China		PCM layer should be located near the cold source and floor surface in the case of the cooling system and heating system, respectively.
Bhamare et al. [27]	Roof	Hot and humid climate of the city of Chennai, India	PCM layer Inclination	Maximum reduction in heat gain savings is obtained for a PCM slab inclination of 2°.

According to the research shown above, including PCM and the air layer in the design of a building is one way to reduce that building's overall demand for energy significantly. On the other hand, the research on the influence of the air layer that was given between the PCM layer, and the wall or roof of the building has not yet been documented. As a result, the usage of an air layer between the PCM layer and the top of a building is investigated in the current research. The novelty of the work is reflected in the findings of this research since a previous study of this kind has not yet been published. Thus, this research aims to explore the influence of optimal air layer thickness on the diurnal changes in indoor temperature and the decrease of heat gain. An experimental investigation is carried out to evaluate the impact of the air layer on the thermal performance of a roof integrating PCM for a prototypic test room located in Surat, India. A square metal enclosure filled with commercial PCM (HS 29) is used to incorporate the PCM layer into the roof of a test room. The air layer thickness varied between 0 to 6 cm, and a fixed PCM layer thickness are considered.

2. Experimental Setup

This section consists of details regarding the experimental setup, material and measurement methods, and methodology for data reduction to investigate the thermal performance of the PCM-integrated roof and the air layer.

2.1. Materials and Testing

It is to be noted that the experiment aims only to determine how effectively PCM and an air layer can minimize the heat load from the roof. Thus, a laboratory-scale test room is developed. However, the dimensions of the experimental room were selected to replicate the study reported by Velraj et al. [25]. Further, it is noted that the thickness of the concrete roof is considered similar to that used in real building conditions. Since all other walls of the model building are kept insulated, the only way for heat entrainment into the room in our experiments is through the roof. With this, considering dominant heat transfer in the vertical direction compared to the lateral direction, the thickness of the roof and PCM layer almost corresponds to real-life scenarios.

The field test was carried out during March and April 2020 using unconditioned single-room test buildings (68 cm × 68 cm × 102 cm) located on the terrace. The test room building is shown in Figure 1. The lightweight fly ash bricks (24 cm × 11 cm × 7 cm) are used to build the test room walls while the conventional concrete slab with a thickness of 12 cm is used as a roof of a building. The outer walls, except for the roof of the room, are insulated by a polyurethane foam (PUF) sheet of a thickness of 3 cm and PUF insulation spray on all the corners of the walls to ensure perfect insulation condition for studying the effect of heat transfer through the roof.

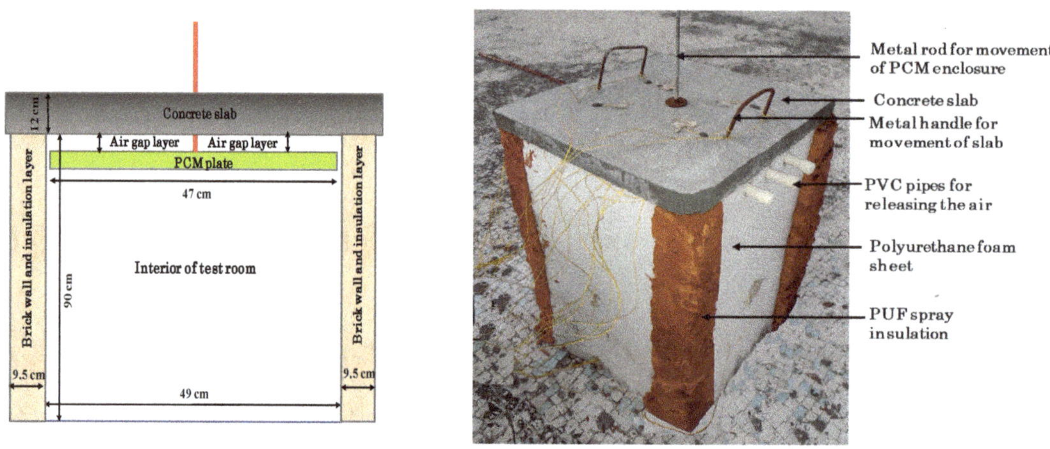

Figure 1. Details of PCM enclosure in the test room.

The PCM is encapsulated in a square metal enclosure having dimensions of 47 cm × 47 cm × 2.8 cm. The enclosure is made of MS steel (0.15 cm thickness). PCM selection was based on the melting temperature and climatic characteristics of the city of Surat, from the study by Bhamare et al. [28], which recommended using a PCM with a melting temperature of 29 °C for hot and humid climatic zones. Hence, the PCM used in this study is the commercially available salt hydrate, HS 29, provided by PLUSS Co Ltd. (Delhi, India). The melting point and latent heat of fusion obtained via a Differential Scanning Calorimeter (DSC) are shown in Appendix A as Figure A1. The observed melting point, latent heat, and specific heat are 29 °C, 217.4 kJ/kg, and 2000 J/kg K, respectively. The density and thermal conductivity of the PCM is taken as 1830 kg/m^3 [25] and 0.478 W/m K [29].

2.2. Instrumentation and Measurement

The temperature of the interior, walls, and PCM enclosure of the test room is measured using K-type thermocouples (total 32 with ±0.1 °C accuracy). Four thermocouples are placed at each (interior and exterior) surface of the concrete slab to measure temperature distribution and heat gain by the concrete slab. Four thermocouples are thus placed on the top side of a concrete slab and four on the bottom side, for a total of eight thermocouples. The heat gain by the concrete slab is calculated based on surface and room temperature. In the same way, eight thermocouples are also embedded in the PCM enclosure. To obtain the temperature gradient at the insulated walls of the test room, another eight thermocouples are used. Thermocouples are positioned at the center of each wall's exterior and interior surfaces. Five thermocouples are placed vertically along the metal rod located at the center of the test room to measure temperature distribution in the interior of the test room. The position of the thermocouples is depicted in Figure A2 of Appendix A. The distance between each thermocouple along the height of a centrally placed rod is 17 cm. To record temperature, a data acquisition system is used, which consists of an RS485 scanner embedded with an intelligent microprocessor. A pyranometer SR2238 (measurement accuracy: ±3% for hourly sums) located outside at a 1.50 m height to reduce the effect of shadows is utilized to collect the solar irradiance. The wind speed is also measured via a weather station installed at the top of the building.

2.3. Climatic Data Deduction

The thermal performance of the test room is established for the reference case (non-PCM) and with PCM having an air layer gap (thickness of 0, 2, 4, and 6 cm). Before conducting the experimentation from 18–19 March 2020, the experiment is set to idle for two days to reach the periodic conditions. The variations in temperature and radiation intensity for different days are shown in Figure 2. It shows hourly averaged values of radiation intensity and ambient temperature on different days. To ease the performance comparison of the configurations with different ambient temperatures and radiation intensity, there is a need for a common climatic parameter by which radiation intensity and ambient temperature can be combined into a single parameter. With this aim, the current study uses the solar air temperature method as a common climatic parameter for combining radiation intensity and ambient temperature. The solar air temperature is the hypothetical outside air temperature that would result in the same temperature distribution and rate of heat transfer through a wall as is seen owing to the influences of the ambient temperature and the incoming solar radiation. In brief, the combined effect of the actual ambient temperature and the incoming solar radiation yields a common characteristic, namely solar air temperature, which can be used to compare various cases. Solar air temperature is calculated based on average roof temperature (T_{roof}), radiation intensity (I_d), ambient temperature (T_o), and wind velocity (V_{wind}) measured during the experimentation [30].

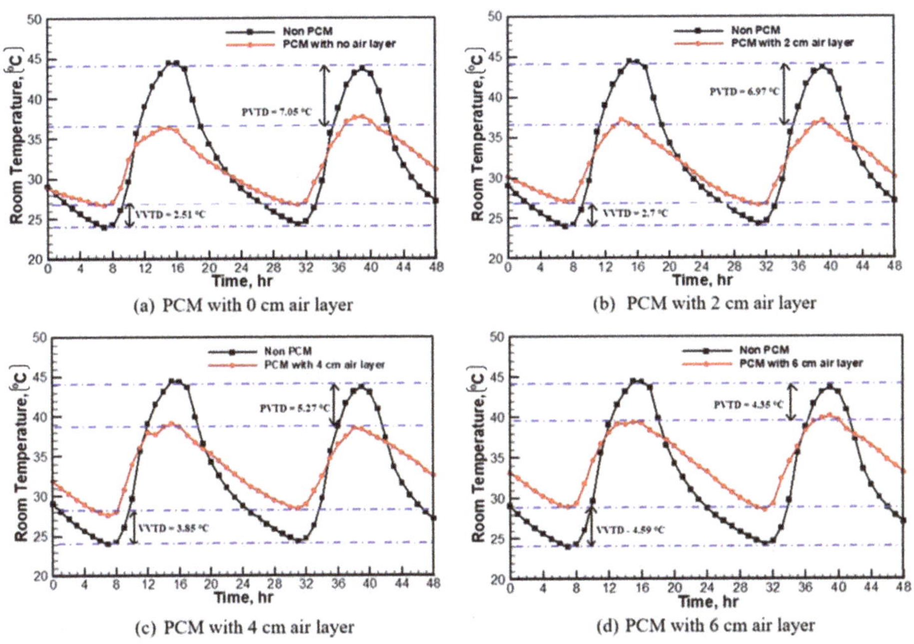

Figure 2. Average room temperature variation in the test room.

Solar air temperature is calculated as [31]

$$T_{sol-air} = \frac{I_d \alpha}{h_{combined}} + T_o \tag{1}$$

$$h_{combined} = h_o + \left[\varepsilon\sigma F_{sky}\left(T_{roof}^2 + T_{sky}^2\right)\left(T_{roof} + T_{sky}\right)\right]\frac{\left(T_{roof} - T_{sky}\right)}{T_{roof} - T_o} \tag{2}$$

where α, ε, and σ are the absorptivity of the surface, the emissivity of the rooftop and Stefan Boltzmann constant, respectively, $h_{combined}$ is the combined radiative and convective heat transfer coefficient, h_o is the ambient heat transfer coefficient, and F_{sky} is shape factor for rooftop slab and sky which is considered as 1.

The convective heat transfer coefficient is calculated as follows [31]

$$h_o = 5.6 + 4.0V_{wind} \text{ for } V_{wind} < 4.88\text{m/s} \tag{3a}$$

$$h_o = 7.2V_{wind}^{0.78} \quad \text{for } V_{wind} > 4.88\text{m/s} \tag{3b}$$

where V_{wind} is wind velocity.

The sky temperature is estimated by [31]

$$T_{sky} = 0.0552T_0^{1.5} \text{ for clear sky} \tag{4}$$

The average solar air temperature is calculated for different configurations using the above formulations and hourly variation of environmental conditions. The obtained average solar air temperatures for non-PCM, PCM with 0 cm, 2 cm, 4 cm, and 6 cm air layers are 31.80 °C, 31.90 °C, 34.65 °C, 32.11 °C, and 34.25 °C respectively. The average solar air temperature represents the extent of climatic conditions experienced by the test room during a particular day of experimentation. The higher average solar air temperature values during the day indicate warm weather conditions, while the lower values correspond to comparatively less warm weather conditions. Thus, an average solar air temperature can be used for comparing the climatic conditions for different days of experimentation. From the average solar air temperatures, one can notice that the highest average solar air temperature is found during the testing of the PCM layer with a 2 cm air layer, whereas the lowest average solar air temperature is shown by non-PCM room testing days.

Further, an index called MKR (Measure of Key Response) [32] is used for the comparative evaluation of the thermal performance of all configurations of PCM. This index, defined in Equation (5), combines crucial TP indicators and allows comparing TP enhancement of roofs incorporating PCM, taking the non-PCM roof as the reference. A high value of this index shows increasing time lag and decreasing solar heat gain and decrement factor, thus improved TP, whereas the lower value implies higher solar heat gain causing an increment in cooling energy consumption and hence poor TP of buildings.

$$\text{MKR} = \frac{f'}{(\phi' + c)Q'} \tag{5}$$

In the above equation, f', ϕ', and Q' respectively represent the time lag, decrement factor, and solar heat gain, all non-dimensional. The value of computational constant c is 0.5. These parameters and their non-dimensional forms are calculated as [32]

$$f = t_{T_{\max\ Ceiling}} - t_{T_{\max\ Roof}} \tag{6}$$

$$f' = \frac{f_{With\_PCM}}{f_{Without\_PCM}} \tag{7}$$

$$\phi = \frac{T_{\max\ Ceiling} - T_{\min\ Ceiling}}{T_{\max\ Roof} - T_{\min\ Roof}} \quad (8)$$

In Equation (6), $t_{T\max\ Ceiling}$ and $t_{T\max\ Roof}$ represent the time required for diurnal temperature waves to reach at ceiling and rooftop slab, respectively.

$$\phi' = \frac{\phi_{With\_PCM}}{\phi_{Without\_PCM}} \quad (9)$$

$$q_{ceiling} = h_i\left(T_{Ceiling} - T_{room}\right) \quad (10)$$

For the non-PCM room, $T_{ceiling}$ is the concrete bottom surface temperature, whereas, for the PCM room, it is the PCM bottom surface temperature. The heat transfer coefficient on the roof's inner side is considered 8.7 W/m^2 K [33]. The total heat gain by the test room for two days is evaluated by summing the hourly heat flux through the ceiling as given by Equation (11),

$$Q_{gain} = \int_0^{48} q_{ceiling}dt \quad (11)$$

$$Q' = \frac{Q_{gain(With\_PCM)}}{Q_{gain(Without\_PCM)}} \quad (12)$$

where $q_{ceiling}$ is the heat flux added from the ceiling surface or released from the room. The values of performance indicators, i.e., f , ϕ , and Q are calculated by the temperature distribution across the roof for all configurations of PCM.

To assess the effectiveness of PCM integrated roof compared with the non-PCM room, the envelope solar heat gain reduction rate can also be calculated as:

$$\eta = \left(1 - \frac{Q_{With\_PCM}}{Q_{Without\_PCM}}\right) \times 100 \quad (13)$$

$$\eta = (1 - Q') \times 100 \quad (14)$$

3. Results and Discussion

The impact of diurnal variation in room temperature for different configurations within the study, namely no- PCM and PCM with 0, 2, 4, and 6 cm air layer thickness, is investigated.

Effect on Room Temperature Variation

The average room temperature variation for different configurations is established during two consecutive days, as shown in Figure 2. From the figure, a diurnal variation in the room temperature on both days for the non-PCM test room is noticed. The room temperature drops to an average minimum of 24.2 °C in the morning, while it increases to an average height of 44.06 °C during the evening. Thus, non-PCM produces conditions that are excessively cooled in the morning i.e., from 1.00 hr to 6.00 hr, and overheated temperatures in the evening from 13.00 hr to 18.00 hr. The application of PCM efficiently eliminates this difference in room temperature. In the case of the PCM integrated test room, the room temperature curve shows minimum variation during morning hours and reaches minimum temperature in the range of 28–29 °C for both testing days. During the two days of room temperature monitoring, the difference between the room temperature for the PCM and the reference room (i.e., the room without PCM) during the morning and the evening is considered respectively, as the average valley and the peak value differences. A high-temperature difference in the valley value indicates that PCM can add more heat during the morning hours, which prevents excessively cooled conditions. In contrast, a high difference in peak temperature value suggests that more incoming heat is absorbed via PCM during the daytime and reduces the increase in the room temperature compared with the non-PCM test room. Hence, the peak value difference is more significant for passive cooling of the test room as it limits the room temperature elevation during the daytime.

The configurations like PCM with 0 cm and 2 cm air layer thickness show peak value temperature difference (PVTD) of 7.05 °C and 6.97 °C, whereas the valley value temperature difference (VVTD) of 2.51 °C and 2.7 °C, respectively. The PVTD of 5.27 °C and 4.35 °C is seen in the case of the PCM layer with 4 cm and 6 cm air layer thickness, respectively. These configurations show VVTD of 3.85 °C and 4.59 °C, respectively. Thus, PCM with 0 cm and 2 cm air layer thickness effectively regulates the diurnal variation in room temperature compared to PCM with 4 cm and 6 cm air layer thickness. Out of these two configurations, PCM with 2 cm air layer thickness is exposed to the highest average solar air temperature, i.e., 34.65 °C, while PCM with no air layer is subjected to the average solar air temperature of 31.90 °C. It indicates that PCM with 2 cm air layer thickness is subjected to more harsh climatic conditions regarding radiation intensity, ambient temperature, and wind velocity compared to the configuration PCM at the bottom. Hence based on maximum peak temperature difference and climatic conditions, PCM with 2 cm air layer thickness is considered the optimum configuration for minimizing the diurnal temperature variation in a test room. The reasons for such results can be stated as follows.

The application of PCM increases the thermal inertia of the roof. Further, effective utilization of the PCM's latent heat also helps reduce the heat load from the roof. This phenomenon is reflected in the comparison of non-PCM and PCM-integrated roofs. The air layer between the roof and the PCM layer acts as insulation. In this way, raising the air layer aids in reducing heat load and so lowering the indoor temperature. However, as air layer thickness increases, PCM may not get enough heat to melt, making it challenging to use latent heat effectively. Thus, there must be an appropriate air layer thickness, which helps reduce heat load through the roof.

The temperature distribution is studied based on temperatures of different surfaces like rooftop surface, roof bottom surface, PCM upper surface, and PCM lower surface. In Figure 3a–e, the heat storage cycle is established for the diurnal period (7.00 hr in the morning to 20.00 hr in the evening). The heat release cycle is established, on the other hand, during the night hours, i.e., from 19.00 to 24.00 hr and 1.00 to 7.00 hr. During the heat storage cycle, for the two days test cycle, the PCM configuration with no air layer, rooftop surface, interface surface, and PCM lower surface acquires the average maximum temperature of 54.89, 44.03, and 38.88 °C, respectively. In contrast, the average minimum temperature attained during the heat release cycle is 19.88, 25.14, and 27.25 °C, respectively. For other configurations, i.e., PCM with 2, 4, and 6 cm air layer, average maximum and minimum temperatures attained by different surfaces during the heat storage and heat release cycle can be obtained from Figure 3. The time lag, decrement factor, and heat gain are obtained from roof temperature distribution based on the hourly variation in surface temperature and its hourly span. The MKR index is obtained using Equation (5) for different configurations and is tabulated in Table 2.

As seen in Table 2, the lowest time lag is registered in the case of a non-PCM room, whereas the maximum time lag is attained in the case of PCM with a 2 cm air layer thickness. Similarly, the lowest decrement factor is seen in the case of PCM with 2 cm air layer thickness, and it is maximum in the case of the room without PCM. In the case of PCM configurations, PCM with 4 cm and 6 cm air layer thickness, there is more diurnal variation in room temperature, which increases the heat gain by these configurations. The heat gain is minimal in PCM with a 2 cm air layer and in PCM with no air layer. Thus, with a lower decrement factor, heat gain, and higher time lag, PCM with 2 cm air layer thickness provides the highest MKR index value compared to other PCM configurations.

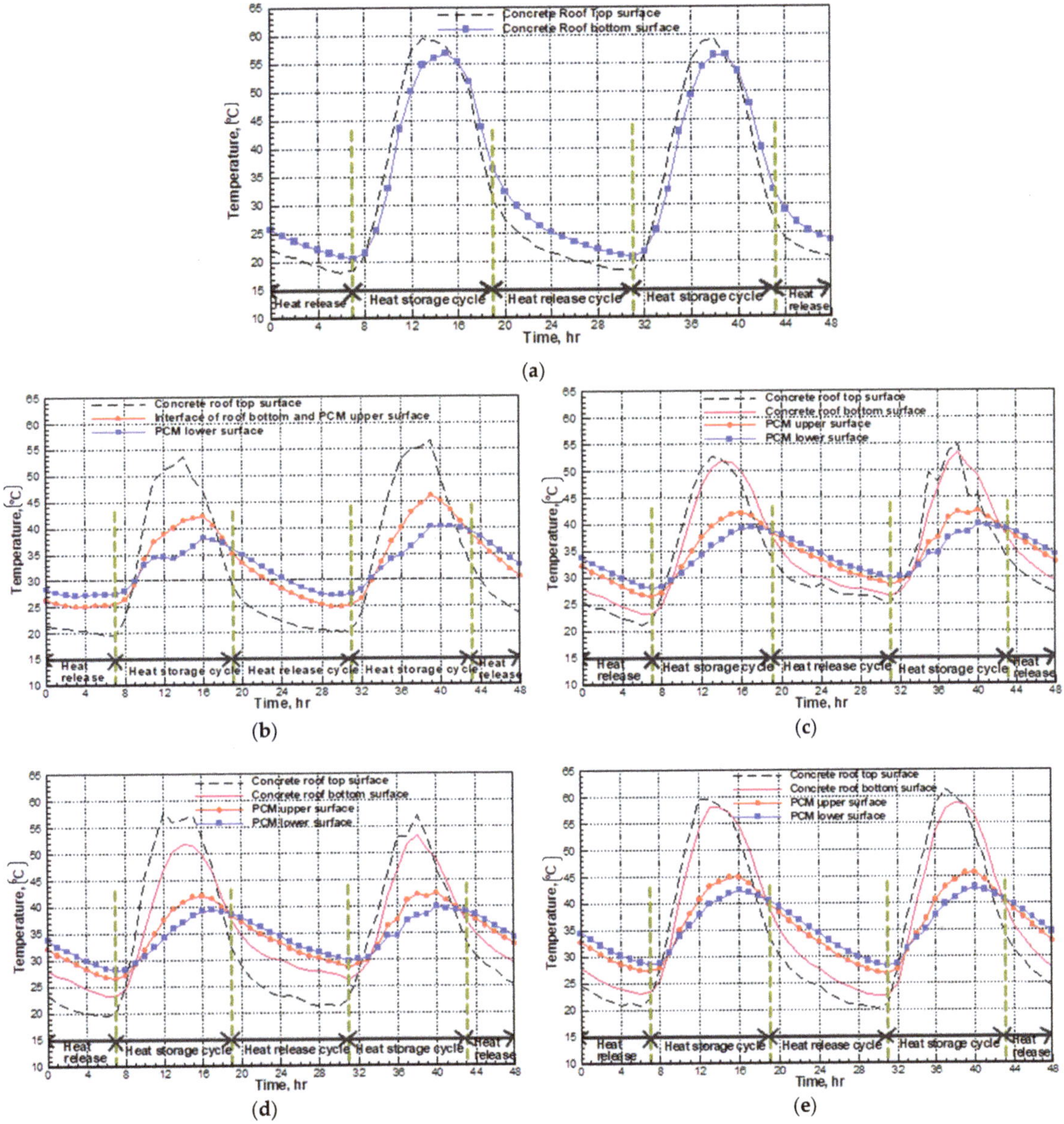

Figure 3. Temperature distribution across the roof slab. (**a**) Non-PCM; (**b**) PCM with 0 cm air layer; (**c**) PCM with 2 cm air layer; (**d**) PCM with 4 cm air layer; (**e**) PCM with 6 cm air layer.

Table 2. MKR index for all configurations of PCM.

	f (hr)	ϕ	Q (Wh/m^2)	f'	ϕ'	Q'	MKR Index
Non-PCM	1.50	0.87	2211.54	-	-	-	-
PCM with 0 cm air layer	2.00	0.36	599.27	1.33	0.41	0.27	5.40
PCM with 2 cm air layer	3.00	0.35	557.45	2.00	0.40	0.25	**8.83**
PCM with 4 cm air layer	2.00	0.38	622.89	1.33	0.43	0.28	5.07
PCM with 6 cm air layer	2.00	0.36	628.09	1.33	0.41	0.28	5.17

Furthermore, in each case, the impact on solar heat gain is assessed for a building with a roof integrated with the PCM. The percentage by which solar heat gain through the envelope is reduced as a result of the presence of the PCM layer is referred to as the envelope solar heat gain reduction rate calculated using Equations (11)–(13).

Table 3 shows the envelope solar heat gain and reduction rate for different configurations.

Table 3. Solar heat gain and reduction rate for all configurations of PCM.

Configuration	Q_{With_PCM} (W hr/m^2)	$Q_{Without\_PCM}$ (W hr/m^2)	Q'	η (%)
PCM with 0 cm air layer	599.27	2211.54	0.27	73
PCM with 2 cm air layer	557.45	2211.54	0.25	75
PCM with 4 cm air layer	622.89	2211.54	0.28	72
PCM with 6 cm air layer	628.09	2211.54	0.28	72

According to Table 3, it can be shown that the solar heat gain reduction rate for PCM falls somewhere in the range of 72–75%. PCM with a 2 cm air layer shows the highest solar heat gain reduction rate of 75% and hence can be treated as optimum air layer thickness for the application of PCM integrated roof.

4. Conclusions

In the present study, field experiments are conducted to analyze air gap thickness's effect on the thermal performance of PCM-integrated test rooms. The investigation is carried out for a non-PCM test room, PCM, with an air layer thickness of 0 cm, 2 cm, 4 cm, and 6 cm for the fixed thickness of the PCM layer. The commercially available material HS 29 is used as PCM. The climatic data during different field tests are used to calculate solar air temperature. The results show that the peak value of room temperature difference between the PCM integrated test room and non-PCM test room is between 4 °C to 7 °C. Thus, it is concluded that the PCM integrated test room reduces the diurnal variations in room temperature when compared with the non-PCM test room.

The effect of the air layer between PCM and the roof is also established with the MKR index. However, the PCM thickness is kept constant and the effect of PCM thickness here is not studied. The PCM with 2 cm air layer thickness shows a lower decrement factor, heat gain, and higher time lag. This reflects the highest MKR index, i.e., 8.83, compared to other configurations. Further, solar heat gain reduction rate is also obtained for all configurations. PCM with an air layer has been demonstrated to have a solar heat gain reduction rate between 72% and 75%. The most effective solar heat gain reduction (75%) is shown with a PCM with an air layer of 2 cm. Hence based on maximum peak temperature difference, MKR index, and climatic conditions, PCM with 2 cm air layer thickness is considered the optimum configuration for adjusting diurnal temperature variation in a test room. Therefore, it is concluded that air layer thickness in the PCM-integrated roof can reduce diurnal temperature swings and readily apply to PCM-integrated passive cooling strategies for buildings.

Author Contributions: Conceptualization, D.K.B. and M.K.R.; methodology, D.K.B.; formal analysis, D.K.B.; writing—original draft preparation, D.K.B.; writing—review and editing, M.K.R., J.B. and M.A.; supervision, J.B. and M.A. All authors have read and agreed to the published version of the manuscript.

Funding: This research received no external funding.

Data Availability Statement: No new data were created, or where data is unavailable due to privacy or ethical restrictions.

Conflicts of Interest: The authors declare no conflict of interest.

Nomenclature

f	Time lag (hr)
f'	Dimensionless time lag
F_{sky}	Shape factor for horizontal surface (=1 for the rooftop slab and sky)
h	Heat transfer coefficient, (W/m^2 K)
I	Solar radiation, (W/m^2)
PCM	Phase Change Materials
q	Heat flux, (W/m^2)
Q	Solar heat gain, (W/m^2)
T	Temperature, (K)
V	Velocity, (m/s)
Vo	Volume, (m^3)
Greek Letters	
α	Absorptivity of surface
ε	Emissivity of rooftop slab
ϕ	Decrement factor
ϕ'	Dimensionless decrement factor
σ	Stefan Boltzmann constant, (W/m^2 K^4)
Subscript	
B	Brick
C	Concrete
combined	Combined radiative and convective conditions
ceiling	Ceiling conditions
d	Direct solar radiation
i	Indoor conditions
o	Ambient or outdoor conditions
P	PCM
roof	Roof conditions
sky	Clear sky conditions
sol-air	Solar air

Appendix A

To measure the melting point, latent heat of fusion, and specific heat of HS29, DSC analysis is carried out. The graph obtained through the DSC is shown in Figure A1.

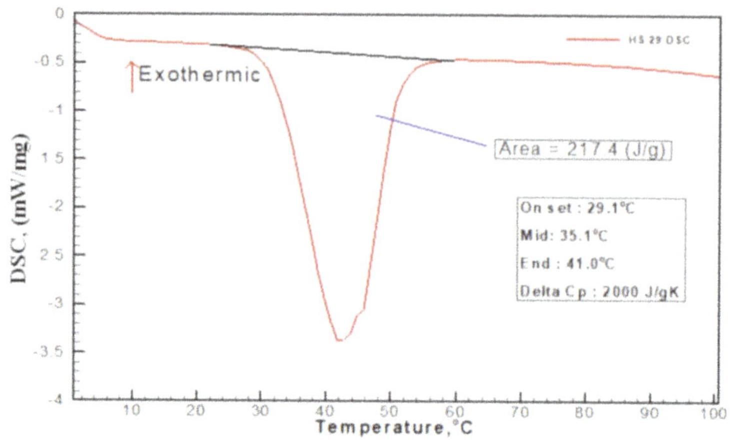

Figure A1. DSC measurement for HS 29.

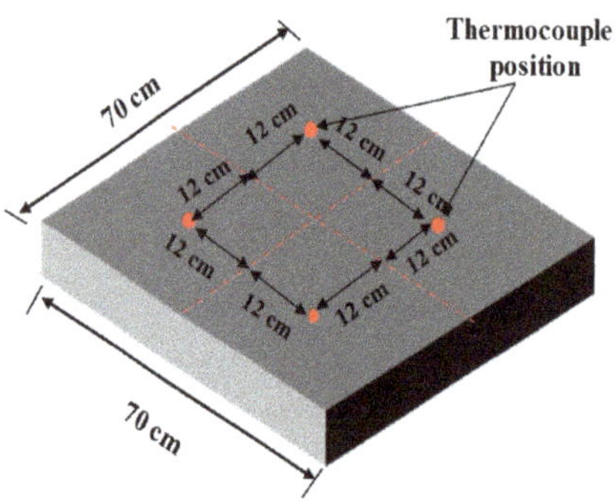

(**a**) Roof surface

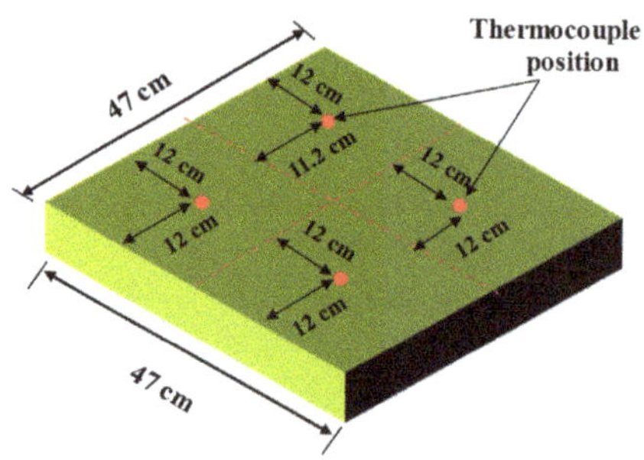

(**b**) PCM enclosure

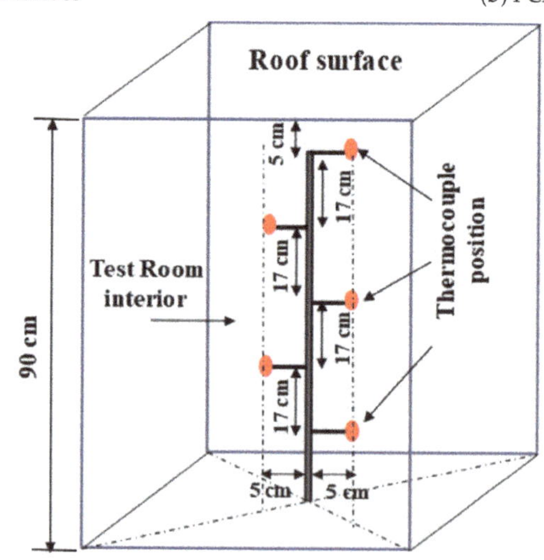

(c) Room interior

Figure A2. Position of thermocouples.

References

1. International Energy Agency. *CO_2 Emissions From Fuel Combustion: Highlights*; IEA Publication: Paris, France, 2012.
2. *2013 April. Annual Energy Outlook 2013 with Projections to 2040*; Energy Information Administration: Washington, DC, USA, 2013.
3. Lei, J.; Yang, J.; Yang, E.-H. Energy performance of building envelopes integrated with phase change materials for cooling load reduction in tropical Singapore. *Appl. Energy* **2016**, *162*, 207–217. [CrossRef]
4. Dehkordi, B.S.; Afrand, M. Energy-saving owing to using PCM into buildings: Considering of hot and cold climate region. *Sustain. Energy Technol. Assess.* **2022**, *52*, 102112.
5. BBland, A.; Khzouz, M.; Statheros, T.; Gkanas, E.I. PCMs for Residential Building Applications: A Short Review Focused on Disadvantages and Proposals for Future Development. *Buildings* **2017**, *7*, 78. [CrossRef]
6. Al-Sanea, S.A.; Zedan, M.; Al-Ajlan, S.A.; Hadi, A.S.A. Heat Transfer Characteristics and Optimum Insulation Thickness for Cavity Walls. *J. Therm. Envel. Build. Sci.* **2003**, *26*, 285–307. [CrossRef]
7. Jamal, B.; Boukendil, M.; Abdelbaki, A.; Zrikem, Z. Numerical simulation of coupled heat transfer through double solid walls separated by an air layer. *Int. J. Therm. Sci.* **2020**, *156*, 106461. [CrossRef]
8. Mahlia, T.; Iqbal, A. Cost benefits analysis and emission reductions of optimum thickness and air gaps for selected insulation materials for building walls in Maldives. *Energy* **2010**, *35*, 2242–2250. [CrossRef]
9. Mavromatidis, L.E.; Bykalyuk, A.; El Mankibi, M.; Michel, P.; Santamouris, M. Numerical estimation of air gaps' influence on the insulating performance of multilayer thermal insulation. *Build. Environ.* **2012**, *49*, 227–237. [CrossRef]
10. Kurt, H. The usage of air gap in the composite wall for energy saving and air pollution. *Environ. Prog. Sustain. Energy* **2010**, *30*, 450–458. [CrossRef]
11. Bhamare, D.K.; Rathod, M.K.; Banerjee, J. Passive cooling techniques for building and their applicability in different climatic zones—The state of art. *Energy Build.* **2019**, *198*, 467–490. [CrossRef]
12. Rathod, M.K.; Banerjee, J. Thermal stability of phase change materials used in latent heat energy storage systems: A review. *Renew. Sustain. Energy Rev.* **2013**, *18*, 246–258. [CrossRef]
13. Saffari, M.; Roe, C.; Finn, D.P. Improving the building energy flexibility using PCM-enhanced envelopes. *Appl. Therm. Eng.* **2022**, *217*, 119092. [CrossRef]
14. Ascione, F.; Bianco, N.; De Masi, R.F.; de' Rossi, F.; Vanoli, G.P. Energy refurbishment of existing buildings through the use of phase change materials: Energy savings and indoor comfort in the cooling season. *Appl. Energy* **2014**, *113*, 990–1007. [CrossRef]
15. Alshuraiaan, B. Efficient utilization of PCM in building envelope in a hot environment condition. *Int. J. Thermofluids* **2022**, *16*, 100205. [CrossRef]
16. Li, D.; Zheng, Y.; Liu, C.; Wu, G. Numerical analysis on thermal performance of roof contained PCM of a single residential building. *Energy Convers. Manag.* **2015**, *100*, 147–156. [CrossRef]
17. Tokuç, A.; Başaran, T.; Yesügey, S.C. An experimental and numerical investigation on the use of phase change materials in building elements: The case of a flat roof in Istanbul. *Energy Build.* **2015**, *102*, 91–104. [CrossRef]
18. Saffari, M.; de Gracia, A.; Ushak, S.; Cabeza, L.F. Economic impact of integrating PCM as passive system in buildings using Finger comfort model. *Energy Build.* **2016**, *112*, 159–172. [CrossRef]
19. Singh, S.; Bhat, V. Performance evaluation of dual phase change material gypsum board for the reduction of temperature swings in a building prototype in composite climate. *Energy Build.* **2018**, *159*, 191–200. [CrossRef]
20. Arıcı, M.; Bilgin, F.; Nižetić, S.; Karabay, H. PCM integrated to external building walls: An optimization study on maximum activation of latent heat. *Appl. Therm. Eng.* **2019**, *165*, 114560. [CrossRef]
21. Jin, X.; Medina, M.; Zhang, X. On the importance of the location of PCMs in building walls for enhanced thermal performance. *Appl. Energy* **2013**, *106*, 72–78. [CrossRef]
22. Khedher, B. Numerical Study of the Thermal Behavior of a Composite Phase Change Material (PCM) Room. *Eng. Technol. Appl. Sci. Res.* **2018**, *8*, 2663–2667. [CrossRef]
23. Lee, K.O.; Medina, M.A.; Raith, E.; Sun, X. Assessing the integration of a thin phase change material (PCM) layer in a residential building wall for heat transfer reduction and management. *Appl. Energy* **2015**, *137*, 699–706. [CrossRef]
24. Lagou, A.; Kylili, A.; Šadauskienė, J.; Fokaides, P.A. Numerical investigation of phase change materials (PCM) optimal melting properties and position in building elements under diverse conditions. *Constr. Build. Mater.* **2019**, *225*, 452–464. [CrossRef]
25. Pasupathy, A.; Velraj, R. Effect of double layer phase change material in building roof for year round thermal management. *Energy Build.* **2007**, *40*, 193–203. [CrossRef]
26. Jin, X.; Zhang, X. Thermal analysis of a double layer phase change material floor. *Appl. Therm. Eng.* **2011**, *31*, 1576–1581. [CrossRef]
27. Bhamare, D.K.; Rathod, M.K.; Banerjee, J. Numerical model for evaluating thermal performance of residential building roof integrated with inclined phase change material (PCM) layer. *J. Build. Eng.* **2019**, *28*, 101018. [CrossRef]
28. Bhamare, D.K.; Rathod, M.K.; Banerjee, J. Proposal of a unique index for selection of optimum phase change material for effective thermal performance of a building envelope. *Sol. Energy* **2021**, *218*, 129–141. [CrossRef]
29. Available online: https://www.pluss.co.in/technical-datasheets/Doc304-TDS-HS-29.pdf (accessed on 7 February 2023).
30. Tchinda, R. Thermal behaviour of solar air heater with compound parabolic concentrator. *Energy Convers. Manag.* **2008**, *49*, 529–540. [CrossRef]

31. EnergyPlus, Engineering References. 2013. Available online: http://apps1.eere.energy.gov/buildings/energyplus/energyplus_documentation.cfm (accessed on 7 February 2023).
32. Bhamare, D.K.; Rathod, M.K.; Banerjee, J. Selection of phase change material and establishment of thermophysical properties of phase change material integrated with roof of a building using Measure of Key Response index: Proposal of a new parameter. *J. Energy Storage* **2020**, *32*, 101812. [CrossRef]
33. Kong, X.; Lu, S.; Li, Y.; Huang, J.; Liu, S. Numerical study on the thermal performance of building wall and roof incorporating phase change material panel for passive cooling application. *Energy Build.* **2014**, *81*, 404–415. [CrossRef]

buildings

MDPI

Article

Nano Engineered Paraffin-Based Phase Change Material for Building Thermal Management

John Paul [1], Mahendran Samykano [1,*], Adarsh Kumar Pandey [2,3,*], Kumaran Kadirgama [1,*] and Vineet Veer Tyagi [4]

1 Faculty of Mechanical & Automotive Engineering Technology, Universiti Malaysia Pahang, Pekan 26600, Pahang, Malaysia
2 Research Centre for Nanomaterials and Energy Technology (RCNMET), School of Engineering & Technology, Sunway University, No. 5, Jalan Universiti, Bandar Sunway, Petaling Jaya 47500, Selangor Darul Ehsan, Malaysia
3 Center for Transdisciplinary Research (CFTR), Saveetha Institute of Medical and Technical Sciences, Saveetha University, Chennai, India
4 School of Energy Management, Shri Mata Vaishno Devi University, Katra 182320, (J&K), India
* Correspondence: mahendran@ump.edu.my (M.S.); adarshp@sunway.edu.my (A.K.P.); adarsh.889@gmail.com (K.K.)

Abstract: Thermal energy storage (TES) and harvesting is an effective technique for optimum building thermal management. Phase-change materials (PCMs) are commonly used for TES applications but are troubled by their degraded thermal conductivity. Recent research progress in latent heat energy storage using PCMs and nano additives provides a viable solution for solar TES. A series of hybrid nano-enhanced phase change materials (HNePCMs) were prepared via two-step synthesis. Hybrid graphene–silver nanofillers were dispersed in commercial paraffin (melting point 25 °C) under different dispersion rates (0.1%, 0.3%, 0.5%). Different characterization techniques, e.g., FESEM, FT-IR, UV-VIS, TGA, XRD, DSC, and Tempos, were used in material characterization. A maximum enhancement of 6.7% in latent heat and 5% in heat storage efficiency was noted for nanocomposites with 0.3 wt% of additives. The nanocomposite with 0.3 Wt% showed great potential in shielding UV rays and showed a reduction of 6.5% in bandgap energy. Furthermore, the thermal conductivity of samples was boosted by a maximum of 90% (from 0.2 W/mK-0.39 W/mK) with 0.3 wt% dispersion of graphene–silver nanofillers. The thermophysical characterization results establish that the synthesized paraffin/graphene–silver hybrid nanocomposites are well suited for building thermal management.

Keywords: phase change materials; nanocomposites; graphene: silver nanofillers; building thermal management

check for updates

Citation: Paul, J.; Samykano, M.; Pandey, A.K.; Kadirgama, K.; Tyagi, V.V. Nano Engineered Paraffin-Based Phase Change Material for Building Thermal Management. *Buildings* **2023**, *13*, 900. https://doi.org/10.3390/buildings13040900

Academic Editor: Ahmed Senouci

Received: 2 February 2023
Revised: 3 March 2023
Accepted: 11 March 2023
Published: 29 March 2023

1. Introduction

In recent decades, a profound rise in sustainable and clean energy technologies arose due to rising global energy demand and environmental challenges [1]. The inconsistency of sustainable energy sources (wind, tidal, and solar energy) reinforces the importance of energy storage [2]. Under the Paris Agreement signed by all United Nations member states in 2015, all its member states are bound to attain a reduction in carbon dioxide (CO_2) emissions mainly by curtailing their fossil fuel usage to 20%. Moreover, the agreement also put forward 17 Sustainable development goals (SDGs). Goal seven among the 17 SDGs stresses providing green and affordable energy around the globe, which literally stresses embracing energy-efficient and renewable energy resources [3]. Approximately 40% of global energy use occurs in the building sector. The general change in living standards and the greater need for heating and cooling in cold and hot climates are the leading causes of rising energy usage [4]. Thermal energy remains an abundant energy source, which is substantially present in nature and is also available as a derivate of numerous energy conversion processes. Thermal energy storage (TES) presents a possible technique to reuse ample heat energy and enhance energy efficiency. TES could be implemented to harvest, store, and even reuse energy as heat whenever needed. Furthermore, it might also provide a viable solution for the existing problem of unmatched energy supply and demand in time and space [5]. Latent heat energy storage (LHES) unveils a higher energy storage density. It also holds a sensitive operating temperature range [6]. Phase change materials (PCMs) were deployed for LHES due to their excellent thermal energy storage capabilities, which resemble a thermal battery [7]. Organic, inorganic, and eutectics are the three main classes of phase change materials. Paraffin wax (PW), an organic PCM, remains the most widely preferred PCM owing to its peculiar attributes: (a) superior heat of fusion, (b) corrosion-free nature, (c) thermal cycling stability, (d) tuneable melting point (20–100 °C), and (e) economic viability. The prominent hurdle in promoting the widespread usage of paraffin as a TES element is its degraded thermal conductivity. Degraded thermal conductivity offers comparatively improved interfacial thermal resistance (among the PCM and dispersed filler), which degrades the charging efficiency of the energy storage material during TES [8]. Installing PCM-enhanced wallboards within the building envelope is the most prevalent technique for building thermal management. Therefore, PCMs can enhance thermal inertia of lightweight structures, resulting in a substantial rise in thermal storage capacity. In order to obtain more significant yearly energy savings, cooling dominant climates demand PCMs with a melting point near 26 °C, whereas heating dominant climates need a melting point closer to 20 °C [9].

Dispersing highly conductive fillers with nano-size provides a viable solution to improving the degraded thermal conductive nature of the PCM matrix. The impact of dispersing thermally conductive nanostructured materials commonly utilized as thermal conductivity enhancers was comprehensively reviewed [10,11]. A meta-analysis of nanofillers was also conducted based on their dimension (0-D,1-D, 2-D, 3-D, and even hybrid materials). Authors also concluded that carbon-based nanomaterials exhibited a higher enrichment in thermal conductivity when compared with metallic/metal oxide nanoparticles [10]. Infusing nanosized carbon-based nanomaterials as fillers is a commonly adopted method to boost the thermal conductivity (λ value) of composites [12]. The chemical exfoliation of graphite flakes generates graphite nanoplatelets, multilayer graphene, and graphene, which possess a higher aspect ratio and ultrahigh intrinsic thermal conductivity ((λ value) > 2000 Wm^{-1} K^{-1}). The loading rate of nanofillers, interfacial thermal resistance, geometrical and dimensional aspects of the dispersants, including size, thickness, and the orientation of sheets, largely influences the enhancement in λ values [13]. The improvement in effective λ value is not pronounced when individual 2D materials are doped onto PCMs. This is primarily because of disorganized geometric contact and interfacial thermal resistance, which is normally elevated when the nanoparticles generate a disordered percolation network [14]. The dispersed fillers could also contribute to unstable λ value during phase transitions [15]. The dispersion of nano-fillers mainly alters the thermal conductivity of the base matrix and even other characteristics, including enthalpy, sub-cooling, phase transition temperature, phase transition duration, density, and viscosity [11]. The excellent thermal conductivity and chemical stability make silver (Ag) another suitable nano-sized dispersant for PCMs. PCM dispersed with mono nanofillers were comprehensively analyzed, and it was understood that the dispersion modifies the thermophysical properties of composites. Moreover, studies on PCMs (especially paraffin) infused with hybrid nanofillers are much fewer or even limited when compared with single nanofillers infused PCMs [10]. PCMs have been widely deployed for thermal management in buildings. The suitability of engineered phase change composites in the thermal management of buildings has been reviewed in detail [16].

George et al. [17] conducted a comparative analysis by dispersing both Polyaniline (PANI) and CuO in paraffin. The rise in thermal conductivity was around 46.7% and 63.5% for PANI and CuO, and the latent heat also logged an increment of 8.25 and 7.8% rise. Habib et al. [18] obtained a boost of 60.5% in thermal conductivity with SWCNT doped into paraffin. Sheng et al. [19] used carbon fiber as a nano additive, and the thermal conductivity was boosted by 148%. The graphene dispersion in paraffin reported by Joseph et al. [20] provided an increment of 60.55 in thermal conductivity. Aslfattahi et al. [21] successfully doped MXene into the paraffin, and the composite registered an improvement of 11.6% in thermal conductivity. Table 1 briefly summarizes the thermophysical properties of recent research works on different paraffin-based nanocomposites. An in-depth literature review by Paul et al. [10,22] highlighted a lack of investigations into the synthesis and thermophysical characterization of low-temperature organic NePCMs used in ambient conditions. Hence, this proved to be an aspiration for researchers in developing low-temperature composite PCMs dispersed with hybrid nanofiller for building thermal management.

Table 1. Thermophysical properties of wax-based nanocomposites.

Nanocomposites	Change in Thermal Conductivity (W/mK)	Change in Enthalpy (J/g)	Nanocomposites	Ref.
Paraffin/Graphene	59.5% increment	12.4% decline	Paraffin/Graphene	[20]
Paraffin/CNT	39.0% increment	0.90% decline	Paraffin/CNT	[23]
Paraffin/PANI	46.7% increment	8.20% increment	Paraffin/PANI	[17]
Paraffin/CuO	63.5% increment	7.80% increment	Paraffin/CuO	[17]
Paraffin/Co-PANI	17.6%increment	8.06% decline	Paraffin/Co-PANI	[24]
Paraffin/Nanographene	741% increment	13.10% decline	Paraffin/Nanographene	[25]
Paraffin/GP: Ag	90% increment	6.70% increment	Paraffin/GP: Ag	PW

In summary, we demonstrate a method to achieve superior λ values with hybrid nano-enhanced phase change materials (HNePCM) formulated via two-step method for the thermal management of buildings and other ambient temperature applications. To the best of our knowledge, the synthesis of HNePCM loaded with 2-D graphene–silver nanofillers in low-temperature paraffin remains a novel work. A maximal enrichment of 90% in thermal conductivity was clocked for the 0.3% dispersion of nano additives. The synergistic effect of hybrid nanofillers, low interfacial thermal resistance, and reagglomeration contribute to an exceptional increment in λ. The λ value of HNePCMs could be tailored within a range of 0.2–0.4 W/mK with nanofiller loadings limited to 0.5 wt%. The composites displayed excellent photo-thermal absorption, latent heat (6.7%), thermal stability (5.7 °C), and reliability. The hybrid nanocomposites could shield UV radiation to a maximum of 100%. The present work provides insights into the exceptional enhancement in the λ value of composites by dispersing hybrid additives, which are currently restricted to PCM-based composites but could also be used to design scalable and higher power density thermal energy harvesting devices for building thermal management applications.

2. Experimental Section

2.1. Materials and Methods

Commercial grade paraffin wax (Plus ice A25 H), purchased from Plus Ice UK, was used as the TES material. Graphene–silver hybrid nanoparticles (with ratio 7:3) was procured from US Research Nanomaterials, Inc, and SDBS was procured from Sigma Aldrich. The entire chemicals used in the synthesis procedure were of analytical reagent grade.

2.2. Synthesis of Paraffin/Graphene: Silver Hybrid Nanocomposites

Two-step synthesis was adopted to synthesize the nanocomposite. The paraffin was heated to ensure a homogeneous liquid state. A calculated quantity of surfactant (SDBS) and liquid PCM mixture was subjected to probe sonication for 40 min. Graphene–silver hybrid nanoflakes were dispersed into the melted wax and further put to probe sonication for 40 min to ensure the homogenous mixing of dispersed additives. Figure 1 details the synthesis route of hybrid nanocomposites. The sonicated mixture was then allowed to cool down to ambient temperature. Figure 2 shows the images of pure and synthesized hybrid nanocomposites.

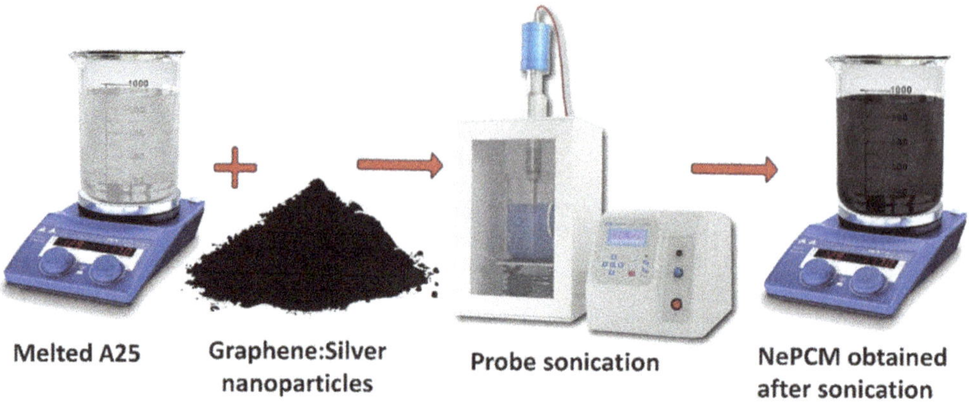

Figure 1. Synthesis route of hybrid nanocomposites.

Figure 2. Images of pure and synthesized hybrid nanocomposites.

2.3. Characterization Techniques

The detailed morphology of Gp: Ag nanofillers was studied under field emission scanning electron microscopy (JSM-IT800, JEOL JAPAN, Tokyo, Japan) and transmission emission scanning electron microscopy (FEI Tecnai G2 20 S-TWIN, FEI Company, Hillsboro, OR, USA). The sample's thermal conductivity was estimated with a thermal property analyzer (model-TEMPOS, SH-3 sensor, Meter Group, Pullman, DC, USA). The latent heat and melting points were calculated using a differential scanning calorimeter analyzer (Model -DSC 822e, DSC-Mettler Toledo, Columbus, OH, USA). The thermal stability analysis of samples was conducted with a TGA (model-TGA 4000, Perkin Elmer, Waltham, MA, USA). The light ab-sorption ability of the HNePCMs was examined via a UV-VIS spectrometer (mod-el-LAMBDA 750, Perkin Elmer, Waltham, MA, USA). The changes in the attached functional group of synthesized HNePCMs were examined with Fourier transform infrared spec-troscopy (model-Spectrum Two FT-IR Spectrometer, Perkin Elmer, Waltham, MA, USA). A thermal imaging camera (model-S60, CAT, Irving, TX, USA) was used to capture infrared images of the samples.

2.4. Uncertainty Analysis

The reliability of the measurements performed was characterized by uncertainty analysis. Moreover, it also reflects the dispersion and accuracy of values measured. The uncertainties involved in direct measurement fall under random errors, named Class

A errors, and are measured by statistical methods. System errors are handled by non-statistical methods and are grouped as Class B errors. The uncertainty associated with the measurements can be estimated by the following equations and the associated terms [26]. The standard uncertainty of measured results is calculated according to specific rules for the A and B uncertainty of a particular measured value, which is commonly referred to as the combined uncertainty.

The mathematical expression gives class A uncertainty.

$$U_A(x) = \sqrt{\sum_{i=1}^{n} \frac{(X_i - X_{avg})^2}{n(n-1)}} \tag{1}$$

where n = total number of readings, X_{avg} = average value of the measurement

Under the same conditions, when x is measured multiple times, the standard uncertainty u(x) of x to be measured is obtained by combining the class A uncertainty $u_A(x)$ and the instrument uncertainty $u_{B2}(x)$.

$$u(x) = \sqrt{u_A{}^2(x) + u_{B2}{}^2(x)} \tag{2}$$

When a single measurement is performed on a physical quantity, the type B uncertainty is composed of the measurement uncertainty $u_{B1}(x)$ and the instrument uncertainty $u_{B2}(x)$.

The measurement uncertainty $u_{B1}(x)$ is caused by estimation, generally 1/10 or 1/5 of the scale value of the instrument.

Instrument uncertainty $u_{B2}(x)$ is determined by the characteristics of the instrument itself and is defined as:

$$u_{B2}(x) = \frac{a}{c} \tag{3}$$

where the confidence factor c can generally be processed with a uniform distribution, and the value assigned for c is $\sqrt{3}$.

For a single measurement x, the standard uncertainty u(x) of x to be measured is obtained by combining the class A uncertainty $u_{B1}(x)$ and the instrument uncertainty $u_{B2}(x)$.

$$u(x) = \sqrt{u_{B1}{}^2(x) + u_{B2}{}^2(x)} \tag{4}$$

$$x \pm u(x) \tag{5}$$

3. Results and Analysis

3.1. Morphological Analysis of Hybrid Nanoparticles

Figure 3 shows the FESEM images of hybrid graphene-silver nanofillers, which also shows the detailed structure of dispersed graphene sheets and silver nanomaterials. A few layers of transparent and thin graphene nanosheets were seen in the FESEM image (Figure 3A). Figure 3A shows that the graphene layers were arranged in a stacked pattern, overlapping each layer with another one in a folded manner [27]. From the FESEM image, the silver nanoparticles can be seen as a filler in between the graphene sheets. FESEM analysis shows that the graphene sheets got intercalated uniformly with silver nanomaterials. The silver nanomaterials lying on top of the graphene layers show a uniform and distinct contrast corresponding to the peripheral transparent graphene flakes (shown in Figure 3A). The TEM image of the HNePCM is given in Figure 3B. In addition, it can also be inferred that the intercalated silver nanomaterials in the graphene sheets may get oriented by interconnecting with each other [28], which eventually assists in linking the graphene flakes. This increases the thermally conducting path in the composite [29]. Standard morphological analysis techniques for low-temperature nanocomposites are not feasible as the phase transition temperature of these samples lies on the lower side. Moreover, the intense beam can readily melt the sample under inspection. Standard FESEM characterization of hybrid nanocomposites is impossible as the highly intense electron beam melts the sample [30]. Energy dispersive x-ray (EDX) assessment was used to ensure the homogeneity of nanofillers and list the dominating elements. Figure 3C,D shows that nanoparticles had 77.79% of carbon (96.92 atomic) and 22.21% of silver (3.02% atomic). This clearly shows that both carbon and silver particles are present in the hybrid nanoparticle.

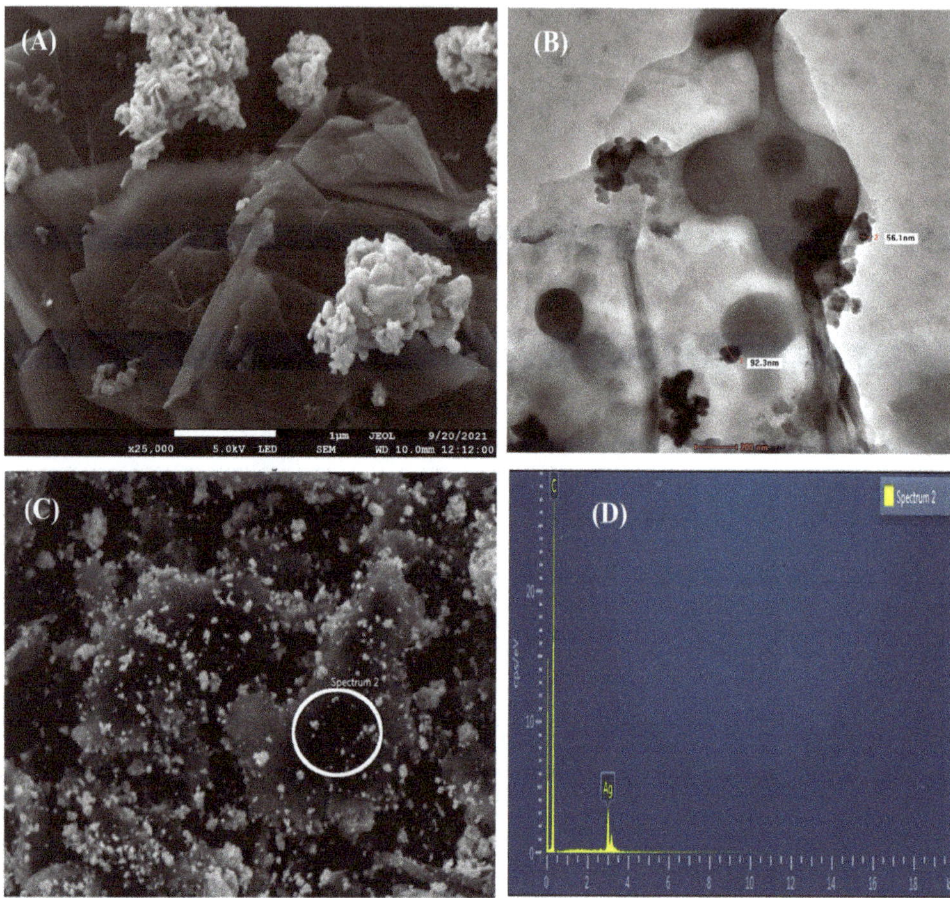

Figure 3. FESEM image of hybrid nanofillers (**A**) magnification 25,000 (**B**) TEM image of the HNePCMs (**C**,**D**) EDX mapping of hybrid nanoparticles.

3.2. *FT-IR Spectrum Analysis of Hybrid Nanocomposites*

The chemical structures of low-temperature pristine wax (A25) and the HNePCMs were studied with FTIR spectroscopy. Figure 4 details the FT-IR spectrum of all synthesized nanocomposites individually. The paraffin wax has an empirical formula of C_nH_{2n+2} and mainly holds three significant peaks of interest [17]. The occurrence of -CH_2- and -CH_3- groups in the composites were evident with the characteristic peak shifts [31]. The peaks within the range of 2800–3000 cm^{-1} indicate a symmetric stretching vibration of the -CH_2 and -CH_3 groups. Similarly, a peak in 1350–1470 region denotes the deformation of alkyl groups (-CH_3- and -CH_2-). Finally, the rocking motion of the -CH2- group is shown by the peak in 720–725 cm^{-1} region [32]. In the FT-IR spectrum, peaks are visible in the following wave numbers 2910 cm^{-1}, 2850 cm^{-1}, 1457 cm^{-1}, and 720 cm^{-1}. Both the peaks at 2910 cm^{-1} and 2850 cm^{-1} lie within the range of 2800–3000 cm^{-1}, showing the symmetric stretching vibration of -CH_3- and -CH_2- groups. The peak 1457 cm^{-1} indicates the deformation of -CH2- and -CH3- groups, and lastly, the corresponding peak at 720 cm^{-1} denotes the rocking vibration of the -CH_2- group as it stays within the range of 720 cm^{-1}– 725 cm^{-1}. Paraffin holds a steady balance of attracting and opposing forces among carbon and hydrogen atoms which is clearly defined by the peaks [33]. Moreover, it is evident from the spectrum that all composites, even with varying loading rates of graphene/silver nanofillers, had similar characteristic peaks at the same wavenumbers (cm^{-1}). The dominant chemical structure of paraffin wax accounts for the uniformity in the FT-IR spectrum of all composites, even with different loading rates of nanofillers [34]. The FT-IR spectrum analysis revealed that no new peaks were present in the synthesized samples. Thus, the FT-IR results establish the absence of chemical reactions that alter the nature of paraffin during the synthesis of samples [35]. As no chemical reaction was present among the additives (paraffin, nanofillers, and surfactant), it can be concluded that the formulated samples were composites.

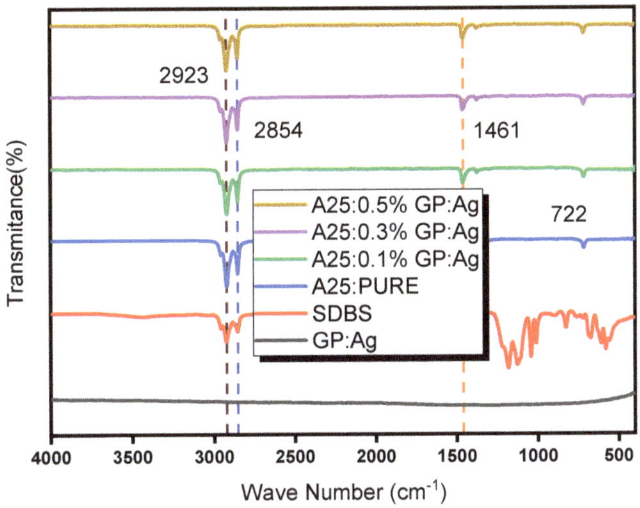

Figure 4. FTIR plot of base, nanoparticles, HNePCMs.

3.3. *UV-Visible Spectrometer Spectrum Analysis of Nanocomposites*

To analyze the light transmission capability of the synthesized hybrid nanocomposite, UV-Vis-NIR transmission spectra of paraffin and nanocomposites were plotted between 200 and 800 nm, as shown in Figure 5. The spectrum curve follows a simple principle which implies that it is impossible to have a material that simultaneously possesses both excellent absorbance and transmissibility. The transmissibility values remain inversely proportional to absorbance values for all samples [17]. The light transmission capability of the hybrid nanocomposite showed a declining trend with the rise in the dispersion of hybrid nanofillers. The improved light absorption of the nanofillers accounts for this trend. In contrast with pure paraffin, hybrid nanocomposites exhibited an entire absorbance of solar radiation throughout the UV-Vis-NIR range. Pure paraffin had an absorbance of 13.22%, and samples with 0.1%, 0.3%, and 0.5% nano fillers exhibited a corresponding absorbance of 97.35%, 99.99%, and 99.75%, respectively. The absorption spectrum was calculated using the solar spectrum data provided by NREL [36]. Moreover, the distinct absorption characteristic of hybrid graphene–silver nanoflakes as the supporting material, enables the nanocomposite to harvest the maximum solar energy. This is because of the firm absorption peaks of the nanocomposite at 245 nm, where most of the solar energy gets distributed (consisting of 7% UV, 50% visible, and 43% infrared) [37]. However, the nanocomposite exhibits total absorption over the entire UV-Vis-NIR range. Solar transmission of nanocomposites in different spectrums are summarized in Table 2. The spectrum analysis results lie in tandem with the physical appearance of paraffin (white) and hybrid nanocomposite (black). The nanosheets function as a perfect black body because of their effective light scattering and trapping ability, enhancing photoabsorption and imparting thermal energy to the wax in the hybrid nanocomposites. With a linear rise in the loading rate of nanofillers, the amount of solar energy absorbed from solar radiation rises, improving the composite's application range, especially in ambient temperature, direct solar thermal applications, such as building thermal management.

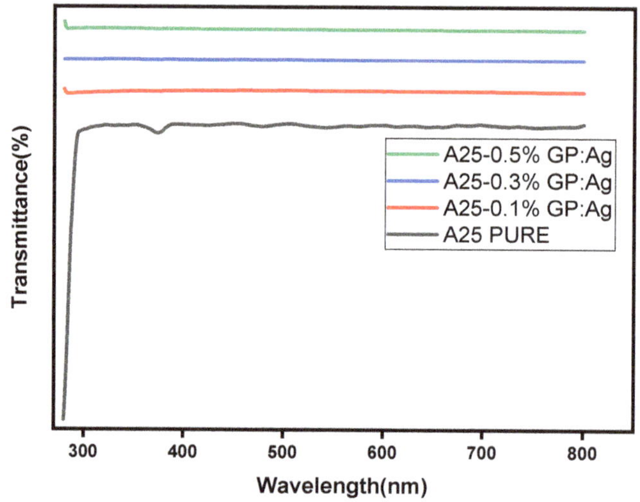

Figure 5. Light transmittance spectrum of samples.

Table 2. Solar transmission of nanocomposites in different spectrums.

SAMPLE	TOTAL	UV	VISIBLE	IR	Reference
SOLAR spectrum	100%	7%	50%	43%	[37]
A25-0.0%	86.78%	6.28%	80.63%	13.09%	Current work
A25-0.1%	2.65%	5.88%	81.73%	12.39%	Current work
A25-0.3%	0.01%	0%	78.03%	21.97%	Current work
A25-0.5%	0.25%	2.51%	76.13%	21.37%	Current work

3.4. Thermal Stability of Paraffin/Graphene: Silver Nanocomposites

The thermal stability of nanocomposites is also pivotal, especially for LHES applications. In this experimental work, the thermal stability of A25 as PCM dispersed with different loading rates of graphene–silver additive was assessed using thermogravimetric analysis (TGA) and derivative thermogravimetry (DTG). Figure 6 displays TGA, and Figure 7 DTG curves of A25 and A25/ graphene–silver nanocomposite with different loading rates of nano additives. The weight loss (5 wt%) and the corresponding temperatures at ($T_{5\%}$) are summarized in Table 3. The temperature at which the maximal weight loss rate (T_{max}) happens is evaluated using DTG curves. The TGA curves show that the base PCM and even the formulated hybrid nanocomposites had a single-step degradation. The samples recorded negligible weight loss in the 30–155 °C. The formulated nanocomposite had excellent thermal stability even at 155 °C, which lies well above the peak melting temperature of pure A25 (25 °C). Higher thermal stability (above the working temperature area) is one of the mandatory prerequisites for a PCM to be deployed for TES. As temperature surpassed 155 °C, A25 in the composite started decomposition and eventually halted at 266 °C, as shown in Figure 7. Furthermore, the heat resistance index [38] (THRI) was calculated by the following mathematical expression.

$$THRI = 0.49 \times [T_{5\%} + 0.6 \times (T_{30\%} - T_{5\%})] \quad (6)$$

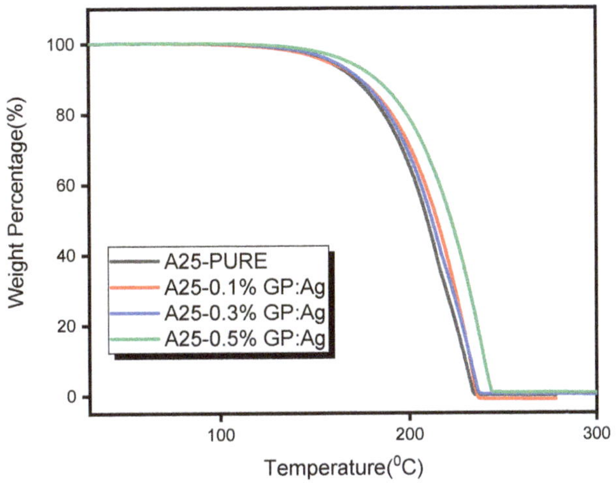

Figure 6. TGA thermograms of A25 and hybrid nanocomposites.

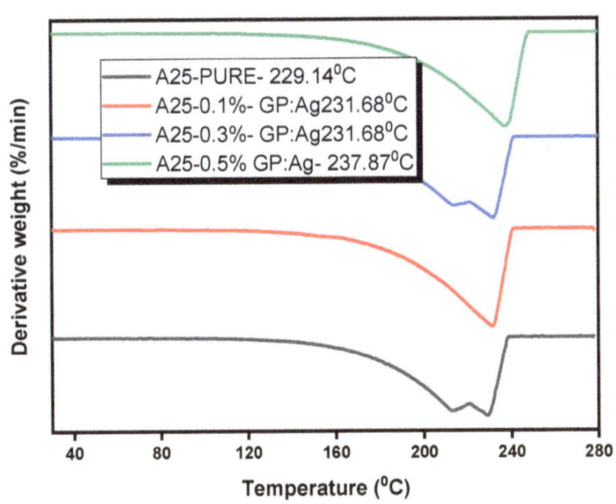

Figure 7. DTG thermograms of A25 and hybrid nanocomposites.

Table 3. Thermal stability parameters of hybrid nanocomposites.

Samples	$T_{5\%}$ (°C)	$T_{30\%}$ (°C)	T_{max} (°C)	Heat Resistance Index (°C)
A25-0.0%	157.37	196.57	229.14	88.63610
A25-0.1%	156.73	201.1	231.68	89.84248
A25-0.3%	159.92	199.11	231.68	89.88266
A25-0.5%	167.96	208.73	237.87	94.28678

The heat resistance index of all samples was tabulated in Table 4. The THRI slightly improved with the addition of nanofillers. This is because nanofillers replace a portion of the PCM matrix. The value almost remains constant for nanocomposites. The temperature readings at which a maximal weight loss rate forA25 and nanocomposite was recorded are 229.14 °C (A25-0%), 231.68 °C (A25-0.1%), 231.68 °C (A25-0.3%), and 237.87 °C (A25-0.5%). These readings could be referred to as the decomposition of paraffin. When compared with pure A25, the decomposition temperatures (the temperature at which maximal weight loss rate) for hybrid nanocomposite were higher than that of A25, which showed the dispersion of Gp. Ag additives enhanced the thermal stability of nanocomposites. The maximum temperature of weight loss of the hybrid nanocomposites was 2–7 °C more than pure A25. In addition, it was also found that the rate of weight loss of the hybrid nanocomposite was comparatively sluggish compared to A25. It is worth mentioning that the synthesized nanocomposites had both improved stability and heat resistance compared to pure A25.

3.5. Effect of Graphene: Silver Dispersion on the Thermal Behavior of Hybrid Nanocomposites

Latent-heat and phase transition temperatures during the phase transition of pristine A25 and HNePCMs were evaluated using a DSC. The numerical values of thermal properties were evaluated using DSC are detailed in Table 4. The tabulated values show that the dispersion of hybrid nanofillers slightly altered the thermal properties of the nanocomposites. The DSC thermograms of all composites had a similar shape which showed that A25 had zero transformation during the synthesis of nanocomposites. Figure 8 illustrates the melting and freezing enthalpies, melting and freezing points of HNePCMs during the phase transition process. Moreover, the enthalpy of melting and freezing of all hybrid composites was found to have increased with the dispersion of hybrid nanofillers. The maximal enhancement in melting latent heat of hybrid nanocomposites was 6.6%, for A25-0.3%, as compared with pure A25. Similarly, the maximal boost in solidification latent heat was 8.4% for the 0.3% sample. The latent heat shows a linear increment with a rise in the dispersion rate of hybrid nano additives. Nanocomposites possess an ultra-thin surface with negligible thickness, resulting in a nearly zero aspect ratio of graphene sheets. Large specific areas and Van Der Waals attraction tend to form aggregate formation. The smaller size of nano additives also means that their molecular density is also high. The rise in latent heat is attributed to a larger surface area of nano additives which further enhances the intermolecular attraction [39].

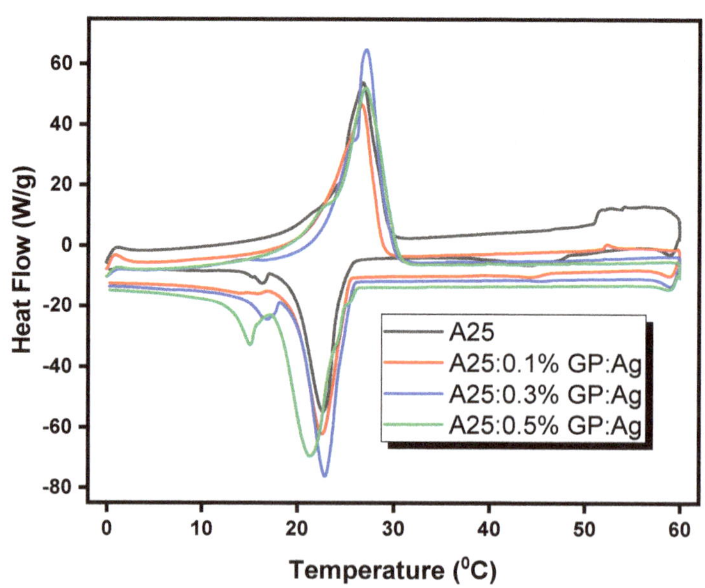

Figure 8. DSC thermogram of samples.

Table 4. Melting and freezing point of composites.

Sample	Enthalpy of Melting ΔH_m (J/g)	Melting Point T_m (°C)	Enthalpy of Freezing ΔH_f (J/g)	Freezing Point T_f (°C)	Degree of Supercooling (°C)	Heat Storage Efficiency γ (%)
A25-0.0%	253.52	26.72	241.98	22.68	4.04	4.55
A25-0.1%	265.75	26.60	253.32	22.58	4.02	4.68
A25-0.3%	270.32	27.07	262.36	22.86	4.21	2.94
A25-0.5%	269.90	27.03	256.40	21.38	5.65	5.00

Furthermore, both the endothermic and exothermic peaks of all synthesized HNePCMs are observed to be narrower/sharper than pristine A25. This is mainly due to the establishment of a thermally conducting network with the dispersion of hybrid nanofillers, providing a rapid thermally conducting path throughout the composite. This eventually promoted the heat transfer rate in the hybrid nanocomposite. Thus, the pace of phase transfer of hybrid nanocomposites will rise, thereby causing a rapid melting and freezing of HNePCMs. It is also noted that both the melting and freezing temperatures show a slight variation mainly due to the rise in λ value of the composites. In this connection, enhancing the loading rate of nanofillers further enhances the heat transfer performance. To assess the phase change performance of synthesized HNePCMs, a typical parameter termed heat storage efficiency (γ) was used. The following mathematical relation is used to estimate (γ).

$$\gamma = \left(1 - \frac{h_f}{h_m}\right) \times 100\% \tag{7}$$

where h_f denotes the enthalpy of freezing and h_m denotes the enthalpy of melting of the samples. The heat loss percentage of entire HNePCMs showed changes within a permissible limit during the cyclic endothermic and exothermic process. The quality loss that occurred while melting during the DSC test accounts for a higher enthalpy of melting for A25 (as compared with freezing enthalpy) and the prepared nanocomposite.

3.6. Thermal Conductivity of Hybrid Nanocomposites

In thermal management applications, the primary function of PCM is to effectively store and release the generated thermal energy using its phase transition ability. The poor λ value of pristine PCMs dampens both the heat storage and release rate, thereby restricting its application range. A PCM having an enhanced thermal conductivity value raises the rate of heat transfer during the phase transition process, which eventually shortens the time for melting and solidification [40]. In this study, among all the synthesized composites, (A25-0.3%) clocked a maximal λ value of 0.395 W/mK (detailed in Table 5). The increment in λ value with hybrid nanofiller infusion could be detailed by two mechanisms, namely the synergy effect and interfacial thermal resistance. The synergy effect (evaluated by synergistic factor η) is strong in the composites, which predominantly forms an effective thermal conductive layer, improving thermal conductivity. The nanoparticle dispersion lowers the interfacial thermal resistance and improves thermal conductivity [41]. The reagglomeration caused by the dispersion of nanofillers also supports the enhancement of λ value.

Table 5. Thermal Conductivity of composites in solid-state.

Sample	Thermal Conductivity of Composites (W/mK)	Enhancement in Thermal Conductivity (%)
A25-0.0%	0.208 ± 0.029	----
A25-0.1%	0.242 ± 0.029	16.44
A25-0.3%	0.395 ± 0.029	89.88
A25-0.5%	0.290 ± 0.029	55.42

The excellent λ value of graphene sheets could be the root cause of a boost in of λ value of hybrid nanocomposite [42]. The nanofillers hold an ultra-thin surface with a negligible thickness, resulting in an almost zero aspect ratio of graphene sheets [43]. A percolating thermally conductive network could be generated in the PCM matrix with graphene sheets having these features. Furthermore, a tunnelling impact that occurs among adjoining graphene sheets also assists in boosting the λ value of samples. The graphene sheets form an entire conductive network by contacting the adjacent nanosheets in percolation, and a substantial increment in the amount of graphene sheets raises the number of conducting networks until the conductivity levels show a decline [44]. Graphene holds an inherent ability to generate an uninterrupted, thermally conductive chain mainly because of its negligible aspect ratio, which improves thermal conductivity [45].

Furthermore, the hybrid nanoparticles connect within the PCM matrix, as graphene sheets exist along with spherical-shaped silver nanofillers [29]. The electrons jump in between adjacent graphene sheets in the matrix with the assistance of silver nanomaterials. Hence, the nanocomposite possesses a stronger energy transport within the matrix [46]. The impact of interfacial thermal resistance between hybrid Gp-Ag nanofillers and PCM matrix cannot be overrated in solid-state nanocomposites. The interfacial thermal boundary resistance could hinder the enrichment of λ value in nanocomposites. This is mainly because the dimensionality of nanofillers is intensified based on the results for 2-D materials, including graphene nanoflakes [47]. The nanocomposites had maximal thermal conductivity at 0.3 Wt%, corresponding to an 89.88% rise. The conductivity further slips to 0.290 W/mK at 0.5 wt%, which shows only an increment of 55.42%. As the nanofiller dispersion crosses the 0.3 Wt% mark, thermal conductivity shows a declining trend. A maximum of 89.88% and a minimum of 16.44% improvement in λ value are obtained with the addition of Gp-Ag nanofillers in wax (shown in Figure 9). The decline in λ value with a rise in nanofiller loading is mainly because (1) the thermal network gets broken down because of the agglomeration of hybrid nanofillers; (2) an increment in phonon-impurity (nanofillers) scattering as the loading rate of nanofillers rises; and (3) finally, a synergic effect of both thermal network crackdown and rise in scattering.

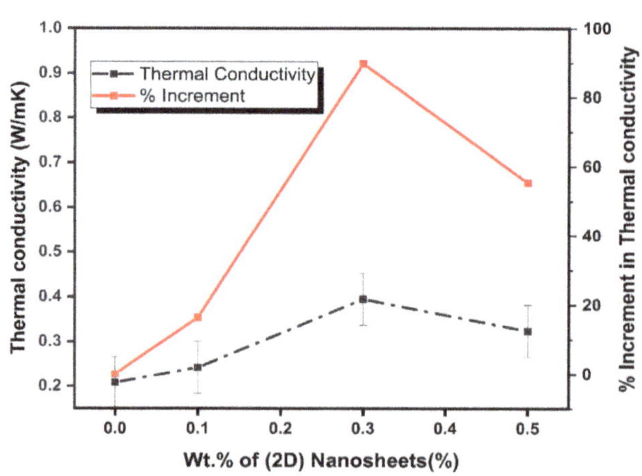

Figure 9. Solid-state thermal conductivity of samples.

3.7. *Infrared Thermal Imaging*

The surface temperature distribution of pure and HNePCMs at various time stamps during the melting process are detailed in Figure 10 (IR thermal images of pristine and HNePCMs). The change in temperature of the samples could be noted from the color pattern, which denotes a lower to higher temperature transition using colors ranging from blue to red. The uniform melting of all samples was noted at a constant temperature. The

HNePCM composite displayed both rapid heat absorption and thermal management when compared with pure A25 PCM under each moment due to dispersed hybrid nanofillers, indicating an incredible temperature and time-regulating ability during the heating process. Figure 10 shows that the HNePCMs logged a higher temperature at every timestamp when compared with pure A25. This indicates that the sample's overall thermal absorption and transfer ability improve with a rise in nanofiller content, and it could be effectively used in building thermal management.

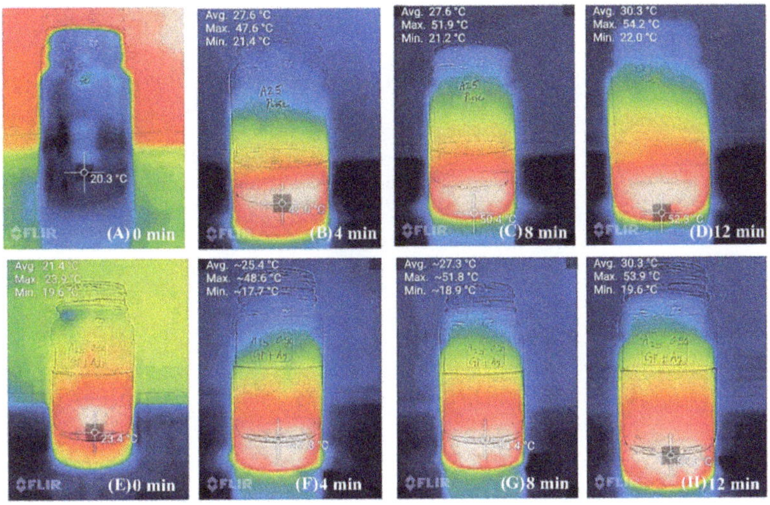

Figure 10. (**A–D**) IR images of pure A25, (**E–H**) IR image of HNePCM.

4. Conclusions

In this experimental work, hybrid nanocomposite phase change materials were produced by dispersing hybrid graphene–silver nanofillers in paraffin wax (A25H) by a two-step method. The primary goal of the experiment was to boost the optical and thermophysical characteristics, particularly thermal conductivity (which is still essential in thermal energy storage applications) for building thermal management. A comparative summary of thermophysical properties of paraffin-based nanocomposites dispersed with different dimensional nanofillers is presented in Table 1. The thermophysical characterization results and its trends clearly state that the synthesized composite is homogenous. To the best of our knowledge, the synthesis of HNePCM loaded with 2-D graphene–silver nanofillers in low-temperature paraffin and its detailed thermophysical characterization remains a novel work. An in depth thermophysical characterization of the novel hybrid nanocomposite at solid state was conducted, and the crucial results are summarized below:

(1) The novel hybrid nano-enhanced composite phase change material logged an increment in enthalpy in comparison with base (6.7%). An increment in enthalpy for nanocomposite is preferable as it can retain more heat, even with the same amount of PCM.

(2) Thermal conductivity was increase by a maximum of 90% with the dispersion of graphene: silver nanofillers (0.3 wt%). Enrichment in thermal conductivity causes a drop in interfacial thermal resistance, and this improves the charging efficiency of HNePCM when used for thermal energy storage.

(3) The pure paraffin had a maximal solar transmissivity of around 87%, and the A25-0.3% sample (0.3% nanofiller loaded) clocked a maximum transmissivity of 0.01%. This result shows that the hybrid nanocomposite could be used in shielding UV rays.

(4) The thermal stability of the composites was also enhanced with the nanofiller infusion. A maximum increment of 5.7 °C was found in THRI results.

(5) Moreover, FT-IR results confirm that the synthesized samples were composites.

The outlined thermophysical characterization results justify an enrichment in thermophysical properties, particularly λ value, optical property, enthalpy, and finally, thermal stability. In addition, the hybrid nanocomposite also had good thermal energy storage capability. The melting and freezing points of the samples lie in the range of 21–27 °C. The fabricated composite could be used for different applications, e.g., the thermal management of PV panels, buildings, etc. This work may provide new insights for fabricating hybrid nanocomposites, presenting broader applications with highly thermally conductive and solar spectrum absorptive capabilities. Finally, with the addition of graphene–silver hybrid nanofillers to low-temperature paraffin, next-generation hybrid composites with improved heat storage and solar absorptivity were synthesized successfully.

Author Contributions: Conceptualization, J.P.; methodology, J.P. and A.K.P.; writing—original draft, J.P. and A.K.P.; funding acquisition, K.K., A.K.P. and M.S.; formal analysis, J.P., A.K.P. and V.V.T.; supervision, K.K., M.S. and A.K.P.; project administration, K.K., M.S. and A.K.P.; writing—review & editing, A.K.P. and V.V.T. All authors have read and agreed to the published version of the manuscript.

Funding: The work is funded by Universiti Malaysia Pahang under RDU213303 and Doctoral Research Scheme. We also received financial aid from Sunway University via Sunway University's Internal Grant Scheme (IGS) 2022 (GRTIN-IGS-RCNMET[S]-15-2022).

Institutional Review Board Statement: Not applicable.

Informed Consent Statement: Not applicable.

Data Availability Statement: The data presented in current work is available within this article.

Conflicts of Interest: The authors declare no conflict of interest.

References

1. Cui, Y.; Zhu, J.; Zhang, F.; Shao, Y.; Xue, Y. Current Status and Future Development of Hybrid PV/T System with PCM Module: 4E (Energy, Exergy, Economic and Environmental) Assessments. *Renew. Sustain. Energy Rev.* **2022**, *158*, 112147. [CrossRef]
2. Du, R.; Wu, Y.; Yang, Y.; Zhai, T.; Zhou, T.; Shang, Q.; Zhu, L.; Shang, C.; Guo, Z. Porosity Engineering of MOF-Based Materials for Electrochemical Energy Storage. *Adv. Energy Mater.* **2021**, *2100154*, 2100154. [CrossRef]
3. Jacob, J.; Pandey, A.K.; Rahim, N.A.; Selvaraj, J.; Samykano, M.; Saidur, R.; Tyagi, V.V. Concentrated Photovoltaic Thermal (CPVT) Systems: Recent Advancements in Clean Energy Applications, Thermal Management and Storage. *J. Energy Storage* **2022**, *45*, 103369. [CrossRef]
4. Ismail, K.A.R.; Lino, F.A.M.; Teggar, M.; Arıcı, M.; Machado, P.L.O.; Alves, T.A.; De Paula, A.C.O.; Benhorma, A. A Comprehensive Review on Phase Change Materials and Applications in Buildings and Components. *ASME Open J. Eng.* **2022**, *1*, 011049 .
5. Cheng, P.; Chen, X.; Gao, H.; Zhang, X.; Tang, Z.; Li, A.; Wang, G. Different Dimensional Nanoadditives for Thermal Conductivity Enhancement of Phase Change Materials: Fundamentals and Applications. *Nano Energy* **2021**, *85*, 105948. [CrossRef]
6. Zhang, N.; Yuan, Y.; Cao, X.; Du, Y.; Zhang, Z.; Gui, Y. Latent Heat Thermal Energy Storage Systems with Solid–Liquid Phase Change Materials: A Review. *Adv. Eng. Mater.* **2018**, *20*, 1–30. [CrossRef]
7. D'Oliveira, E.J.; Pereira, S.C.C.; Groulx, D.; Azimov, U. Thermophysical Properties of Nano-Enhanced Phase Change Materials for Domestic Heating Applications. *J. Energy Storage* **2022**, *46*, 103794. [CrossRef]
8. Magendran, S.S.; Khan, F.S.A.; Mubarak, N.M.; Vaka, M.; Walvekar, R.; Khalid, M.; Abdullah, E.C.; Nizamuddin, S.; Karri, R.R. Synthesis of Organic Phase Change Materials (PCM) for Energy Storage Applications: A Review. *Nano-Struct. Nano-Objects* **2019**, *20*, 100399. [CrossRef]
9. Martín, M.; Villalba, A.; Fernández, A.I.; Barreneche, C. Energy & Buildings Development of New Nano-Enhanced Phase Change Materials (NEPCM) to Improve Energy Efficiency in Buildings: Lab-Scale Characterization. *Energy Build.* **2019**, *192*, 75–83. [CrossRef]
10. Paul, J.; Kadirgama, K.; Samykano, M.; Pandey, A.K.; Tyagi, V.V. A Comprehensive Review on Thermophysical Properties and Solar Thermal Applications of Organic Nano Composite Phase Change Materials. *J. Energy Storage* **2022**, *45*, 103415. [CrossRef]
11. Leong, K.Y.; Abdul Rahman, M.R.; Gurunathan, B.A. Nano-Enhanced Phase Change Materials: A Review of Thermo-Physical Properties, Applications and Challenges. *J. Energy Storage* **2019**, *21*, 18–31. [CrossRef]
12. Chen, L.; Zou, R.; Xia, W.; Liu, Z.; Shang, Y.; Zhu, J.; Wang, Y.; Lin, J.; Xia, D.; Cao, A. Electro- and Photodriven Phase Change Composites Based on Wax-Infiltrated Carbon Nanotube Sponges. *ACS Nano* **2012**, *6*, 10884–10892. [CrossRef]
13. Ma, T.; Liu, Z.; Wen, J.; Gao, Y.; Ren, X.; Chen, H.; Jin, C.; Ma, X.L.; Xu, N.; Cheng, H.M.; et al. Tailoring the Thermal and Electrical Transport Properties of Graphene Films by Grain Size Engineering. *Nat. Commun.* **2017**, *8*, 1–9. [CrossRef] [PubMed]

14. Kargar, F.; Barani, Z.; Salgado, R.; Debnath, B.; Lewis, J.S.; Aytan, E.; Lake, R.K.; Balandin, A.A. Thermal Percolation Threshold and Thermal Properties of Composites with High Loading of Graphene and Boron Nitride Fillers. *ACS Appl. Mater. Interfaces* **2018**, *10*, 37555–37565. [CrossRef] [PubMed]
15. Zheng, R.; Gao, J.; Wang, J.; Chen, G. Reversible Temperature Regulation of Electrical and Thermal Conductivity Using Liquid-Solid Phase Transitions. *Nat. Commun.* **2011**, *2*, 289. [CrossRef]
16. Suresh, C.; Kumar Hotta, T.; Saha, S.K. Phase Change Material Incorporation Techniques in Building Envelopes for Enhancing the Building Thermal Comfort-A Review. *Energy Build.* **2022**, *268*, 112225. [CrossRef]
17. George, M.; Pandey, A.K.; Abd, N.; Tyagi, V.V.; Shahabuddin, S.; Saidur, R. A Novel Polyaniline (PANI)/Para Ffi n Wax Nano Composite Phase Change Material: Superior Transition Heat Storage Capacity, Thermal Conductivity and Thermal Reliability. *Sol. Energy* **2020**, *204*, 448–458. [CrossRef]
18. Habib, N.A.; Ali, A.J.; Chaichan, M.T.; Kareem, M. Carbon Nanotubes/Paraffin Wax Nanocomposite for Improving the Performance of a Solar Air Heating System. *Therm. Sci. Eng. Prog.* **2021**, *23*, 100877. [CrossRef]
19. Sheng, N.; Rao, Z.; Zhu, C.; Habazaki, H. Honeycomb Carbon Fibers Strengthened Composite Phase Change Materials for Superior Thermal Energy Storage. *Appl. Therm. Eng.* **2020**, *164*, 114493. [CrossRef]
20. Joseph, M.; Sajith, V. Graphene Enhanced Paraffin Nanocomposite Based Hybrid Cooling System for Thermal Management of Electronics. *Appl. Therm. Eng.* **2019**, *163*, 114342. [CrossRef]
21. Aslfattahi, N.; Saidur, R.; Arifutzzaman, A.; Sadri, R.; Bimbo, N.; Sabri, M.F.M.; Maughan, P.A.; Bouscarrat, L.; Dawson, R.J.; Said, S.M.; et al. Experimental Investigation of Energy Storage Properties and Thermal Conductivity of a Novel Organic Phase Change Material/MXene as A New Class of Nanocomposites. *J. Energy Storage* **2020**, *27*, 101115. [CrossRef]
22. Paul, J.; Pandey, A.K.; Mishra, Y.N.; Said, Z.; Mishra, Y.K.; Ma, Z.; Jacob, J.; Kadirgama, K.; Samykano, M.; Tyagi, V.V. Nano-Enhanced Organic Form Stable PCMs for Medium Temperature Solar Thermal Energy Harvesting: Recent Progresses, Challenges, and Opportunities. *Renew. Sustain. Energy Rev.* **2022**, *161*, 112321. [CrossRef]
23. Wang, J.; Xie, H.; Xin, Z. Thermal Properties of Paraffin Based Composites Containing Multi-Walled Carbon Nanotubes. *Thermochim. Acta* **2009**, *488*, 39–42. [CrossRef]
24. B, K.; Pandey, A.K.; Shahabuddin, S.; George, M.; Sharma, K.; Samykano, M.; Tyagi, V.V.; Saidur, R. Synthesis and Characterization of Conducting Polyaniline@cobalt-Paraffin Wax Nanocomposite as Nano-Phase Change Material: Enhanced Thermophysical Properties. *Renew. Energy* **2021**, *173*, 1057–1069. [CrossRef]
25. Li, M. A Nano-Graphite/Paraffin Phase Change Material with High Thermal Conductivity. *Appl. Energy* **2013**, *106*, 25–30. [CrossRef]
26. Sun, L.; Diao, R.; Yang, F.; Lin, B. Analysis of the Thermal Performance of the Embedded Composite Phase Change Energy Storage Wall. *ACS Omega* **2020**, *5*, 17005–17021. [CrossRef] [PubMed]
27. Wang, S.; Wang, C.; Ji, X. Towards Understanding the Salt-Intercalation Exfoliation of Graphite into Graphene. *RSC Adv.* **2017**, *7*, 52252–52260. [CrossRef]
28. Ramesh, B.P.; Blau, W.J.; Tyagi, P.K.; Misra, D.S.; Ali, N.; Gracio, J.; Cabral, G.; Titus, E. Thermogravimetric Analysis of Cobalt-Filled Carbon Nanotubes Deposited by Chemical Vapour Deposition. *Thin Solid Films* **2006**, *494*, 128–132. [CrossRef]
29. Cheng, C.; Li, D. Solvated Graphenes: An Emerging Class of Functional Soft Materials. *Adv. Mater.* **2013**, *25*, 13–30. [CrossRef] [PubMed]
30. Jacob, J.; Pandey, A.K.; Rahim, N.A.; Selvaraj, J.; Paul, J.; Samykano, M.; Saidur, R. Quantifying Thermophysical Properties, Characterization, and Thermal Cycle Testing of Nano-Enhanced Organic Eutectic Phase Change Materials for Thermal Energy Storage Applications. *Sol. Energy Mater. Sol. Cells* **2022**, *248*, 112008. [CrossRef]
31. Aslfattahi, N.; Saidur, R.; Arifutzzaman, A.; Abdelrazik, A.S.; Samylingam, L.; Sabri, M.F.M.; Sidik, N.A.C. Improved Thermo-Physical Properties and Energy Efficiency of Hybrid PCM/Graphene-Silver Nanocomposite in a Hybrid CPV/Thermal Solar System. *J. Therm. Anal. Calorim.* **2022**, *147*, 1125–1142. [CrossRef]
32. Mekaddem, N.; Ali, S.B.; Fois, M.; Hannachi, A. Paraffin/Expanded Perlite/Plaster as Thermal Energy Storage Composite. *Energy Procedia* **2019**, *157*, 1118–1129. [CrossRef]
33. Allahyarzadeh, V.; Montazer, M.; Nejad, N.H.; Samadi, N. In Situ Synthesis of Nano Silver on Polyester Using NaOH/Nano TiO 2. *J. Appl. Polym. Sci.* **2013**, *129*, 892–900. [CrossRef]
34. Kim, C.H.; Joo, C.K.; Chun, H.J.; Yoo, B.R.; Noh, D.I.; Shim, Y.B. Instrumental Studies on Silicone Oil Adsorption to the Surface of Intraocular Lenses. *Appl. Surf. Sci.* **2012**, *262*, 146–152. [CrossRef]
35. Mansur, H.S.; Oréfice, R.L.; Mansur, A.A.P. Characterization of Poly(Vinyl Alcohol)/Poly(Ethylene Glycol) Hydrogels and PVA-Derived Hybrids by Small-Angle X-Ray Scattering and FTIR Spectroscopy. *Polymer* **2004**, *45*, 7193–7202. [CrossRef]
36. Gueymard, C.A. The Sun's Total and Spectral Irradiance for Solar Energy Applications and Solar Radiation Models. *Sol. Energy* **2004**, *76*, 423–453. [CrossRef]
37. Wang, J.; Li, Y.; Deng, L.; Wei, N.; Weng, Y.; Dong, S.; Qi, D.; Qiu, J.; Chen, X.; Wu, T. High-Performance Photothermal Conversion of Narrow-Bandgap Ti2O3Nanoparticles. *Adv. Mater.* **2017**, *29*, 1–6. [CrossRef]
38. Yang, Z.; Mao, Z.; Xiang, B.; Zhang, J. Construction of a Binary Channel Efficient Cooling Composites with Reflective and Phase-Change Properties. *Compos. Part B Eng.* **2019**, *178*, 107517. [CrossRef]
39. Shaikh, S.; Lafdi, K.; Hallinan, K. Carbon Nanoadditives to Enhance Latent Energy Storage of Phase Change Materials. *J. Appl. Phys.* **2008**, *103*, 094302. [CrossRef]

40. Arshad, A.; Jabbal, M.; Yan, Y. Preparation and Characteristics Evaluation of Mono and Hybrid Nano-Enhanced Phase Change Materials (NePCMs) for Thermal Management of Microelectronics. *Energy Convers. Manag.* **2020**, *205*, 112444. [CrossRef]
41. Qu, Y.; Wang, S.; Zhou, D.; Tian, Y. Experimental Study on Thermal Conductivity of Paraffin-Based Shape-Stabilized Phase Change Material with Hybrid Carbon Nano-Additives. *Renew. Energy* **2020**, *146*, 2637–2645. [CrossRef]
42. Balandin, A.A.; Ghosh, S.; Bao, W.; Calizo, I.; Teweldebrhan, D.; Miao, F.; Lau, C.N. Superior Thermal Conductivity of Single-Layer Graphene. *Nano Lett.* **2008**, *8*, 902–907. [CrossRef] [PubMed]
43. Shahil, K.M.F.; Balandin, A.A. Graphene-Multilayer Graphene Nanocomposites as Highly Efficient Thermal Interface Materials. *Nano Lett.* **2012**, *12*, 861–867. [CrossRef] [PubMed]
44. Sadasivuni, K.K.; Ponnamma, D.; Kim, J.; Thomas, S. Graphene-Based Polymer Nanocomposites in Electronics. In *Graphene-Based Polymer Nanocomposites in Electronics*; Sadasivuni, K.K., Ponnamma, D., Kim, J., Sabu, T., Eds.; Springer International Publishing: Berlin/Heidelberg, Germany, 2015; pp. 25–45. ISBN 9783319138756.
45. Barkoula, N.M.; Alcock, B.; Cabrera, N.O.; Peijs, T. Flame-Retardancy Properties of Intumescent Ammonium Poly(Phosphate) and Mineral Filler Magnesium Hydroxide in Combination with Graphene. *Polym. Polym. Compos.* **2008**, *16*, 101–113.
46. Baby, T.T.; Ramaprabhu, S. Investigation of Thermal and Electrical Conductivity of Graphene Based Nanofluids. *J. Appl. Phys.* **2010**, *108*, 124308. [CrossRef]
47. Warzoha, R.J.; Fleischer, A.S. Heat Flow at Nanoparticle Interfaces. *Nano Energy* **2014**, *6*, 137–158. [CrossRef]

buildings

MDPI

Article

Hot Water Generation for Domestic Use in Residential Buildings via PCM Integrated U-Tube Based Solar Thermal Collector: A 4-E Analysis

Sudhir Kumar Pathak [1], V. V. Tyagi [1,*], K. Chopra [2], A. K. Pandey [3,4] and Ahmet Sari [5,6]

1 School of Energy Management, Shri Mata Vaishno Devi University, Katra 182320, India; 20dem001@smvdu.ac.in
2 School of Mechanical Engineering, Shri Mata Vaishno Devi University, Katra 182320, India; kapil.chopra@smvdu.ac.in
3 Research Centre for Nano-Materials and Energy Technology (RCNMET), School of Engineering and Technology, Sunway University, No. 5, Jalan Universiti, Bandar Sunway, Petaling Jaya 47500, Selangor Darul Ehsan, Malaysia
4 Center for Transdisciplinary Research (CFTR), Saveetha Institute of Medical and Technical Sciences, Saveetha University, Chennai 602105, India
5 Department of Metallurgical and Material Engineering, Karadeniz Technical University, 61080 Trabzon, Turkey
6 Center of Research Excellence in Renewable Energy (CORERE), King Fahd University of Petroleum & Minerals, Dhahran 4000, Saudi Arabia
* Correspondence: v.tyagi@smvdu.ac.in

Abstract: In recent years, building energy consumption has increased every day due to population growth and an increased human desire for a healthy and pleasant lifestyle, and this is responsible for a crisis of energy shortages worldwide. Therefore, use of solar water heating (SWH) systems in buildings for hot water demand is the prime need of the hour to maintain sustainability. The novelty of this work was in developing a phase change material (stearic acid)-filled U-tube based evacuated tube solar collector (collector A). In addition, another collector B, left without energy storage material, was considered a reference unit for comparing the energy and exergy outputs. The study's main aim was to examine the energy, exergy, enviro- and exergoeconomic analysis of newly developed water heating systems. The findings of study revealed that the maximum daily energy outputs of collector A were found to be 85.86% (simultaneous mode) and 84.27% (midday charging mode) at a high mass flow rate (0.5 LPM), and exergy outputs were 19.41% and 21.35%, respectively, at a low flow rate. The thermal output of collector A was higher than that of collector B. The per liter cost of hot water produced from collector A with PCMs was found to be INR 0.1261 and INR 0.1276, respectively, under both modes, which is less compared with the electric geyser (0.325 INR). The levelized energy cost, net present worth, and the payback time of the developed collector A obtained were 4.61 INR/kWh, INR 49710, and 4.49 years (simultaneous), and 4.67 INR/kWh, INR 48130, and 4.64 years (mid-day charging), respectively. Furthermore, the amount of CO_2 mitigation from the energy and exergy perspective for collector A was found to be 24.30 and 23.76 tCO_2/lifetime and 5.31, 5.58 tCO_2/lifetime, respectively.

Keywords: energy; exergy; economic; enviroeconomic; residential water heating; PCM; SWH

Citation: Pathak, S.K.; Tyagi, V.V.; Chopra, K.; Pandey, A.K.; Sari, A. Hot Water Generation for Domestic Use in Residential Buildings via PCM Integrated U-Tube Based Solar Thermal Collector: A 4-E Analysis. *Buildings* **2023**, *13*, 1212. https://doi.org/10.3390/buildings13051212

Academic Editor: Kian Jon Chua

Received: 28 March 2023
Revised: 17 April 2023
Accepted: 20 April 2023
Published: 4 May 2023

1. Introduction

In recent years, building energy consumption has increased every day due to population growth and an increased human desire for a healthy and pleasant lifestyle, and this is responsible for a crisis of energy shortages worldwide. The building construction segment accounts for nearly one-third of global CO_2 emissions and energy consumption. In addition, nearly one-third of materials are consumed in buildings and produce waste [1,2].

A significant amount of energy consumed in buildings is needed for water heating applications, especially in cold climatic conditions. Nobody now denies the necessity to minimize energy consumption in buildings, seeking better energy output and more integration of renewable sources of energy and, eventually, making our buildings sustainable due to the significance of energy consumption in buildings [3,4]. Therefore, the installation of solar thermal collectors (FPCs and ETCs) that are primarily focused on hot water production applications in buildings is a priority. Their installation can be a successful move toward the goal of net-zero and sustainable buildings (see Figure 1) [5,6].

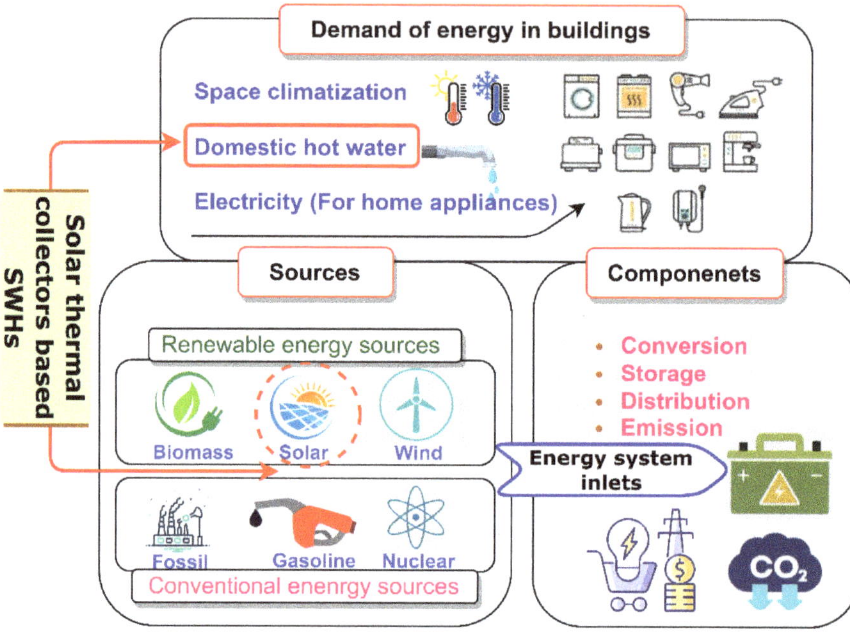

Figure 1. Promotion of solar collectors in the various energy schemes in buildings.

The progressive development of solar thermal collectors (STCs) provides a significant amount of domestic hot water, which varies as per the climatic zones and type of collectors. Flat plate collectors are the most common devices with wide options in design modifications, but vacuum tube collectors have a high thermal performance and temperature range due to vacuum insulation. Therefore, vacuum tube collectors around the world are more preferred or being installed in residential buildings for hot water needs. Generally, residential water heating systems use storage tanks with thermal stratification to improve energy efficiency [7,8].

1.1. Hot Water Consumption in Buildings

The consumption profile for domestic hot water is complex and has changed significantly over time. The most important factors in the different studies were discovered to be the graphical location, weather, occupant number, occupant behavior towards hot water usage, lifestyle, and social and economic conditions. Daily and hourly hot water consumption profiles by households are not similar in all countries [9] Figure 2A shows the consumption of hot water per household and person in different composite climatic conditions (country-wise), and Figure 2B shows the consumption of hot water on an hourly basis per household for different seasons.

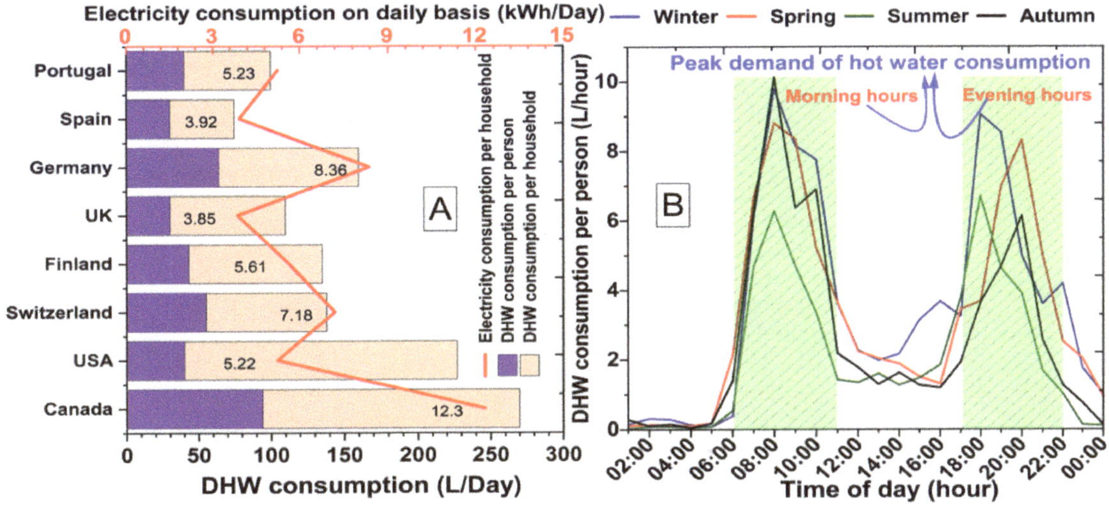

Figure 2. (**A**) Hot water uses per home and person in different countries (**B**) Domestic hot water use pattern per household in different seasons [10].

Researchers are growing concerned about solar water heating (SWH) systems due to several nations' and international agencies' long-term planning of low-carbon energy. Various studies were conducted with the purpose of analyzing the impact of PCM integration with SWH systems instead of using it as a separate storage unit. Chopra et al. [11] evaluated the 4E analysis of the HP-ETC based SWH system for Jammu's diverse climate. The system was planned and constructed for a family of six occupants. The experiments were conducted at six fluid flow rates. The max. average energy and exergy outputs were 72% and 5.2% at 20 LPH, and 55% and 1.25% at 60 LPH, respectively. At 20 and 60 LPH, the max. and min. average outlet temperatures were determined to be 76.4 and 45 °C, respectively. The per liter price of hot water at the specified temperature was determined to be 0.12 INR/L, compared with 0.40 and 0.26 INR/L for electric and gas geysers, respectively. The payback period for the SWH was four years, which was a much shorter time frame. Naghavi et al. [12] evaluated the thermal output of the PCM-filled heat pipe based ETC. They used paraffin wax as the PCM. They found that the proposed system's thermal output fluctuates by 8% as a result of varying environmental circumstances. They also observed that the mass flow rate significantly influenced the overall system performance. In addition, the use of fins with a heat pipe overcame the problem of the PCM's poor conductivity and prevented the heat pipe from overheating. Xue [13] examined the output of a water-in-glass ET-SWH combined with PCMs and found it to be superior to the performance of a traditional type with a comparable collector area. Abokersh et al. [14] applied PCM using a similar technique while additionally adding aluminum fins to the inner and outer surfaces of the glass absorber and U-tube. They observed that, despite the fact that adding an aluminum fin increased heat retention capacity, it reduced efficiency by 14% compared with a system filled with only PCM and no aluminum fin. O'Neil and Sobhansarbandi [15] compared the heating efficiency of a U-tube ETSC to that of a heat pipe based ETSC that is commercially available. The ETSC based on the U-pipe obtained a 31 °C higher peak water temperature than the ETSC based on the heat pipe. It will result in a 13% increase in thermal output in the absence of storage medium. The finned U-tube-based ETSC with PCM yielded water tank peak temperatures of 47 °C and 38.7 °C, while the heat pipe based ETSC yielded peak water temps of 60 °C and 36.7 °C, respectively. The maximum temperatures for fin and tank water for the U-tube based ETSC with nanofluid were 63.9 and 43.3 °C, respectively. Li et al. [16] analyzed the heat produced by a PCM integrated ETC based on U-tubes. U-type ETCs with fin-filled PCMs increase the amount of time hot water is supplied and decrease the maximum exit temperature of the HTF. When the PCM melting temperature reached 323 K, the peak HTF output temperature dropped by 7.4 K, the PCM liquid percentage was high, and the ETC's thermal and storage efficiencies were 50.72% and 19.20%, respectively.

PCM adds an extra 160 min to the delivery duration of hot water over 308 K. Researchers found that adding fins to PCM loaded with ETC increased overall performance. Huang et al. [17] analyzed the heat produced by a novel vacuum tube collector loaded with PCM. Radial metal fins were attached to the U-tube to increase efficiency via high heat transmission. The U-tube's PCM was paraffin. The findings showed that, when PCM was added to the ETC, the water temperature was higher at night than without a PCM system. Not only that, but a higher output flow rate resulted in cooler water. The thermal output of U-tube finned direct flow collectors with and without PCM was compared by Essa et al. [18]. It was the first time a collector used PCM (paraffin wax). The PCM's energy storage advantage was realized at low flow rates, allowing for a full phase variation. When the complete phase changed, the system performed at its best. The 0.25 LPM (0.81 m^2 aperture area) PCM ETC system with integrated energy storage was 21.9% more efficient than a standard collector. Because of a phase shift that occurs in the PCM between its absorbing and releasing modes, this enhancement was possible.

1.2. Research Gaps and Objective of the Study

Latent heat storage is crucial for heating and cooling systems. The main aim of heat storage with solar water heaters is to overcome the issue of fluctuations in the demand and supply of hot water, which is a common issue in solar thermal systems due to the fluctuating nature of solar energy. The main function of phase change materials (PCMs) is to absorb and release thermal energy at a constant temperature (isothermal), which significantly reduces the additional space requirement. As per the literature, it was observed that various modifications in designs and integration of latent heat storage materials with ETSC had been carried out in previous studies. Undoubtedly, integration of energy storage materials offers various merits, such as improving thermal performance and supply of hot water for an extended period. Still, a few gaps have not been addressed in the previous studies.

Nowadays, monetary savings and increasing the energy efficiency of SWH systems used in residential buildings are both clear objectives of researchers, and researchers have discussed the thermal performance of SWH systems only from an energy and exergy perspective. Nobody has discussed the thermal output of stearic acid (PCM) filled with U-tube based ETSCs from enviro- and exergoeconomic perspectives, which are dynamic factors for the growth of energy storage based SWH systems. This work is novel in examining the energy, exergy, enviro- and exergoeconomic analysis of newly designed U-tube based ETSCs with and without PCMs, which are operated under dual modes (simultaneous and midday charging) at three mass flow rates. In addition, economic and environmental analysis makes it clear whether using PCMs with an ETC based SWH system is financially and environmentally viable for commercial utility, because, for any solar thermal system, it is important to analyze sustainability's economic and environmental aspects.

2. Selection and Addition of PCM with Proposed ETC-SWH

A phase change material is a "latent" heat storage substance that melts or solidifies (phase transition) isothermally. By dissolving their chemical bonds endothermically during the solid-to-liquid phase change, PCMs store thermal energy as latent heat, which releases exothermically during cooling back to the solid. It is crucial to evaluate many essential factors to select PCMs to be employed as energy storage materials in SWH systems for water heating applications in buildings [19]. The selection of an appropriate PCM relies on several recommendations. However, PCMs must exhibit thermo–physical, kinetic, chemical, technological, and economic characteristics. These include an adequate melting temperature range, high specific heat, low-density fluctuation during phase transition, chemical stability, nontoxicity, and cost. Previous studies mainly employed paraffin wax as latent heat storage, although its lower thermal conductivity influences the heat transfer phenomenon. The current research uses stearic acid (K = 0.32–0.34 W/m K) as a PCM due to its superior thermophysical properties over paraffin wax. Stearic acid has the following thermophysical properties: melting temperature (T_m = 60 °C), latent heat (191 kJ/kg), density (1.08–1.17 gm/cm^3), and specific heat (2–2.45 kJ/kg K) [20]. In this experimental study, the PCM was filled inside the annular gap of the evacuated tube between the U-tube and the absorber surface to act as a latent heat storage (LHS) medium. The 75% volume (2.3 kg) of the total capacity of the evacuated tube was then immediately filled with liquified PCM using a beaker. Figure 3 shows the design analysis and PCM integration with ETSC.

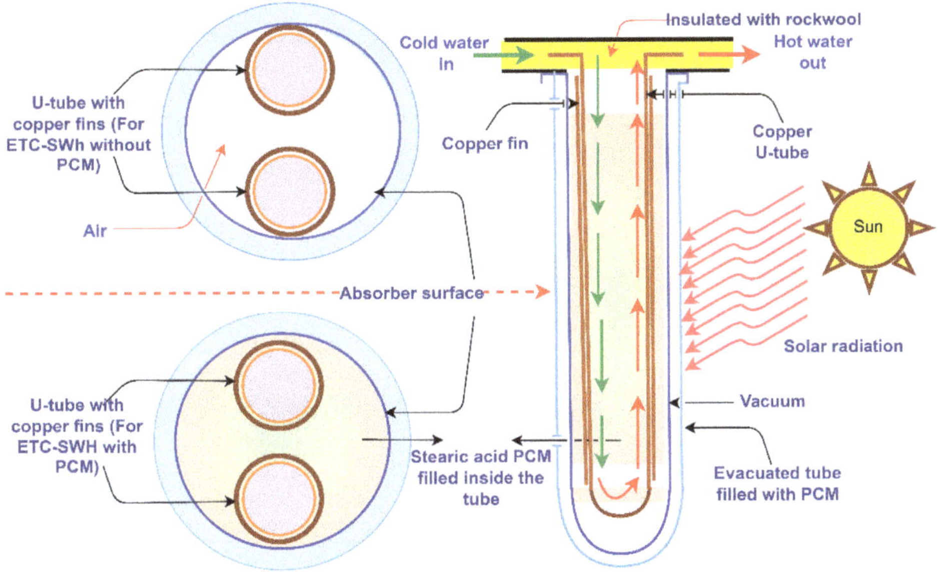

Figure 3. Schematic arrangement of the designed PCM integrated ETSC.

Chopra et al. [21,22] evaluated stearic acid for 0–1500 heat cycles. Spectroscopy and differential scanning calorimetry were used to analyze both samples. DSC analysis was performed at 40–85 °C. Stearic acid was chemically analyzed using Fourier transform infrared spectroscopy. The thermogravimetric analysis determined the heat stability of cycled and uncycled PCM samples. After the following characterization tests, they found that both 0 and 1500 cycled stearic acid samples were thermally degraded at 297 °C and 291 °C, respectively. X-ray diffraction analysis employed Cu (K = 1.5406) powder diffraction meters. DSC analysis of stearic acid samples' 0th and 1500th cycles was completed to evaluate latent heat storage capacity. After 1500 cycles, the stearic acid sample melted at 67.10 °C and froze at 65.74 °C. Stearic acid's melting and freezing latent heat capacities were 228.65 and 233.87 J/g, respectively, after 1500 cycles, compared with 244.21 and 244.14 J/g, respectively, at 0 cycles. After 1500 cycles, stearic acid's melting temperature remained 0.2 °C and its latent heat storage capacity declined to 6.37%. These results reveal that stearic acid provides reliable latent heat energy storage life for solar thermal applications.

3. Experimental Setup and Methodology

This experiment mainly assessed the 4E analysis of a newly designed U-tube based ETSC with and without PCM used for constant hot water supply in buildings. Therefore, the study underwent three tests (Test 1 to Test 3) for composite climatic conditions of Jammu, India. The test setup was installed at the Shri Mata Vaishno Devi University, Katra rooftop at the optimized position (45°) to capture the maximum solar radiation during the daytime. All the data were recorded from 8:00 a.m. to 9:00 p.m. Furthermore, the total amount of hot water produced during the night was measured and stored in the insulated tank.

3.1. Test Setup

Figure 4 depicts a schematic with photographs of the proposed experimental design. The test setup had two collectors, A and B. Collector B was used as a reference collector without phase change material. In contrast, collector A was integrated with PCM and operated under dual modes (simultaneous and midday charging modes). Newly designed collector's thermal performance and other economic/environmental output parameters were compared with collector B. The system comprised two insulated storage tanks for collecting the exit hot water. The inlet water was flowed to both collectors from a large tank (1000 L capacity). Two separate rotameters equipped with each collector were used to determine the mass flow rate of the water. The water was circulated in the system and collected in the storage tanks. Each collector had a 0.49 m^2 heat-collecting surface. A solar power meter measured the global solar radiation that fell on the aperture area. Table 1 displays the specific structural features and materials of the U-tube based ET-SWH.

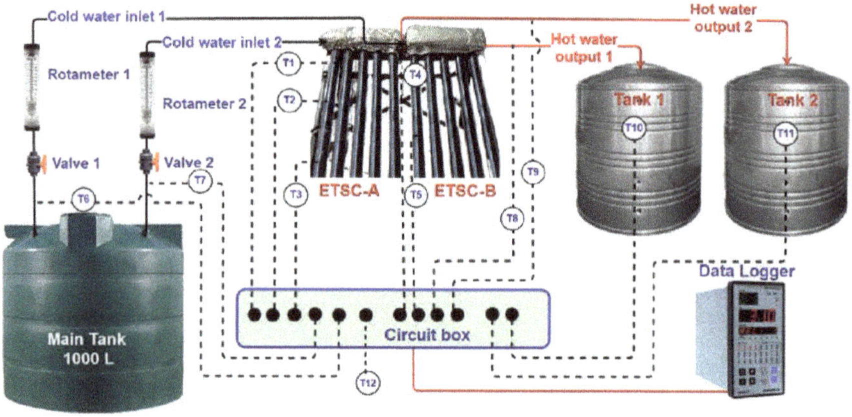

Figure 4. Schematic arrangement of the experimental test ring (collector A and collector B).

Table 1. Detailed specifications of the test setup components.

Component	Properties	Value
Evacuated tube	Outer/inner diameters	58 mm and 47 mm
	Thickness of tube	1800 mm
	Length of tube	1800 m
	Specific heat	980 (J/kg K)
	Thermal conductivity	1.14 (W/m K)
	Transmissivity	0.92
Absorber coating	Thickness of coating	0.1 mm
Three target ALN/SS/Cu	Absorptivity	0.92
	Emissivity	0.08
Heat exchanger	Length of tube	1700 mm
(U-tube of copper)	Outer/inner dia. of tube	8 mm and 7.5 mm
	Fin material	Copper

Table 1. *Cont.*

Component	Properties	Value
	Fin thickness	0.5 mm
	Fin length	1600 mm
	Thermal conductivity	385 W/m K
	Density of copper	8830 kg/m^3
Collector	Aperture area	0.490 m^2

In the test setup, 12 resistance temperature detector (RTD) sensors were used to measure the temperature at defined locations (see Figure 4). Three RTD sensors were installed in an evacuated tube of collector A, and two RTD sensors were placed inside an evacuated tube of collector B to measure the temperature of latent heat storage material and air. The digital data logging device Masibus 85x++ linked with the RTD sensors. Where five-minute difference temperature data were recorded and saved, a digital solar power meter monitored the solar energy from 8:00 a.m. to 6:00 p.m. every 30 min. Rockwool was used for insulating the header portion of the U-tube based ETSC to reduce the heat losses at the top.

3.2. *Research Methodology*

The experimental procedure adopted for this study is discussed in detail in this subsequent section. Before starting the experiment, ensure all the connections are working correctly. The experiments were performed on two U-tube based evacuated tube solar collectors A and B, with three mass flow rates (0.167, 0.34, and 0.5 LPM) under the composite climatic conditions of Jammu, India. The experiment was run from 8:00 a.m. to 9:00 p.m. for all three test days.

Each test was performed for a specific mass flow rate. First, uncover both collectors at 8:00 a.m. sharp; the absorber surface captures the solar energy fall on both collectors. The evacuated glass tube consists of a metal absorber with a selective coating having low thermal conductivity and high absorptivity. The low emissivity of the selective coating reduces the amount of heat lost from the absorber surface by radiation. It allows the absorber to maintain the higher temperature and transfer more heat to the water flowing through the U-tube. In collector A, the PCM filled inside the tubes absorbs a part of the solar energy, and the remaining is transferred to the U-tube for heating the water. In collector B, the total amount of solar energy absorbed by the absorber surface is transferred to heat the working fluid. Collector A was tested under dual operation modes (simultaneous and midday charging modes). In simultaneous mode, the system was operated from early morning, 8:00 a.m., and inlet water was supplied to the collector in parallel for heating. In midday charging mode, collector A with PCM was uncovered from 8:00 a.m. for complete charging of the PCM. The input water was supplied to a collector at 12:00 a.m. for heating. The reference collector B was operated from 8:00 a.m. to 6:00 p.m. The thermal performance and economic/environmental outputs of the reference collector were compared with collector A with PCM. The same procedure was repeated for the other two mass flow rates (see Figure 5).

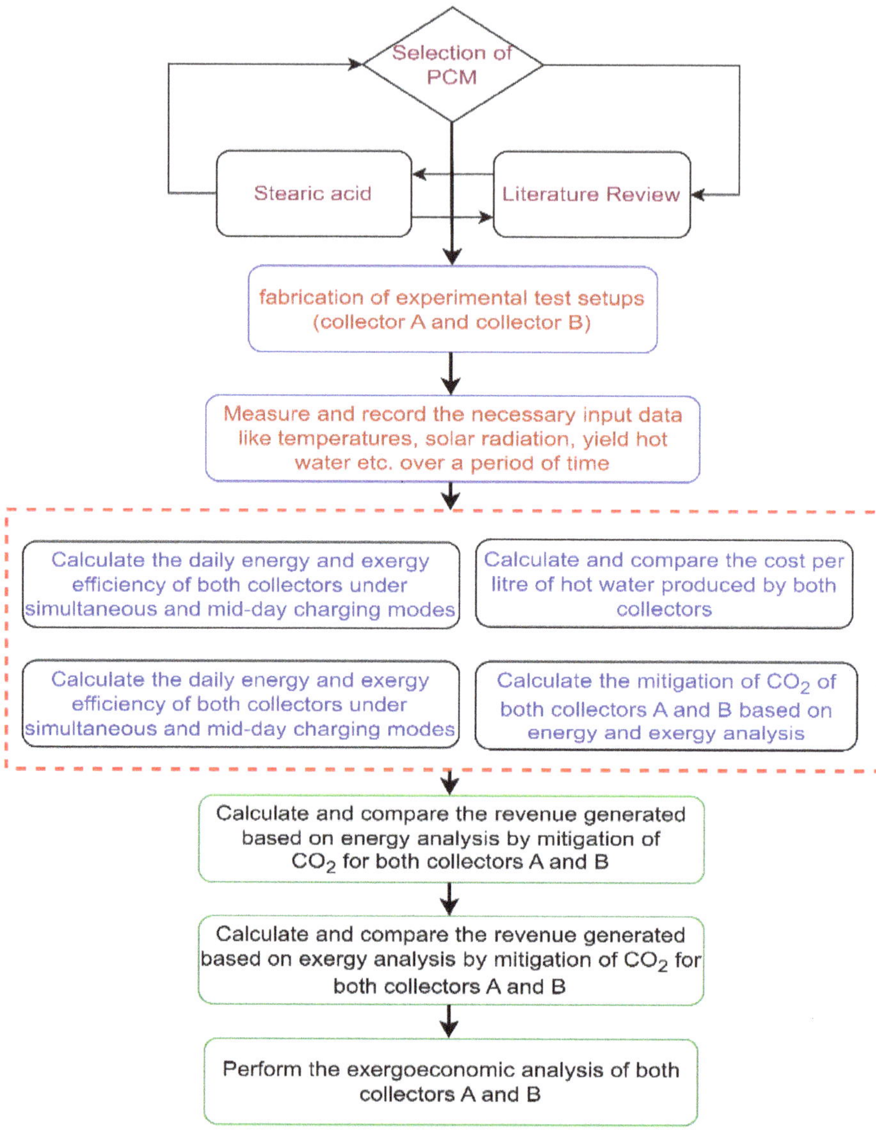

Figure 5. A flow chart of processes involved in the experimental procedure of the current study.

4. Thermodynamic Analysis of U-Tube Based ETSC

4.1. Energy Analysis

The energy study of any solar thermal system is conducted with the help of the 1st law of thermodynamics to analyze its energy efficiency. The irreversibility of the processes was not considered during the energy analysis.

For an open system moving between states 1 and 2, the steady-state flow energy equation is written as

$$\dot{Q} - \dot{W} = \sum \dot{m}_o \left(\psi_o + \frac{V_o^2}{2} + g\lambda_o \right) - \sum \dot{m}_i \left(\psi_i + \frac{V_i^2}{2} + g\lambda_i \right) \tag{1}$$

The SFEE can be simplified by eliminating the change in kinetic and potential energy changes. In addition, there is no work produced by the system [23]:

$$\dot{Q} = \sum \dot{m}_o \psi_o - \sum \dot{m}_i \psi_i \tag{2}$$

Therefore, the useful amount of heat absorbed by the water can be calculated by the below given expression:

$$\dot{Q}_u = \left[\dot{m}_{HTF} \times C_{HTF} \times (T_{o,HTF} - T_{i,HTF})\right] \tag{3}$$

The total useful amount of heat gained by the water for a time strap can be calculated by the equation:

$$Q_u = \sum_{t=0}^{k} \dot{Q}_u \times \Delta\tau = \left[\dot{m}_{HTF} \times C_{HTF} \sum_{t=0}^{k} (T_{o,HTF} - T_{i,HTF}) \times \Delta\tau\right] \tag{4}$$

The instantaneous solar energy fall on the area of the collector can be evaluated as

$$\dot{Q}_i = I_G \times A_{aperture} \tag{5}$$

The total incident solar energy for a full day can be obtained from the given expression:

$$Q_i = \sum_{t=0}^{k} I_G \times A_{aperture} \times \Delta\tau \tag{6}$$

The solar thermal collector's overall energy efficiency is the ratio of total usable heat gained by water to the overall solar energy fall on the collector per day. The daily energy efficiency (DEE) of the collector can be represented as:

$$\eta_{DEE} = \left[\frac{\dot{m}_{HTF} \times C_{HTF}}{A_{aperture}} \times \sum_{t=0}^{k} \frac{(T_{o,HTF} - T_{i,HTF})}{I_G}\right] \tag{7}$$

Both solar collectors, collector A with PCM and collector B without PCM, are operated under the same ecological conditions. Therefore, the energy enhancement ratio (EER) can be calculated as [24]:

$$EER = \left[\frac{Q_{u(A)} - Q_{u(B)}}{Q_{u(A)}}\right] \tag{8}$$

4.2. Exergy Analysis

Exergy analysis is an analytical method based on the second law of thermodynamics that provides an alternative and informative way of logically and meaningfully evaluating and comparing processes and systems.

The inlet exergy ($\dot{E}x_i$) and outlet exergy ($\dot{E}x_o$) of the newly designed solar collectors (A and B) can be calculated as [25]:

$$\dot{E}x_i = \dot{m}_{HTF} \times C_{HTF}\left[T_{i,HTF} - T_{atm} \times \left\{1 + ln\left(\frac{T_{i,HTF}}{T_{atm}}\right)\right\}\right] \tag{9}$$

$$\dot{E}x_o = \dot{m}_{HTF} \times C_{HTF}\left[T_{o,HTF} - T_{atm} \times \left\{1 + ln\left(\frac{T_{o,HTF}}{T_{atm}}\right)\right\}\right] \tag{10}$$

The exergy loss ($\dot{E}x_{loss}$) from the collectors A and B from the absorber surface to atmospheric air can be obtained as [22]:

$$\dot{E}x_{loss} = -U_{loss} \times A_{aperture}\left[(T_{absorber} - T_{atm}) - T_{atm} \times ln\left(\frac{T_{HTF}}{T_{atm}}\right)\right] \tag{11}$$

The amount of incident exergy ($\dot{E}x_{inc}$) can be written as [26]:

$$\dot{E}x_{inc} = I_G \times A_{aperture} \times \left[1 + \frac{1}{3}\left(\frac{T_{atm}}{T_{sun}}\right)^4 - \frac{4}{3}\left(\frac{T_{atm}}{T_{sun}}\right)\right] \quad (12)$$

Therefore, the overall exergy output of the newly developed SWH systems (collector A and collector B) is the ratio of gain of exergy to the total fall exergy input, and it is expressed as:

$$\eta_{DEXE} = \left[\frac{\dot{m}_{HTF} \times C_{HTF} \times \left\{(T_{o,HTF} - T_{i,HTF}) - T_{atm} \times ln\left(\frac{T_{o,HTF}}{T_{i,HTF}}\right)\right\}}{I_G \times A_{aperture} \times \left[1 + \frac{1}{3}\left(\frac{T_{atm}}{T_{sun}}\right)^4 - \frac{4}{3}\left(\frac{T_{atm}}{T_{sun}}\right)\right]}\right] \quad (13)$$

4.3. Numerical Models for the Generation of Energy and Exergy

The energy and exergy generation models were developed by considering the effect of these parameters: energy and exergy efficiency, solar radiation, and degradation in thermal output of newly developed collectors A and B for time. The prediction of energy and exergy from both collectors throughout the year can be obtained by the equations [27]

$$Q_{u,EGY} = \left(365 \times I_G \times A_{aperture} \times \eta_{DEE} \times \Delta\tau\right) \times (1 - d_R)^n \quad (14)$$

$$Q_{u,EXGY} = \left(365 \times I_G \times A_{aperture} \times \eta_{DEXE} \times \Delta\tau\right) \times (1 - d_R)^n \quad (15)$$

4.4. Economic and Environmental Analysis

The existing SWH system faced the main issues of low thermal performance and inability to produce hot water in off-sunshine/nighttime. Therefore, the main aim of this study was to overcome these two issues with existing systems by integrating the thermal energy storage unit. However, although the integration of stearic acid as a PCM with SWH increases the energy efficiency of existing systems, it is not enough to determine their economic viability and commercial applicability. Therefore, it is also required to analyze the economic analysis of the stearic acid (PCM) integrated SWH system. Therefore, a detailed economic analysis of the newly developed SWH system has been conducted, and its results compared with a conventional SWH system of the same configuration.

The levelized cost of water heating (LCWH) is an important parameter of economic study, as it measures collector A and collector B based on their prices. It provides the economic viability of a newly developed SWH system. It shows the cost of hot water production per kg by SWH. The following equations can be used for the economic analysis of SWH systems [28,29]:

$$LCWH = \frac{\sum_{n=0}^{N-1}\left[\frac{C_n}{(1+i)^n}\right]}{\sum_{n=0}^{N-1}\left[\frac{(E_o)_n}{(1+i)^n}\right]} \quad (16)$$

where i is known as the rate of interest, which can be calculated as:

$$i = EQ_R[RR_i + (i_m \times DE_R)] \quad (17)$$

Whereas the total price of solar collectors in a particular year can be calculated by using the given expression

$$C_n = (IE_o)_{n=0} + LC_n + OM_n + T_{an} \quad (18)$$

The initial expenditure on SWH can be calculated by using the following mathematical expression:

$$IE_o = C_o \times (1 - DE_R) \quad (19)$$

In the above-written equation, the loan cost for a particular financial year can be calculated as:

$$LC_n = (C_o)_{n=0} \times DE_R \times \left\{ \frac{1}{N} + \left(\frac{N-n}{N} \right) \times i_m \right\} \tag{20}$$

The net present value (NPV) is the difference between the present value of cash inflows and outflows over a certain lifespan of solar thermal collectors. SWH's profitability is examined using NPV in capital costing and investment planning. The net present value (NPV) of the newly designed collectors A and B can be calculated using the following equations [27]:

$$NPV = \sum_{n=0}^{N-1} \frac{CF_n}{(1+i)^n} - C_0 \tag{21}$$

The profitability index $(Pr)_n$ represents the NPV of the funded project compared with the total initial cost. It indicates the discounted per cent return of the initially invested price, and a number greater than zero means a profit in investment. Mathematically, it can be written as [29,30]:

$$(Pr)_n = \left[\frac{\sum_{n=0}^{N-1} \frac{CF_n}{(1+i)^n} - C_0}{C_0} \right] \tag{22}$$

The payback period represents a return on investment for time after recovering the total expenditure cost on the newly developed SWH system. It is mainly used to guess the economic sustainability of capital assets. Moreover, the payback period of the developed system must be less than the total lifespan of the solar collector. The following mathematical expression calculates the payback period considering the recovery on investment and overall useful thermal energy gain from the developed collector [8]:

$$PB_p = \frac{C_0}{AR_n - C_n} \tag{23}$$

The enviroeconomic analysis based on CO_2 emission has been discussed in this section. The newly developed evacuated tube solar collector working on solar energy, which is clean, reduces CO_2 emissions. The thermal power plant emits almost 980 gCO_2 to generate one unit (kWh) of electricity. The amount of CO_2 emissions increases to 2 kg of Co_2/kWh. Therefore, saving on CO_2 emissions (β_{e,CO_2}) can be calculated as [23]:

$$\beta_{e,CO_2} = \frac{\propto_{CO_2} \times N \times Q_{u,EGY}}{1000} \tag{24}$$

$$\beta_{ex,CO_2} = \frac{\propto_{CO_2} \times N \times Q_{u,EXGY}}{1000} \tag{25}$$

4.5. Experimental Error Analysis

To determine the accuracy of recorded data, the error analysis of measurements is investigated. A substantial amount, in general, cannot be described clearly but depends on parameters that can be assessed directly and are referred to as errors [31]. This study determines how errors related to each computed variable influence the measured amount value. It can be written as:

$$U_z = \sqrt{\sum_i \left(\frac{\delta Y}{\delta X_i} \right)^2 U_{xi}^2} \tag{26}$$

For example, the temperature sensors ±1 °C, solar power meter ±1.5%, and rotameter ±2% of the standard deviation. The error in exergy and energy outputs at different mass flow rates for both collectors A and B is presented in Table 2.

Table 2. Error analysis results in daily energy and exergy outputs of both collectors.

Type of System	Flow Rate (LPM)	Daily Energy Efficiency (%)	Error (%)	Daily Exergy Efficiency (%)	Error (%)
Collector B (reference system)	0.167	59.62 ± 1.91	1.13	18.52 ± 0.0113	1.13
	0.34	69.40 ± 2.18	1.51	16.12 ± 0.0144	1.44
	0.50	75.08 ± 3.01	2.26	14.43 ± 0.0197	1.97
Collector A with PCM (simultaneous mode)	0.167	72.15 ± 1.84	1.32	19.41 ± 0.0122	1.21
	0.34	79.64 ± 2.44	1.94	17.40 ± 0.0153	1.53
	0.50	85.75 ± 2.85	2.45	15.48 ± 0.0193	1.93
Collector A with PCM (midday charging mode)	0.167	70.93 ± 1.51	1.07	21.35 ± 0.0101	1.01
	0.34	77.86 ± 1.93	1.51	17.83 ± 0.0143	1.43
	0.50	84.27 ± 2.42	2.03	16.06 ± 0.0175	1.75

5. Results and Discussion

The thermal output, and economic and environmental results of both collectors A and B operated under dual modes are discussed in this section. The experiments were conducted at three mass flow rates (0.167, 0.34, and 0.50 LPM) for composite climatic conditions of Jammu, India. Each experiment was performed on a separate day.

5.1. Effect of Low Mass Flow Rate (0.167 LPM) on Various Parameters

Figure 6a–d shows the changes in solar radiation, ambient temperature, and inlet and outlet water temperatures in Test 1 for collectors A and B under simultaneous and midday charging modes. Figure 6a shows that solar radiation gradually increased from 8:00 a.m. to the afternoon (12:00 a.m.–1:00 p.m.) with minimum fluctuations. The maximum solar intensity was obtained as 1053 and 1061 W/m^2 for both simultaneous and midday charging modes, respectively, in Test 1. After that, the solar radiation slowly decreased with small fluctuations due to changes in weather conditions. The fluctuation in solar intensity was found to be minimal during both days for Test 1. Therefore, the energy and exergy outputs of the proposed collectors A and B can be easily compared for Test 1 under both modes. The ambient temperature variation was observed to be 25–31 °C and 28–33 °C for both collectors at 0.167 LPM under simultaneous and midday charging modes, respectively. The temperature change with respect to time increased during the day and decreased in the evening hours. The maximum ambient temperature varied between 3 and 5 °C for Test 1 under both modes. From Figure 6b, it was found that the maximum outlet water temperature from collector B without storage reached 73.28 °C at 12:40 p.m. After that, it underway a decrement due to a drop in solar radiation till 6:00 p.m. evening. The average outlet and inlet water temperatures of collector B operated under simultaneous mode were 56.10 °C and 32.6 °C, respectively. The average temperature change between collector B's inlet and outlet water was 23.52 °C.

Moreover, the maximum outlet water temperature from collector A with storage operated under simultaneous mode reached 64 °C at 2:40 p.m. After that it started decreasing (see Figure 6c). Due to the PCM integration, fluctuations in hot water output were minimized, and its supply was extended for few more hours because of thermal energy storage. The average outlet and inlet water temperatures of collector A were 51.30 °C and 32.11 °C, respectively, with a temperature difference of 19.19 °C.

Furthermore, from Figure 6d, it was observed that the maximum outlet water temperature of collector A with storage operated under midday charging mode reached 82.68 °C at 12:15 p.m. because of the complete charging of the PCM filled inside the evacuated tubes of a collector. After that, the outer water temperature of collector A with PCM linearly decreased until 9:00 p.m. The average outlet and inlet water temperatures of collector A operated under midday charging mode were 58.08 °C and 34.96 °C, respectively, with a temperature difference of 23.11 °C. The constant hot water obtained from collector A with storage is mainly utilized in medium-temperature household applications in residential buildings.

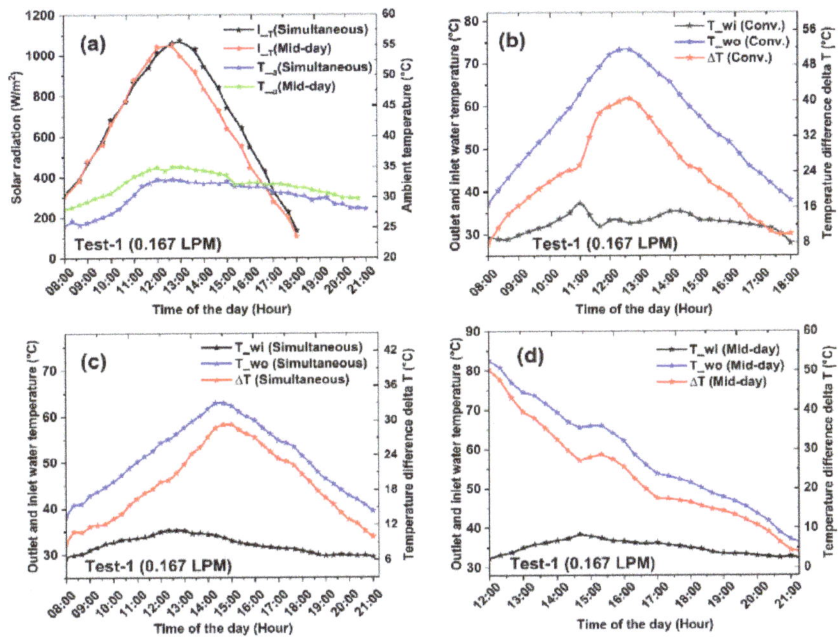

Figure 6. (**a**) Variation of solar radiation and ambient temperature for both modes in Test 1 and variation of outlet and inlet water temperature, (**b**) for collector B without PCM, (**c**) for collector A with PCM under simultaneous mode, and (**d**) for a collector with PCM under midday charging mode.

5.2. Effect of Medium Mass Flow Rate (0.34 LPM) on Various Parameters

Figure 7a–d shows the change in solar radiation, ambient temperature, and outlet and inlet water temperatures of both collector A and B when operated in Test 2 (0.34 LPM) under simultaneous and midday charging modes.

Test 2 was conducted on clear sunny days with minimum solar intensity changes. For both days, the solar intensity rose gradually from 8:00 a.m. to 12:35 a.m. and touched its maximum peaks of 1058 W/m^2 (12:30 p.m.) and 1060 W/m^2 (12:15 p.m.), respectively. After 12:35 p.m., solar intensity steadily decreased to its lowest (6:00 p.m.). In the late noon hours, there were slight variations in solar intensity due to changes in environmental conditions. Figure 7a shows the ambient temperature variation. The ambient temperature was 24–32 °C for simultaneous mode and 27–33 °C for midday charging mode for collector A.

It is observed from Figure 7b that the maximum outlet water temperature obtained was 66 °C at 12:40 p.m. After that, it dropped over time with slight variations. The reference collector's average daily outlet and inlet water temperatures were 52.53 °C and 34.70 °C, respectively, with a temperature change of 18.20 °C. The maximum outlet temperature of collector A with PCM operated under simultaneous mode at 0.34 LPM reached 56.5 °C at 3:10 p.m., and the daily average outlet/inlet water temperatures were recorded as 47.05 °C and 32.60 °C, respectively, with a temperature difference of 14.46 °C (see Figure 7c).

Furthermore, Figure 7d shows that collector A's maximum outlet water temperature with PCM operated under midday charging mode was 75 °C at 12:15 p.m. After that, the temperature of hot water started decreasing linearly till 9:00 p.m. The average daily outlet and inlet water temperatures of collector A with PCM were 50.01 °C and 31.65 °C, respectively.

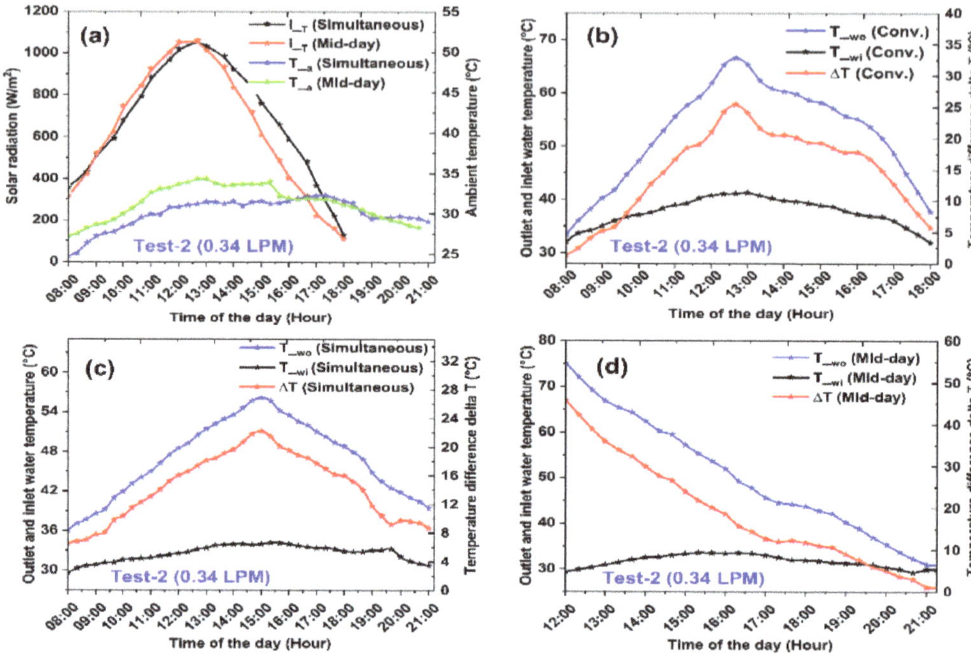

Figure 7. (**a**) Variation of solar intensity and ambient temperature for both collectors under dual modes and change in outlet and inlet water temperatures, (**b**) for collector B without PCM, (**c**) for collector A with PCM under simultaneous mode, and (**d**) for collector A with PCM under midday charging mode for Test 2.

5.3. Effect of High Mass Flow Rate (0.5 LPM) on Various Parameters

The variation of solar intensity, ambient temperature, outlet and inlet water temperatures of both collectors A and B operated under simultaneous and midday charging modes during the increased mass flow rate of water from 0.34 LPM to 0.5 LPM for Test 3 is shown in Figure 8a–d.

Similar to Test 1, Test 3 shows the change in solar intensity and ambient temperature. Solar radiation for both days increased linearly with minor fluctuations and reached maximum values of 1045 W/m^2 at 12:30 p.m. and 1028 W/m^2 at 12:45 p.m., respectively. The ambient temperature was 25–32 °C for simultaneous mode and 27–33 °C for midday charging mode. The variation in solar radiation and ambient temperature was less, which means that Test 3 operated under clear sunny days for both modes. It was ideal for comparing the thermal output of both collectors for a fixed mass flow rate.

At this specific flow rate, collector B without storage attained the maximum temperature of 56 °C (1:10 p.m.). The average outlet and inlet water temperatures of collector B were recorded as 46.93 °C and 35.22 °C, respectively, with a temperature difference of 11.70 °C. Collector B with PCM operated under simultaneous mode attained a maximum temperature of 51.10 °C at 2:20 p.m. (see Figure 8c). Its outlet and inlet water temperatures were recorded as 44.59 °C and 32.18 °C, respectively.

Similarly, Figure 8d shows that the maximum outlet temperature of collector A with PCM operated under midday charging mode was 70 °C at 2:15 p.m. After that, the hot water temperature dropped linearly till late evening. Its average outlet and inlet water temperatures were recorded as 47.64 °C and 33.15 °C, respectively. Thus, it can be revealed that the reference system reached the outlet's maximal temperature before collector A with PCM for simultaneous mode. This was because the PCM filled inside the tubes of collector A required a certain amount of time to charge. The maximum change in water temperature between the outlet and inlet of both collectors was maximum at a low flow rate. This occurred because the contact period was decreased between the U-tube and the flowing fluid. At this flow rate, the temperature difference between the input and exit of collector A with PCM was more than 5 °C until 9:00 p.m. on the same day. Still, the reference system delivered water with this temperature difference until 6:00 p.m. Therefore, collector A with PCM supplied hot water at this temperature differential for about 3 h more than the reference system. The key need for SWH systems used in buildings for daily household usage, i.e., cleaning, bathing and washing, is to ensure a constant hot water supply during overcast and night periods. This is made feasible by installing a larger-than-necessary collector and storing the extra hot water in a storage tank. However, this raises the initial cost as well as the heat loss from the storage tank. Therefore, the PCM integrated U-tube based ETSCs developed in this research provide hot water in rainy/cloudy or off-sunshine periods and improve the system's performance. In contrast, the reference system (collector B) cannot provide hot water at these times.

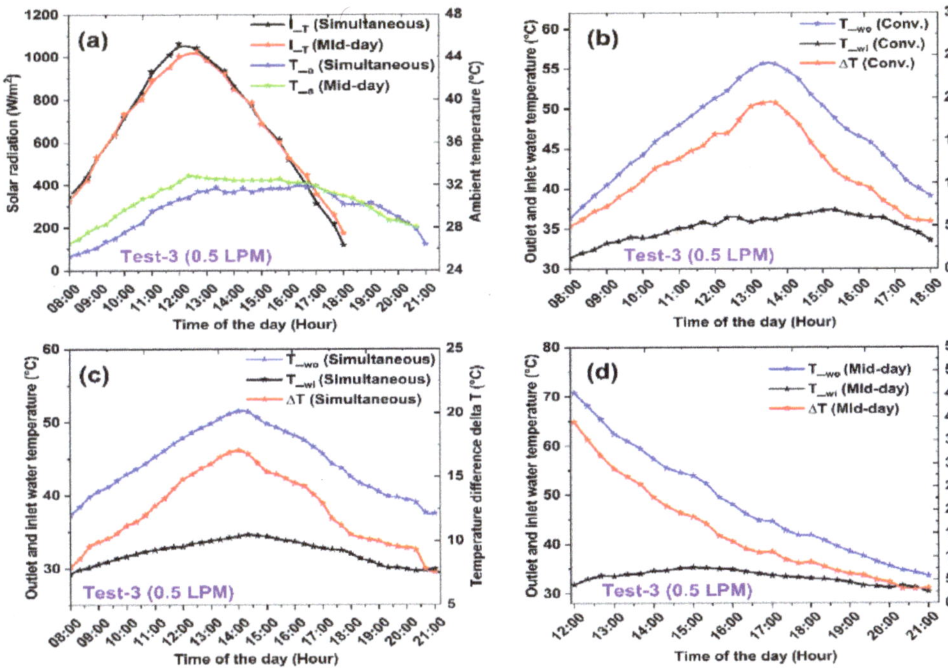

Figure 8. (**a**) Variation of solar intensity and ambient temperature for both simultaneous and midday charging modes in Test 3 and variation of outlet and inlet water temperature, (**b**) for reference collector B without PCM, (**c**) for collector A with PCM under simultaneous mode, and (**d**) for collector A with PCM under midday charging mode for Test 3.

5.4. Effect of Mass Flow Rates on the Temperature of Air and PCM Inside the Tubes of Collectors

Figure 9a–c depicts the fluctuation of internal air temperature and phase change material temperature for collectors A and B operated under simultaneous and midday charging modes at different runs (Test 1 to Test 3). The air temperature inside the tube for collector B from 8:00 a.m. to 12:00 p.m. for three test days under a simultaneous mode of operation varied between 50 °C and 120 °C for all test days at 0.167, 0.34 and 0.50 LPM, respectively.

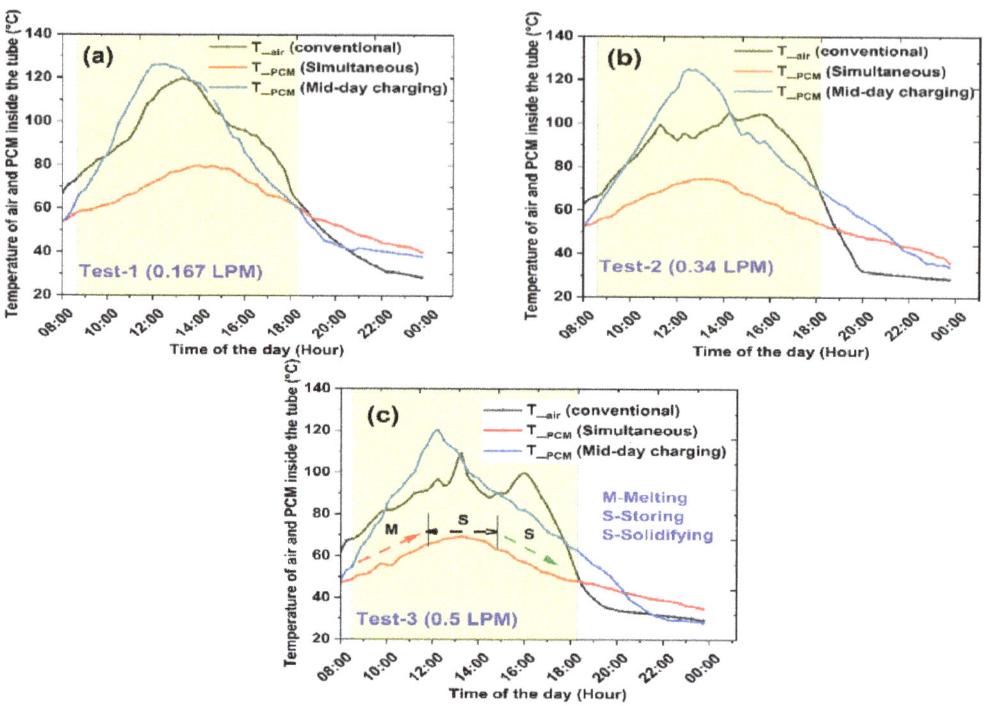

Figure 9. Variation of air and PCM temperature inside the evacuated tubes of both collectors at (**a**) 0.167 LPM, (**b**) 0.34 LPM, and (**c**) 0.5 LPM mass flow rates under simultaneous and midday charging modes.

The average temperature of the PCM loaded within the evacuated tubes of collector-A fluctuated with time for different mass flow rates when operated in simultaneous and midday charging modes. The melting temperature of the chosen PCM (stearic acid) is 57–60 °C [32,33]. From Figure 9a–c, it is observed that the temperature of the PCM inside the U-tube heat exchanger approached the melting point after two hours, from 8:00 a.m., and the PCM begins to melt (charge) from solid to liquid. The thermal energy absorbed by the PCM is affected by the mass flow rate of a working fluid. The charging period of PCM began early in the morning (sunrise) and continued until the temperature of the PCM reached a steady value. The stability period of the PCM temperature was attained between 12:30 p.m. and 03:00 p.m. for all mass flow rates in simultaneous mode. PCM charging was high at the low flow rate compared with medium and high flow rates. During the steady PCM temperature phase, the working fluid directly absorbed the quantity of solar energy. It was discovered that the discharge process for all flow rates began after the stable phase and lasted until the complete heat absorbed by the PCM was released. Similar results were reported by Mellouli et al. [34].

Figure 9a–c illustrates the average PCM temperature within the tubes of collector A with PCM at varied water mass flow rates in midday charging mode. In this mode, the PCM within the tubes of collector A absorbed all solar energy hitting the surface area. From 8:00 a.m. until 12:00 a.m., the PCM was charged entirely; after that, the cold water flowed via collector A to operate under midday charging mode. The average PCM temperature inside the tube of collector A reached 131 °C, 125 °C, and 119 °C between 11:45 a.m. and 12:00 a.m. at respective mass flow rates of 0.167, 0.34, and 0.50 LPM, respectively. Midday charging mode increased collector A heat transfer losses. However, all flow rates had higher outlet water temperatures than simultaneous mode had.

5.5. Effect of Mass Flow Rates on the Temperature of Stored Water in Both Collectors

Figure 10a–c shows stored water temperature fluctuation over time of collectors A and B operated under dual modes. Both collectors A and B's outlet hot water was kept in two well-insulated tanks to reduce nighttime heat loss. The researchers employed 50 mm rock wool to insulate both tanks. The collector B's stored water temperature fluctuation over time for all three mass flow rates was 35–60 °C. The stored water temperature fluctuated during the day due to changes in the outlet hot water. After 6:00 p.m., the stored water temperature of collector B stabilized and declined due to tank thermal energy loss. At 6:00 a.m., after 22 h of operation, collector B's insulated tank water temperature was 49.7 °C, 45.7 °C, and 41.5 °C for 0.167, 0.34, and 0.50 LPM, respectively. The water temperature profiles in storage tanks varied between 38 and 53 °C with regard to time for collector A operating in simultaneous mode on all test days (Tests 1–3). At a low flow rate (0.167 LPM), the stored water temperature was the highest compared with other flow rates, but heat loss from the storage tank to the environment was also considerable. On each day of the tests, the water temperature in the storage tank grew linearly, peaked, and then decreased until the following day at 6:00 a.m. Water stored in collector A's insulated tank (at 6:00 a.m., 22 h after the start of operation) was 46 °C, 44 °C, and 41.7 °C for the respective mass flow rates during the simultaneous mode of operation.

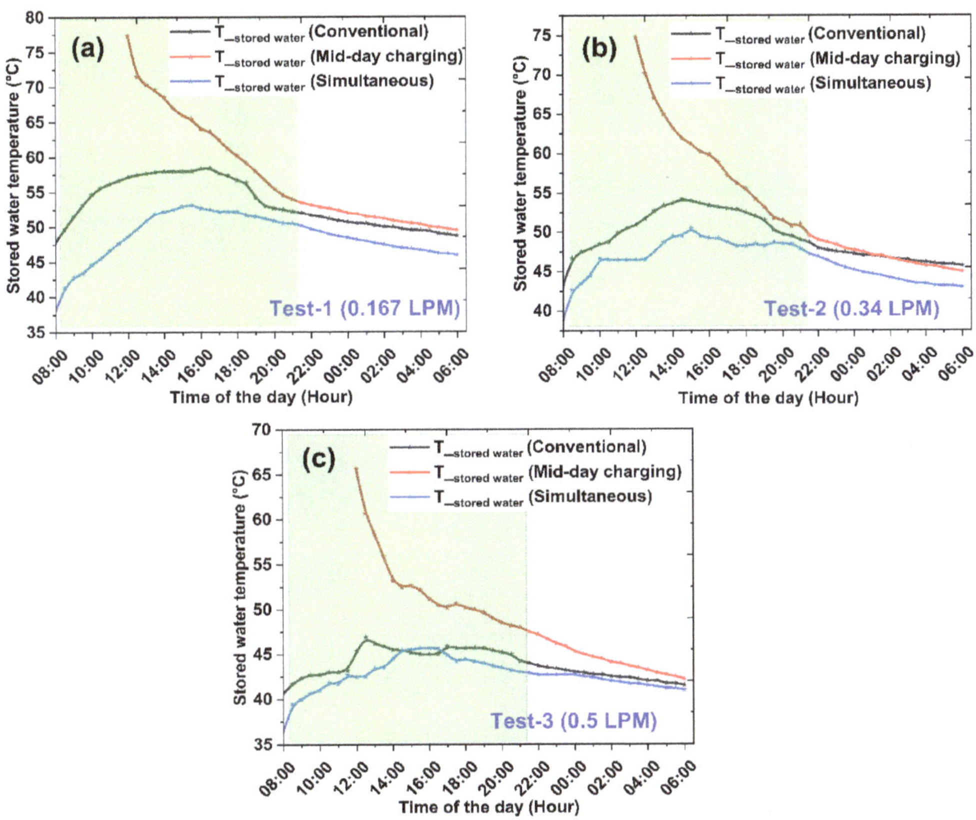

Figure 10. Variation of stored water temperature into the insulated tank for both collectors at (**a**) 0.167 LPM, (**b**) 0.34 LPM, and (**c**) 0.5 LPM mass flow rates under simultaneous and midday charging modes.

In midday charging mode, collector A with PCM operation (flow of water) began at 12:00 a.m. and continued until PCM heat was released. The water temperature was highest at the beginning of the process. The temporal variation of the stored water temperature profiles was insignificant on all test days. At 6:00 a.m., 18 h after the operation, the water temperature was 49.5 °C, 45 °C, and 42.1 °C for mass flow rates at 0.167, 0.34, and 0.50 LPM, respectively. On all test days, it was noticed that the stored water temperature for both collectors exceeded 40 °C, making it suitable for domestic hot water applications.

5.6. Effect of Mass Flow Rates on Average Daily Energy and Exergy Output of Both Collectors

Figure 11a–c reveals the daily energy and exergy outputs of both collectors A and B operated under simultaneous and midday charging modes for different test runs (Test 1–Test 3). For collector B without PCM, it was found that both useful and incident energy quantities were 6.97, 8.81, and 9.31 MJ, and 11.69, 12.68, and 12.40 MJ for respective mass flow rates of 0.167, 0.34, and 0.50 LPM. In addition, collector B's daily energy and exergy outputs were 59.62%, 69.40%, and 75.08%, and 18.52%, 16.12%, and 14.43%, respectively. Table 3 shows the thermal performance parameters of both collectors A and B at different flow rates under simultaneous and midday charging modes.

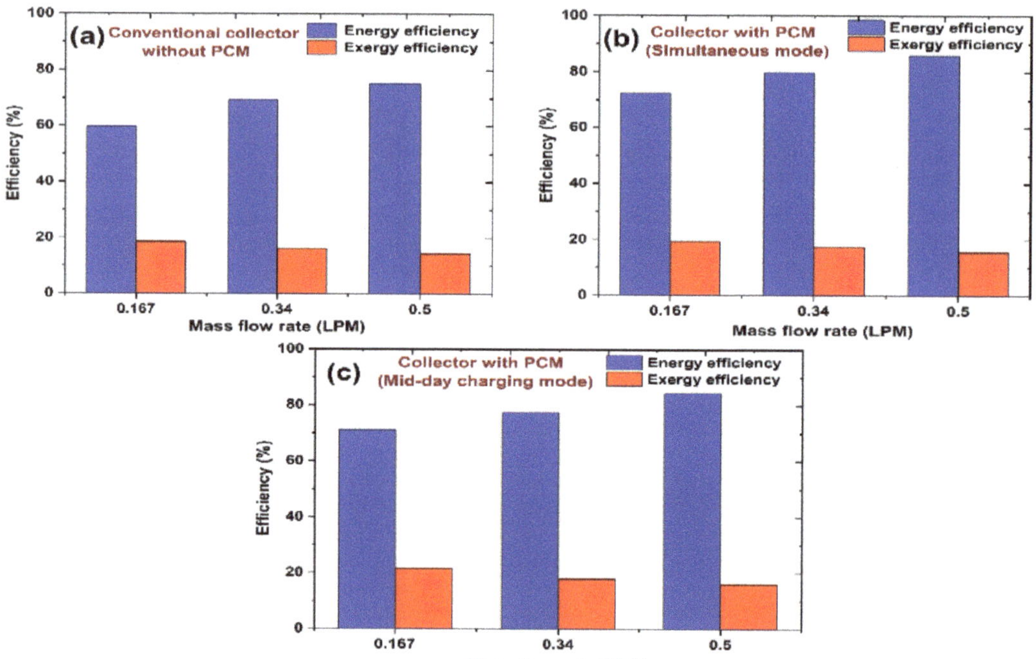

Figure 11. Average daily energy and exergy outputs (**a**) for collector B without PCM under simultaneous mode and (**b**,**c**) for collector A with PCM under simultaneous and midday charging modes.

Collector A with PCM operated under simultaneous mode had the useful energy outputs of 8.44, 10.10, and 10.65 MJ for respective mass flow rates. The input energy valuesremained the same as for conventional collector B. The daily energy and exergy outputs of collector A with PCM operated under simultaneous mode were 72.15%, 79.64%, and 85.86%, and 19.41%, 17.40%, and 15.48% at 0.167, 0.34, and 0.50 LPM, respectively. Similarly, collector A with PCM was operated under midday charging mode and then used, and incident energy outputs were 8.82, 9.31, and 10.39 MJ, and 12.42, 12.04, and 12.33 MJ for their respective mass flow rates. The daily energy and exergy outputs of collector A with PCM under midday charging mode were 70.97%, 77.35%, and 84.27%, and 21.35%, 17.83%, and 16.06%, respectively. Therefore, from the results, it was summarized that the PCM-filled collector A operated under simultaneous mode produces more useful energy for all test days compared with the midday charging mode and collector B (without PCM). It generally happens because, in collector A operated under midday charging mode, the thermal losses are more and system operating hours are less than the simultaneous mode. The results show that the energy and exergy efficiencies enhancement ratio improve collector A's energy gain over collector B. Consequently, regardless of the flow rates used, collector A with PCM operated under simultaneous mode is more efficient regarding usable energy and exergy outputs than the reference system. In addition, collector A, with PCM outputs of useable energy and exergy, continued long after the sun had set.

Table 3. Thermal output parameters of both collectors at different flow rates under simultaneous and midday charging modes.

Type of the Collector	Mass Flow Rate (LPM)	Daily Incident Energy (MJ)	Daily Useful Energy (MJ)	Daily Energy Efficiency (%)	Daily Exergy Efficiency (%)
Collector B without PCM (reference collector)	0.167	11.69	6.97	59.62	18.52
	0.34	12.68	8.81	69.4	16.12
	0.5	12.40	9.31	75.08	14.43
Collector A with PCM (simultaneous mode)	0.167	11.69	8.44	72.15	19.41
	0.34	12.68	10.10	79.64	17.4
	0.5	12.40	10.65	85.56	15.48
Collector A with PCM (midday charging mode)	0.167	12.42	8.82	70.97	21.35
	0.34	12.04	9.31	77.35	17.83
	0.5	12.33	10.39	84.27	16.06

5.7. *Effect on Economic, Enviro- and Exergoeconomic Outputs with and without Use of PCM*

In this section, the economic, enviro- and exergoeconomic analysis results are discussed in detail. The current study's economic analysis was conducted using the input parameters based on India's market economic survey (see Table 4). This section evaluates the annual cost of hot water production, cost per liter of hot water, and payback time of both collectors operated under dual modes at different flow rates. In addition, economic factors determined for the electric geyser (EG) are compared with these values. The proposed systems are anticipated to be in operation for 15 years, and the collector's useable energy production will decrease by 1% each year.

Table 4. Technical input parameters as per Indian market survey used in economic analysis.

Input Economic Parameters	Value
Total aperture area (m^2)	0.49
Capital cost of collector per unit collector area (Rs/m^2-aperture area)	32,000
Total initial cost of solar collectors (Rs)	15,810.6
Debt–equity ratio	90%
Nominal interest rate	8%
Debt term (years)	15
Percentage of O&M cost	1.00%
Electricity price in first year (Rs/kWh)	6.50
Inflation rate in electricity	5.0%
Inflation rate	4.5%
Rate of discount	5.0%
Rate of degradation	1.00%

Figure 12a depicts the variation of bank loan cost, operation and maintenance cost, auxiliary energy, and ETC-SWH annual cost over the years for collector B. It has been found that the loan cost decreases for years because of decreasing interest amounts. It was calculated that the interest amount would be highest in the first year (INR 1871) and lowest in the fifteenth year (INR 882). Notably, 90% of the total investment is taken from the bank with an interest rate and other terms and conditions. In comparison, the investor invests the remaining amount. The O&M cost increases with the number of years because this is the function of the inflation rate [35]. It has been found that approximately INR 3057 will be required for the O&M of the proposed SWH system after 15 years. In comparison, the auxiliary energy cost is the fuel cost for running the SWH system during cloudy/rainy days. This cost also increases with the years due to inflation in electricity prices, which is assumed to be 5%, as depicted in Table 4.

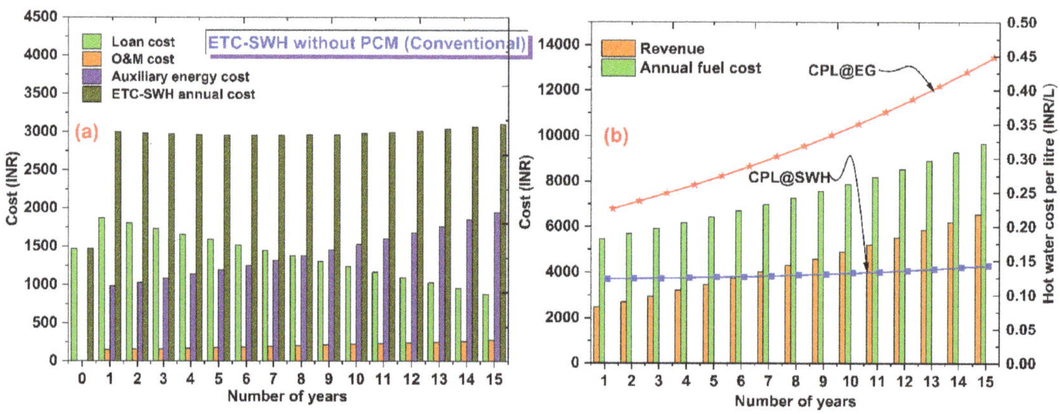

Figure 12. (**a**) Variation of the loan cost, O&M cost, auxiliary energy and ETC-SWH annual cost, and (**b**) CPL of EG and collector B with generated revenue and annual fuel cost.

Thus, due to increasing O&M and auxiliary energy costs, SWH annual costs increase with the passage of years. Approximately INR 45,999 must be paid after 15 years of working on the proposed conventional SWH system to accomplish the heat requirement. From Figure 12b, it is noticed that the revenue is the advantage gained from implementing the proposed conventional SWH system above EG. The income is growing over time due to the continually rising yearly fuel cost of electricity. It has been discovered that, by using a suggested conventional SWH system instead of an EG, over INR 65,720 can be saved after 15 years of operation. The cost per liter (CPL) of both the water heating systems collector B and EG increases over time due to increased electricity costs. The CPL for collector B was found to be 0.1314 INR/L on average and 0.325 INR/L for EG. According to the input economic parameters assessed, the levelized energy cost (LEC), net present worth (NPW), and payback time (PBT) were determined to be 4.82 INR/kWh, 41,290 INR, and 5.05 years, respectively, for collector B without PCM. Similar to conventional collector B, the variation of the loan cost, O&M cost, auxiliary energy, and ETC-SWH annual cost of collector A with PCM operated under simultaneous and midday charging modes are presented in Figures 13a and 14a. It was calculated that the interest amount would be highest in the first year (INR 2010 for simultaneous mode and INR 2011 for midday charging mode) and lowest in the fifteenth year (INR 948 for both modes). It has been found that approximately INR 3284 and INR 3286 will be required for the O&M of collector A with PCM operated under simultaneous and midday charging modes after 15 years. The annual costs of collector A with PCM increase over years. INR 51,230 and INR 50,683 must be paid after 15 years of working on the proposed system under simultaneous and midday charging modes to accomplish the heat requirement.

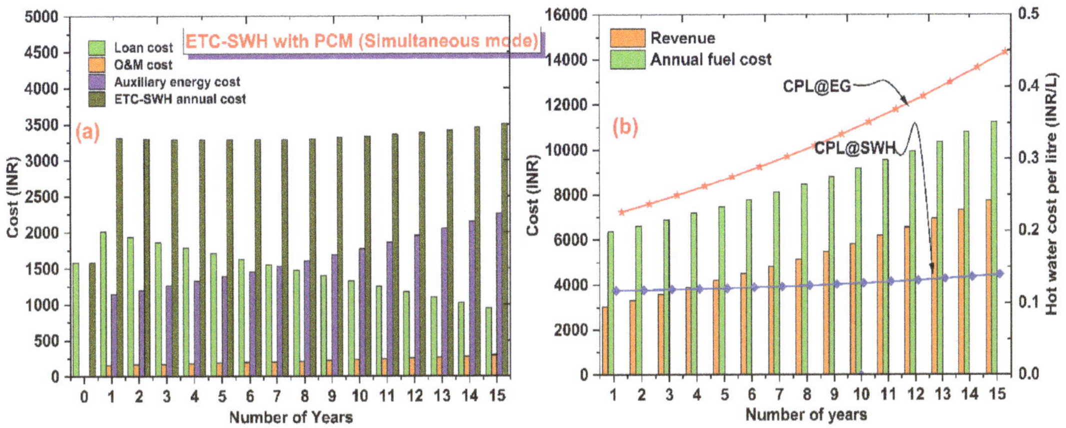

Figure 13. (**a**) Variation of loan cost, O&M cost, auxiliary energy, and ETC-SWH annual cost, and (**b**) CPL of EG and collector A operated under simultaneous mode with generated revenue and annual fuel cost.

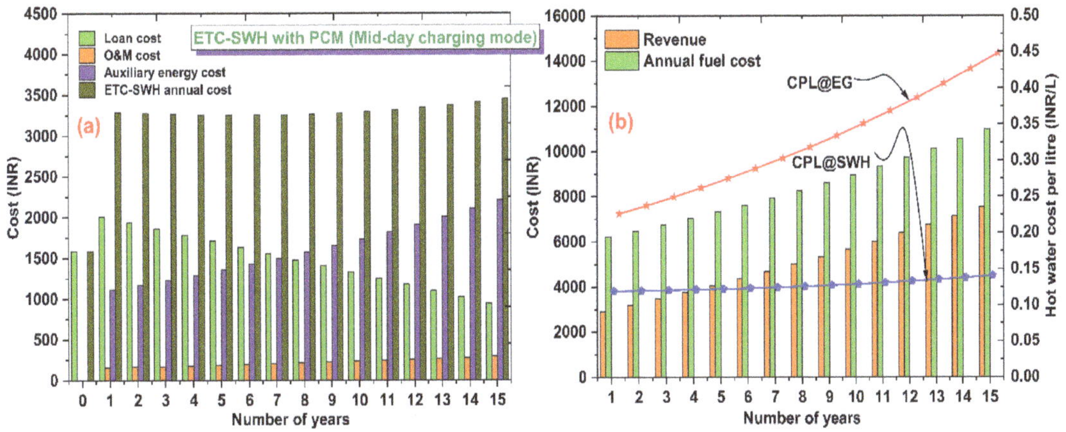

Figure 14. (**a**) Variation of loan cost, O&M cost, auxiliary energy, and ETC-SWH annual cost, and (**b**) CPL of EG and collector A operated under midday charging mode with generated revenue and annual fuel cost.

Figures 13b and 14b show that collector A's revenue with PCM was INR 78,615 and INR 76,252 when operated under simultaneous and midday charging modes, respectively. The cost per liter of hot water generation by the PCM integrated collector A was INR 0.1261 and INR 0.1276 under dual modes, whereas the CPL of EG was found to be INR 0.325. The levelized energy cost (LEC), net present value (NPW), and the payback time (PBT) for collector A with PCM under simultaneous and midday charging modes were calculated to be 4.61 and 4.67 INR/kWh, INR 49,710 and INR 48,130 and 4.49 and 4.64 years, and INR 41,290, respectively, based on the input economic factors evaluated. The values of these criteria indicate that deploying the suggested solar collector system is highly profitable economically.

Figure 15 depicts the quantity of CO_2 mitigation by the proposed SWH system for electric geysers to reduce carbon dioxide emissions into the atmosphere for various tests. The amount of CO_2 mitigation by the developed SWH system was calculated to be 20.88, 24.30, and 23.76 tCO_2/lifetime for a conventional collector, collector A with PCM (simultaneous mode), and collector A with PCM (midday charging mode), respectively, based on useable energy output. Conversely, it was determined that the suggested system mitigates 4.95, 5.31, and 5.58 tCO_2/lifetime for a conventional collector, collector A with PCM (simultaneous mode), and collector A with PCM (midday charging mode), respectively, based on useable exergy output. These calculations demonstrate that the suggested system is advantageous in terms of energy and exergetic efficiency and economics, and reduces CO_2 emissions into the atmosphere.

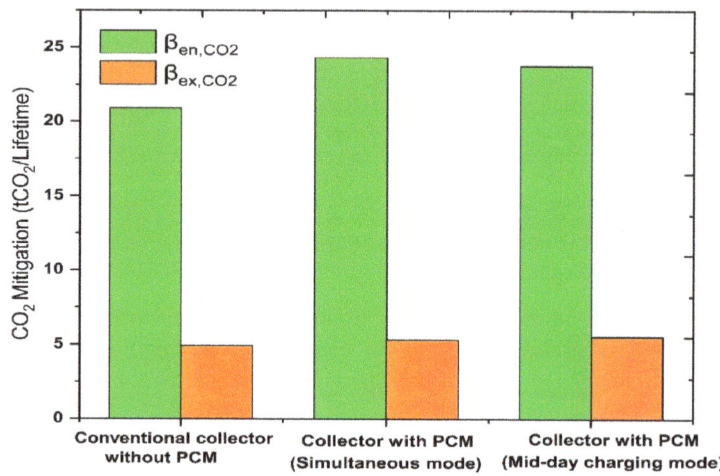

Figure 15. CO_2 mitigation from energetic and exergetic outputs of both collectors A and B.

The proposed SWH system's energoeconomic, exergoeconomic, and economic characteristics are calculated and compared with EG and presented in Tables 5–7 for conventional collector, and collector A with PCM operated under simultaneous and midday charging modes, and also shows the impact of operating life and bank interest rates on these variables. These numbers have been derived from typical daily hot water outputs, usable energy, and exergy gain. It has been found that the uniform yearly cost rises with an increase in the bank interest rate; hence the CPL value of the suggested system grows for a given lifespan. However, the change in interest rate has zero impact on the CPL price of electric geysers. There is also a rise in the average yearly cost and CPL value of the developed SWH systems and EG as the expected lifetime of the system is extended. The energoeconomic and exergoeconomic benefits decline as the standard cost per year rises. From an energoeconomic, exergoeconomic, and economic vantage point, it can be said that the suggested arrangement is preferable in residential water heating applications [36].

Table 5. Energoeconomic and exergoeconomic variations with bank interest rate and life of the conventional collector without PCM.

N (Years)	Interest Rate	Uniform Annual Cost (INR)		Energoeconomic Factor Based on Energy (kJ/INR)		Exergoeconomic Factor Based on Energy (kJ/INR)	
		Proposed ETC	EG	Proposed ETC	EG	Proposed ETC	EG
10	6	3097	6604	809.81	379.76	192.28	90.17
	7	3156		794.67		188.68	
	8	3216		779.85		185.16	
12	6	2959	6899	847.58	363.53	201.25	86.31
	7	3019		830.73		197.25	
	8	3080		814.28		193.34	

Table 5. *Cont.*

N (Years)	Interest Rate	Uniform Annual Cost (INR)		Energoeconomic Factor Based on Energy (kJ/INR)		Exergoeconomic Factor Based on Energy (kJ/INR)	
		Proposed ETC	EG	Proposed ETC	EG	Proposed ETC	EG
15	6 7 8	2869 2930 2992	7373	874.17 855.97 838.23	340.16	207.56 203.24 199.03	80.76

Table 6. Energoeconomic, exergoeconomic, and economic variation with bank interest rate and life of the collector with PCM (simultaneous mode).

N (Years)	Interest Rate	Uniform Annual Cost (INR)		Energoeconomic Factor Based on Energy (kJ/INR)		Exergoeconomic Factor Based on Energy (kJ/INR)	
		Proposed ETC	EG	Proposed ETC	EG	Proposed ETC	EG
10	6 7 8	3437 3501 3565	7687	849.28 833.76 818.79	379.73	184.78 181.40 178.14	82.62
12	6 7 8	3294 3360 3426	8030	886.15 868.75 852.01	363.51	192.80 189.01 185.37	79.09
15	6 7 8	3208 3275 3341	8582	909.91 891.29 873.69	340.13	197.97 193.92 190.09	74.00

Table 7. Energoeconomic and exergoeconomic variation with bank interest rate and life of the collector with PCM (midday charging mode).

N (Years)	Interest Rate	Uniform Annual Cost (INR)		Energoeconomic Factor Based on Energy (kJ/INR)		Exergoeconomic Factor Based on Energy (kJ/INR)	
		Proposed ETC	EG	Proposed ETC	EG	Proposed ETC	EG
10	6 7 8	3405 3470 3534	7512	837.88 822.19 807.30	379.79	196.38 192.70 189.21	89.01
12	6 7 8	3261 3327 3392	7848	874.88 857.52 841.09	363.53	205.05 200.99 197.14	85.20
15	6 7 8	3172 3238 3304	8388	899.43 881.09 863.49	340.12	210.81 206.51 202.39	79.72

6. Conclusions and Future Recommendations

6.1. Conclusions

This experimental work performed a comprehensive 4E analysis of newly designed U-tube based ETSCs with and without PCM at three flow rates under the composite climate of Jammu, India. In this study, stearic acid was selected as the PCM from deep literature analysis and filled inside the evacuated tubes of collector A, while collector B was left empty without PCM, and was known as the reference collector. Integrating PCM with a collector eliminates the need for additional space/arrangement for energy storage material and provides constant hot water during cloudy or nighttime conditions with increased thermal performance. It will also reduce the carbon footprint compared with EG. The conclusions have been made from the experimental investigation:

- Integrating copper fin with U-tube of collector A would increase the heat transfer rate between the U-tube and phase change material;
- The energy and exergy outputs of collector A, operated under dual modes, were higher than collector B;
- The maximum energy efficiency of collector A was 85.86% (simultaneous mode) and 84.27% (midday charging mode), whereas for collector B it was 75.08% at 0.5 LPM. In contrast, the exergy efficiencies were found to be 19.41% and 21.35% for collector A, and 18.52% for collector B at a low flow rate (0.167 LPM) for the same modes;
- The energy enhancement ratio of collector A was obtained in the 14.35–21.01% range compared with collector B. In contrast, the exergy enhancement ratio of collector A was obtained in the range of 7.2–15.25% compared with collector B;
- The CPL of collector A was found to be INR 0.1261 for simultaneous mode and INR 0.1276 for midday charging mode, which is much less than the CPL of EG (INR 0.325);
- The levelized energy cost, net present worth, and the payback time of collector A were found to be 4.61 INR/kWh, INR 49710, and 4.49 years (S), and 4.67 INR/kWh, INR 48130, and 4.64 years (M), respectively;
- The amount of CO_2 mitigation from the energy and exergy perspectives for collector A was 24.30 for simultaneous mode, 23.76 tCO_2/lifetime for midday charging mode, and 5.31 and 5.58 tCO_2/lifetime for the same modes, respectively.

6.2. Future Recommendations

Optimization and experimental investigations using nano-enhanced PCMs are advised for future system improvement. To develop more effective finned storage systems, scientists will need to consider the PCMs' cycle performance, long-term stability, and overall environmental impact. Additionally, the heat transmission model of PCMs should be studied in the future with various PCM kinds, porous materials, PCM filling ratio, and seasonal changes taken into account.

Author Contributions: S.K.P.: writing—original draft, investigation, methodology, acquisition of data, review and editing, V.V.T.: writing, reviewing and editing, supervision, K.C.: methodology, writing, review and editing, A.K.P.: formal analysis, reviewing and editing, A.S.: writing- reviewing and editing. All authors have read and agreed to the published version of the manuscript.

Funding: This work was supported by Shri Mata Vaishno Devi University, Katra (J&K) through the Research Project with Unique ID as RP-141 (Japan International Cooperation Agency Project for AUN/SEED-NET and University of Malaya Malaysia).

Data Availability Statement: Data should be made available on request.

Conflicts of Interest: The authors declare that they have no known competing financial interests or personal relationships that could have appeared to influence the work reported in this paper.

Nomenclature

$A_{aperture}$—Aperture area of the collector (m^2)
C_{HTF}—Specific heat of working fluid (J/kg·K)
CF_n—Cash flow cost (INR)
C_o—Initial cost (INR)
DE_R—Debt–equity ratio
d_R—Depreciation rate (%)
Ex_i—Exergy at the inlet (J)
β_{en,CO_2}—Saving on CO_2 emission from energy
β_{ex,CO_2}—Saving on CO_2 emission from exergy
IE_o—Initial expenditure cost (INR)
V_i and V_o—Velocity at inlet and outlet (m/s)
$\psi_i$and ψ_o—Energy function at inlet and outlet
LC_n—Loan cost for a particular financial year

Ex_o—Exergy at the outlet (J)
Ex_{loss}—Exergy loss (J)
Ex_{inc}—Exergy incident on aperture area (J)
I_G—Incident solar radiation (W/m^2)
$\dot{Q}_u$—Useful amount of heat (J)
$Q_{u,EXGY}$—Useful exergy from predicted model
$Q_{u,EGY}$—Useful energy from predicted model
$m_{\dot{H}TF}$—Mass flow rate of working fluid (LPM)
PB_P—Payback period (years)
i—Interest (%)
i_R—Reverse interest rate (%)
n—Number of years
$T_{i,HTF}$—Working fluid inlet temperature (°C)
$T_{o,HTF}$—Working fluid outlet temperature (°C)
T_{atm}—Atmospheric temperature (°C)
$\Delta\tau$—Time strap (sec)
T_{sun}—Temperature of sun (°C)
U_{loss}—Overall heat transfer loss
η_{DEE}—Daily energy efficiency (%)
η_{DEXE}—Daily exergy efficiency (%)

Abbreviations

CFD—Computational fluid dynamics
HTF—Heat transfer fluid
INR—Indian rupees
LCWH—Levelized cost of water heating
NPV—Net present worth
EER—Efficiency enhancement ratio
LHS—Latent heat storage
SWH—Solar water heater
ETC—Evacuated tube collector
ETSC—Evacuated tube solar collector
FPC—Flat plate collector
PCM—Phase change material
TES—Thermal energy storage

References

1. Soares, N.; Costa, J.J.; Gaspar, A.R.; Santos, P. Review of passive PCM latent heat thermal energy storage systems towards buildings' energy efficiency. *Energy Build.* **2013**, *59*, 82–103. [CrossRef]
2. Yao, S.; Yang, L.; Shi, S.; Zhou, Y.; Long, M.; Zhang, W.; Cai, S.; Huang, C.; Liu, T.; Zou, B. A Two-in-One Annealing Enables Dopant Free Block Copolymer Based Organic Solar Cells with over 16% Efficiency. *Chin. J. Chem.* **2023**, *41*, 672–678. [CrossRef]
3. Souayfane, F.; Fardoun, F.; Biwole, P.-H. Phase change materials (PCM) for cooling applications in buildings: A review. *Energy Build.* **2016**, *129*, 396–431. [CrossRef]
4. Fan, R.; Li, Y.; Bu, T.; Sun, K.; Zhou, Y.; Shi, J.; Zhang, X. Comparative study of solar hot air heating systems with phase change materials in plateau areas. *Energy Build.* **2020**, *224*, 110265. [CrossRef]
5. Aoul, K.T.; Hasan, A.; Riaz, H. Solar water heating systems for different buildings under a hot climate; parametric optimization and economic analysis. *Sustain. Build.* **2018**, *3*, 3. [CrossRef]
6. Chopra, K.; Tyagi, V.; Pathak, A.K.; Pandey, A.; Sari, A. Experimental performance evaluation of a novel designed phase change material integrated manifold heat pipe evacuated tube solar collector system. *Energy Convers. Manag.* **2019**, *198*, 111896. [CrossRef]
7. Pathak, S.K.; Tyagi, V.; Chopra, K.; Sharma, R.K. Recent development in thermal performance of solar water heating (SWH) systems. *Mater. Today Proc.* **2022**, *63*, 778–785. [CrossRef]
8. Pathak, S.K.; Tyagi, V.; Chopra, K.; Rejikumar, R.; Pandey, A. Integration of emerging PCMs and nano-enhanced PCMs with different solar water heating systems for sustainable energy future: A systematic review. *Sol. Energy Mater. Sol. Cells* **2023**, *254*, 112237. [CrossRef]
9. Baneshi, M.; Bahreini, S.A. Impacts of hot water consumption pattern on optimum sizing and techno-economic aspects of residential hybrid solar water heating systems. *Sustain. Energy Technol. Assess.* **2018**, *30*, 139–149. [CrossRef]

10. Fuentes, E.; Arce, L.; Salom, J. A review of domestic hot water consumption profiles for application in systems and buildings energy performance analysis. *Renew. Sustain. Energy Rev.* **2018**, *81*, 1530–1547. [CrossRef]
11. Chopra, K.; Tyagi, V.V.; Pandey, A.K. Thermodynamic and techno-economic analysis of heat pipe ETC water heating system for Indian composite climate. *J. Therm. Anal. Calorim.* **2019**, *139*, 1395–1407. [CrossRef]
12. Naghavi, M.; Ong, K.; Badruddin, I.; Mehrali, M.; Metselaar, H. Thermal performance of a compact design heat pipe solar collector with latent heat storage in charging/discharging modes. *Energy* **2017**, *127*, 101–115. [CrossRef]
13. Xue, H.S. Experimental investigation of a domestic solar water heater with solar collector coupled phase-change energy storage. *Renew. Energy* **2016**, *86*, 257–261. [CrossRef]
14. Abokersh, M.H.; El-Morsi, M.; Sharaf, O.; Abdelrahman, W. On-demand operation of a compact solar water heater based on U-pipe evacuated tube solar collector combined with phase change material. *Sol. Energy* **2017**, *155*, 1130–1147. [CrossRef]
15. O'neil, T.J.; Sobhansarbandi, S. Thermal performance investigation of energy storage based U-pipe evacuated tube solar collector: An experimental study. *Sustain. Energy Technol. Assess.* **2022**, *52*, 102146. [CrossRef]
16. Li, Y.; Liang, X.; Song, W.; Li, T.; Wang, D.; Liu, Y. Optimization and thermal performance of U-type evacuated tube solar collector filled with phase change material. *Energy Rep.* **2022**, *8*, 6126–6138. [CrossRef]
17. Huang, K.; Feng, G.; Zhang, J. Experimental and numerical study on phase change material floor in solar water heating system with a new design. *Sol. Energy* **2014**, *105*, 126–138. [CrossRef]
18. Essa, M.A.; Asal, M.; Saleh, M.A.; Shaltout, R. A comparative study of the performance of a novel helical direct flow U-Tube evacuated tube collector. *Renew. Energy* **2020**, *163*, 2068–2080. [CrossRef]
19. Nemati, H.; Souriaee, V.; Habibi, M.; Vafai, K. Design and Taguchi-based optimization of the latent heat thermal storage in the form of structured porous-coated pipe. *Energy* **2023**, *263*, 125947. [CrossRef]
20. Ao, C.; Yan, S.; Zhao, S.; Hu, W.; Zhao, L.; Wu, Y. Stearic acid/expanded graphite composite phase change material with high thermal conductivity for thermal energy storage. *Energy Rep.* **2022**, *8*, 4834–4843. [CrossRef]
21. Chopra, K.; Pathak, A.K.; Tyagi, V.; Pandey, A.; Anand, S.; Sari, A. Thermal performance of phase change material integrated heat pipe evacuated tube solar collector system: An experimental assessment. *Energy Convers. Manag.* **2019**, *203*, 112205. [CrossRef]
22. Chopra, K.; Tyagi, V.; Pandey, A.; Popli, S.; Singh, G.; Sharma, R.; Sari, A. Effect of simultaneous & consecutive melting/solidification of phase change material on domestic solar water heating system. *Renew. Energy* **2022**, *188*, 329–348. [CrossRef]
23. Pathak, S.K.; Tyagi, V.; Chopra, K.; Kalidasan, B.; Pandey, A.; Goel, V.; Saxena, A.; Ma, Z. Energy, exergy, economic and environmental analyses of solar air heating systems with and without thermal energy storage for sustainable development: A systematic review. *J. Energy Storage* **2023**, *59*, 106521. [CrossRef]
24. Essa, M.A.; Rofaiel, I.Y.; Ahmed, M.A. Experimental and Theoretical Analysis for the Performance of Evacuated Tube Collector Integrated with Helical Finned Heat Pipes using PCM Energy Storage. *Energy* **2020**, *206*, 118166. [CrossRef]
25. Pandey, A.; Laghari, I.A.; Kumar, R.R.; Chopra, K.; Samykano, M.; Abusorrah, A.M.; Sharma, K.; Tyagi, V. Energy, exergy, exergoeconomic and enviroeconomic (4-E) assessment of solar water heater with/without phase change material for building and other applications: A comprehensive review. *Sustain. Energy Technol. Assess.* **2021**, *45*, 101139. [CrossRef]
26. Petela, R. Exergy of undiluted thermal radiation. *Sol. Energy* **2003**, *74*, 469–488. [CrossRef]
27. Chopra, K.; Tyagi, V.; Pathak, S.K.; Tripathi, P.A.; Sharma, R.; Singh, G.; Pandey, A. Thermal and chemical reliability of paraffin wax and its impact on thermal performance and economic analysis of solar water heater. *Energy Sustain. Dev.* **2023**, *73*, 39–53. [CrossRef]
28. Rezvani, S.; Bahri, P.; Urmee, T.; Baverstock, G.; Moore, A. Techno-economic and reliability assessment of solar water heaters in Australia based on Monte Carlo analysis. *Renew. Energy* **2017**, *105*, 774–785. [CrossRef]
29. Tewari, K.; Dev, R. *Exergy, Environmental and Economic Analysis of Modified Domestic Solar Water Heater with Glass-to-Glass PV Module*; Elsevier Ltd.: Amsterdam, The Netherlands, 2019; Volume 170. [CrossRef]
30. Hossain, M.S.; Pandey, A.K.; Tunio, M.A.; Selvaraj, J.; Hoque, K.E.; Rahim, N.A. Thermal and economic analysis of low-cost modified flat-plate solar water heater with parallel two-side serpentine flow. *J. Therm. Anal. Calorim.* **2015**, *123*, 793–806. [CrossRef]
31. Abo-Elfadl, S.; Yousef, M.S.; Hassan, H. Energy, exergy, and enviroeconomic assessment of double and single pass solar air heaters having a new design absorber. *Process. Saf. Environ. Prot.* **2020**, *149*, 451–464. [CrossRef]
32. Alwaeli, A.H.A.; Kazem, H.A.; Chaichan, M.T.; Sopian, K. Experimental investigation of using nano-PCM/nanofluid on a photovoltaic thermal system (PVT): Technical and economic study. *Therm. Sci. Eng. Prog.* **2019**, *11*, 213–230. [CrossRef]
33. Murali, G.; Mayilsamy, K.; Arjunan, T.V. An Experimental Study of PCM-Incorporated Thermosyphon Solar Water Heating System. *Int. J. Green Energy* **2014**, *12*, 978–986. [CrossRef]
34. Mellouli, S.; Alqahtani, T.; Algarni, S. Parametric Analysis of a Solar Water Heater Integrated with PCM for Load Shifting. *Energies* **2022**, *15*, 8741. [CrossRef]

35. Azari, P.; Lavasani, A.M.; Rahbar, N.; Yazdi, M.E. Performance enhancement of a solar still using a V-groove solar air collector—Experimental study with energy, exergy, enviroeconomic, and exergoeconomic analysis. *Environ. Sci. Pollut. Res.* **2021**, *28*, 65525–65548. [CrossRef] [PubMed]
36. Pawar, V.R.; Sobhansarbandi, S. Heat transfer enhancement of a PCM-porous metal based heat pipe evacuated tube solar collector: An experimental study. *Sol. Energy* **2023**, *251*, 106–118. [CrossRef]

buildings

MDPI

Article

Optimizing PCM Integrated Wall and Roof for Energy Saving in Building under Various Climatic Conditions of Mediterranean Region

Sana Dardouri [1], Ekrem Tunçbilek [2,3], Othmen Khaldi [4], Müslüm Arıcı [2,3,*] and Jalila Sghaier [1]

1 Research Laboratory of Thermal and Thermodynamic of Industrial Processes (LRTTPI), National Engineering School of Monastir, University of Monastir, Monastir 5000, Tunisia
2 Mechanical Engineering Department, Engineering Faculty, Kocaeli University, Umuttepe Campus, Kocaeli 41001, Turkey
3 International Joint Laboratory on Low-Carbon and New-Energy Nexus Research and Development, Kocaeli University, Kocaeli 41001, Turkey
4 Laboratory of Materials Organization and Properties LMOP (LR99ES17), University of Tunis El Manar, Tunis 2092, Tunisia
* Correspondence: muslumarici@gmail.com

Abstract: Energy conservation in buildings has been the focus of many studies since nearly one-third of global energy consumption is due to buildings. Phase change material (PCM) technology promises to be an attractive solution for energy saving in buildings since it is a passive and effective technology, as demonstrated in the literature. Therefore, this study focuses on the energy-saving performance of PCM-integrated buildings located in a Mediterranean climate to reveal their energy-saving potential. PCM is integrated both in external or internal south walls and roofs of buildings under four different climatic conditions. EnergyPlus, which is a well-known building simulation software, is adopted for building thermal analyses. The effects of melting temperature, location of PCM layer in the wall, thickness of PCM layer, type of envelope (wall or roof), and PCM double-layer system in the wall are investigated. The corresponding energy savings and CO_2 emission reductions are obtained for the considered cases. The results showed that up to 41.6% reduction in energy demand can be obtained depending on the PCM application. Besides, PCM with a low melting temperature (21 °C) favored heating energy savings, while PCM with a high melting temperature (29 °C) favored cooling energy savings. Moreover, the double-layer PCM system provided higher energy savings than the single-layer PCM system, especially in warm and arid regions (Sousse and Tozeur).

Keywords: phase change materials; building performance; climatic condition; PCM melting temperature; CO_2 reduction; PCM double layer

check for updates

Citation: Dardouri, S.; Tunçbilek, E.; Khaldi, O.; Arıcı, M.; Sghaier, J. Optimizing PCM Integrated Wall and Roof for Energy Saving in Building under Various Climatic Conditions of Mediterranean Region. *Buildings* **2023**, *13*, 806. https://doi.org/10.3390/buildings13030806

Academic Editor: Cinzia Buratti

Received: 12 January 2023
Revised: 6 March 2023
Accepted: 11 March 2023
Published: 18 March 2023

1. Introduction

Global energy consumption is gradually increasing with the continuous development of technology, although considerable efforts have been put into increasing energy efficiency and thereby decreasing energy demand. Considering the fact that most fossil fuels are used as primary energy sources to supply energy to the energy networks, an inevitable shortage of these sources is obvious in the upcoming years. In addition to that, the high amount of greenhouse gas emitted in the process of generating energy is another important factor that has to be focused on when it is considered that these emissions lead to the greenhouse gas effect. In this context, the development of new energy sources is a major issue for the scientific community, and several works and research studies have been carried out in this field with the aim to improve energy efficiency, gain energy saving, and limit greenhouse gas emissions [1–3].

In Tunisia, the energy consumed in the building sector is approximately 36% of the national total energy consumption (Figure 1). Obviously, this percentage tends to increase in the upcoming years considering the technological development, rising population, and increasing comfort expectations of people [4,5]. Transportation and industry followed the building sector's energy consumption with 31% and 27% shares, respectively, in the national final energy consumption. The building sector indeed has great potential for energy savings since a majority of the buildings lack even proper thermal insulation, and previous works in the literature showed that energy saving in buildings can reach considerable rates [6–8]. In addition, the Social Commission for Western Asia (ECWAS), the German Development Agency (GIZ), and the UN's Economic and Social Council have recently evaluated the energy situation of buildings in Tunisia [9]. The study evaluated the diverse energy consumption items (air conditioning, heating, lighting, cooling, office equipment, etc.) in public, private, and office buildings in detail. These findings represented a database of electricity consumption estimations in Tunisia's buildings. For instance, in 2017, the total energy consumption for the residential sector reached 5176.2 GWh (13.07 kWh/m^2 over the year), a radical increase of 27% as compared to the energy consumption observed in 2014. Among the evaluated parameters, cooling is classified as the first on the list (30.3% of the total), followed by air conditioning (22.2% against 3.4% in 2004) and television sets (17.2% against 21.16% in 2004) [9].

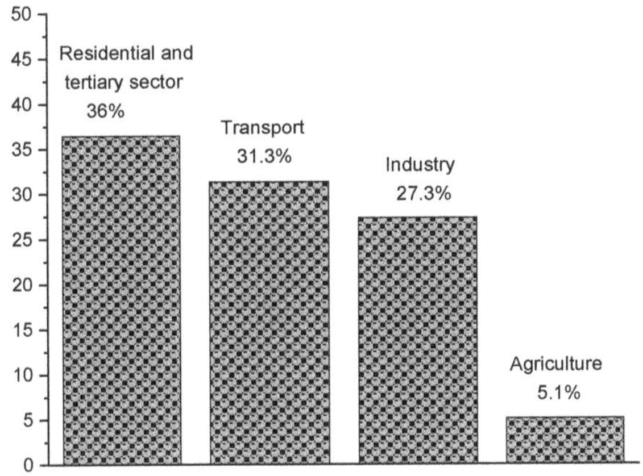

Figure 1. Final energy consumption in 2020 with respect to sector in Tunisia [9].

Thus, it is crucial to decrease energy consumption in the residential sector by improving both energy efficiency as well as saving measures. In this context, phase change thermal energy storage represents an innovative and promising way for energy storage technology. This can be realized particularly with phase change materials (PCM), which have the potential to store latent heat in materials such as paraffin waxes, eutectics, and salt hydrates. The application of PCMs in building envelopes can reduce space heating/cooling energy demand and be considered a promising passive cooling technology in buildings that have poor energy efficiency and thermal comfort [10].

Recently, a good amount of research has been dedicated to the utilization of PCM as well as nano-enhanced PCM applications in buildings [11–16]. It is obvious from the literature review that the PCM technique is successfully integrated into the walls [11,13], windows [14,15], floors [16,17], and roofs [18,19] to reduce the energy requirement of the buildings [20]. For instance, Chelliah et al. [13] studied the energy-saving potential of air-conditioning by evaluating PCM-filled terracotta bricks in the external walls. They reported that there is an optimum PCM with a certain melting temperature and that up to EUR 74 in cost savings can be achieved. Hamidi et al. [21] integrated PCM into hollow brick walls and indicated that the percentage of energy saved is about 56% in the northeast Mediterranean zone, while no energy-saving effect is observed for the southeast zone. Besides, for the northeast zone, a 26 °C optimum melting temperature for PCM is suggested as it gives the best thermal performance. Kenzhekhanov et al. [22] demonstrated the promising effect of the PCM location as well as the PCM amount on the energy savings and discomfort hours of a building located in a subarctic climate. They reported that increasing the thickness and amount of PCM in the layer promotes more energy savings. Besides, up to 10,000 kWh and 500 h of annual energy savings and discomfort hour reductions are respectively achieved with PCM integration.

Applications related to PCM building and analysis of energy efficiency, energy saving, or thermal comfort are simulated and analyzed in several works. However, studies on the position, thickness, and type of PCM in the envelope should be performed and detailed carefully prior to the application of innovative materials to buildings [23,24]. For example, Hu [25] studied thermal and building envelope responses to light under several climate situations. The total energy consumed was reduced by 19% for the roof with a thermochromic coating (TC roof)—PCM, which in turn reduced CO_2 emissions by 5%. So, as the PCM roof thickness increases, the energy savings increase and reach 29%, while CO_2 emission reductions become 8%. Additionally, as mentioned in [26], PCM with a melting temperature close to 24 °C was successfully integrated into the building wall to maintain the indoor temperature of the building. This work shows that comfortable temperatures can be maintained for the whole year without the contribution of any mechanical air-conditioning system. These results agree well with several numerical investigations in Morocco, in which the improvement of the thermal efficiency of PCM-integrated double external walls was considered [27]. A previous work by Hagenau [28] investigated the role of PCM and the related dynamic energy in order to evaluate building envelope performance under various Danish climate zones, and the researchers reported that energy savings in the range of 32–36% can be achieved. Considering the abovementioned works, it is important to note that using PCM in the walls and roofs increases the cost of construction. Therefore, it is obvious to search for the most favorable location for PCM integration to gain the best possible energy conservation. In this context, a dynamic model based on hysteresis was used to find the best location of PCM thickness in the wall and roof, as well as its appropriate position for heat flux reduction [29].

From the literature review, it is clear that PCM integration into the building's external walls promises to be a good candidate for energy saving. Besides, it is also known that melting temperature, location, and thickness of PCM are important parameters affecting TPCM energy-saving performance. The optimum melting temperature and location can be different depending on the outdoor boundary conditions. Therefore, in this study, an optimization of PCM properties is conducted to gain the highest possible energy saving from PCM application to the building's external envelope. In this respect, different PCM melting temperatures, PCM locations (in the wall and roof), and PCM thicknesses are studied based on thermal analyses. Furthermore, these thermal analyses are conducted for four different climatic regions of Tunisia so an effective guideline can be presented with relevant findings. The main targets of this study can be listed as (i) exploring the effect of PCM features (melting temperature, location, and thickness) on energy saving; (ii) revealing the energy saving potentials by utilizing PCM in different climatic regions of Tunisia; (iii) the impact of using double-layer PCM in the external wall on the energy saving; and (iv) providing a practical guideline that can be valid for different climatic regions of Tunisia for building envelope applications of PCM.

2. Methodology

2.1. Model Description

A building with standard wall, roof, and floor structures is considered to be used in thermal analyses. The building is one story, has two standard-sized windows on the south side, and the window-to-wall ratio on that side is approximately 40%. The floor area is 72 m^2, and the building has a flat roof. The building is conditioned by using an air-cooled heat pump for both heating and cooling purposes. In order to ensure thermal comfort conditions inside the building, the heating and cooling set temperatures are taken as 21 °C and 24 °C, respectively. Besides, the heating and cooling setback temperatures are considered for preventing excessive cooling (in winter) and heating (in summer) when the building is not occupied.

2.2. Numerical Simulation

In this work, Design Builder is employed as the graphical interface for EnergyPlus software. EnergyPlus is a software that is capable of performing building thermal simulation, and its development was funded by the U.S. Department of Energy's (DOE) Building Technologies Office (BTO). Based on a definition of a building in terms of the physical design of the building, associated mechanical systems, etc., the software simulates the outdoor and indoor conditions to calculate the heating and cooling loads required to maintain thermal control setpoints. EnergyPlus includes several modules to estimate building energy requirements. It does not have its own graphical user interface, but several interfaces are available for inputting data and processing simulation results (Figure 2).

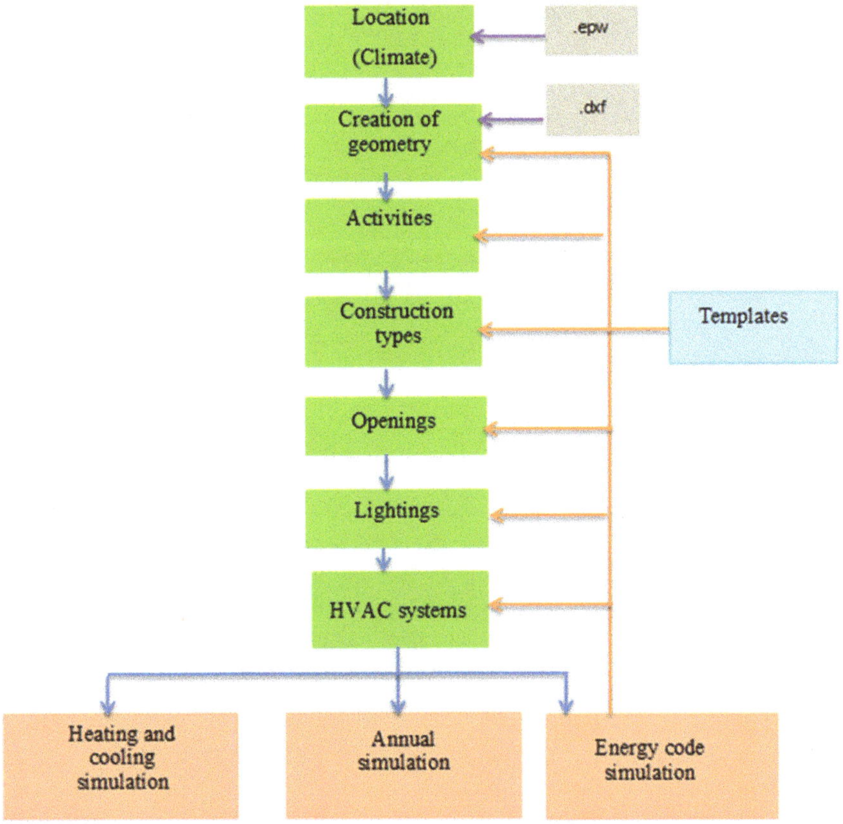

Figure 2. EnergyPlus program flow chart.

The EnergyPlus PCM model based on the one-dimensional conduction finite-difference (CondFD), which was subject to various verification studies and has been validated by several researchers [30–34], is adapted as the solution algorithm. This algorithm gives the user the possibility to choose between an implicit finite difference scheme and the Crank–Nicholson method. In this study, an implicit finite difference scheme coupled with the enthalpy–temperature equation is used to evaluate phase change and find the temperature field (Equations (1) and (3)). All elements were discretized using a node spacing Δx, which depends on the time step (Δt), the space discretization constant (c), and the thermal diffusivity of the material (α) (Equation (2)).

$$\rho C_p \Delta x \frac{T_j^{n+1} - T_j^n}{\Delta t} = \left(\frac{\lambda_{j+1}^{n+1} + \lambda_j^{n+1}}{2}\right)\left(\frac{T_{j+1}^{n+1} - T_j^{n+1}}{\Delta x}\right) + \left(\frac{\lambda_{j-1}^{n+1} + \lambda_j^{n+1}}{2}\right)\left(\frac{T_{j-1}^{n+1} - T_j^{n+1}}{\Delta x}\right) \quad (1)$$

$$\Delta x = \sqrt{c\alpha\Delta t} \quad (2)$$

$$C_p = \frac{h_j^{new} - h_j^{old}}{T_j^{new} - T_j^{old}} \quad \text{where} \quad h = h(T) \quad (3)$$

where T stands for temperature and λ stands for thermal conductivity of the wall material. Besides, ρ and C_p represent the density and specific heat capacity of the material, respectively. Enthalpy is denoted by h. In the thermal analyses, 60 time steps per hour (Δt = 1 min) are taken, while the space discretization constant (c) in the node spacing equation is taken as 0.5.

2.3. Climatic Conditions

Climatic conditions have a great impact on building thermal efficiency and performance. The scope is focused on the region of Tunisia, due to its strategic location. As shown in Figure 3, the climate of Tunisia is divided into five climatic zones, namely, arid, upper humid, Saharan, lower semi-arid, and lower humid. According to the Köppen–Geiger classification [35], a great difference between the north and the rest of the country occurs due to the chain of the Tunisian ridge, which separates the zones subjected to the Mediterranean climate as humid (for Bizerte and Tabarka), lower semi-arid (for Sousse) and the hot desert climate typical of the Sahara (for Tozeur) (Table 1). The weather data for each city are exported to the EnergyPlus climate file database. The geographic and meteorological data of these regions, as well as the city description, are given in Table 1.

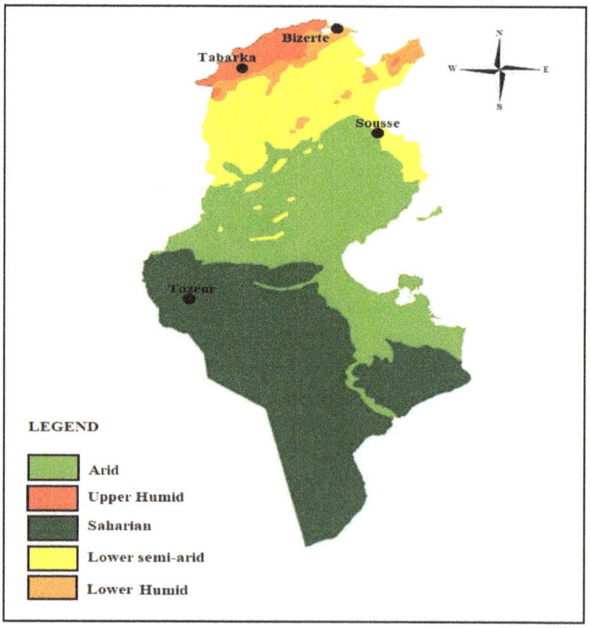

Figure 3. Map of climatic zones in Tunisia.

Table 1. Considered cities located in different climatic zones.

Region (City)	Climate Classification	Latitude	Longitude	Elevation [m]
Sousse	Lower semi-arid	35°40′ N	10°45′ E	24
Bizerte	Lower humid	37°15′ N	9°48′ E	5
Tabarka	Upper humid	36°57′ N	8°45′ E	4.69
Tozeur	Saharian	33°55′ N	8°06′ E	42

With a cold and rainy winter season, the heating loads are dominant in Tabarka, and snow is possible during this period of the year with an average temperature of 7.4 °C. However, summer is usually hot and dry, with an average temperature of 28.2 °C. Besides, Sousse is characterized by a hot, semi-arid climate, which makes it an all-season resort with hot, dry summers and warm, wet, mild winters. Bizerte is characterized by a warm-summer Mediterranean climate, with mild and rainy winters. Summers are hot and dry, and together with the Mediterranean Sea, summers are cooler and more humid than in the interior of Tunisia. For Bizerte and Sousse, both heating and cooling are required. Conversely, Tozeur suffers from hot desert climate conditions. Summer is exceptionally hot, with daily temperatures frequently exceeding 45 °C in the shadows, and the sirocco (Mediterranean wind) can force temperatures to exceed 50 °C. In winter, it can sometimes turn freezing at night, and just before sunrise, the temperature drops below 0 °C. Depending on the incident solar radiation, the highest solar radiation is observed in April, May, June, July, and August in Sousse, Tozeur, Bizerte, and Tabarka, respectively, which in turn improves the solar heat gain in buildings through the building envelope.

The outdoor temperatures (T_{air}) as well as the wind speed (w) are given in Ref. [36], while the collected solar radiation data are available in Ref. [37]. Therefore, the corresponding annual climate data are collected from these sources and depicted in Figure 4.

2.4. PCM Properties

In this work, the Infinite R™ PCMs were chosen, which are used in diverse commercial applications such as energy storage with diverse melting temperatures of 18, 21, 23, 25, and 29 °C, referred to as PCM18, PCM21, PCM23, PCM25, and PCM29, respectively. The thermophysical properties of the Infinite R™ PCMs are given in Table 2.

Table 2. Thermophysical properties of PCM.

	PCM18	PCM21	PCM23	PCM25	PCM29
Latent heat during the entire phase change (kJ/kg)	200	200	200	200	200
Peak melting temperature for melting curve (°C)	19	22	24	26	30
Peak melting temperature for freezing curve (°C)	17	20	22	24	28
Liquid state thermal conductivity (W/(mK))	0.54	0.54	0.54	0.54	0.54
Solid state thermal conductivity (W/(mK))	1.09	1.09	1.09	1.09	1.09
Liquid state density (kg/m^3)	1540	1540	1540	1540	1540
Solid state density (kg/m^3)	1540	1540	1540	1540	1540
Liquid state specific heat (J/(kgK))	3140	3140	3140	3140	3140
Solid state specific heat (J/(kgK))	3140	3140	3140	3140	3140

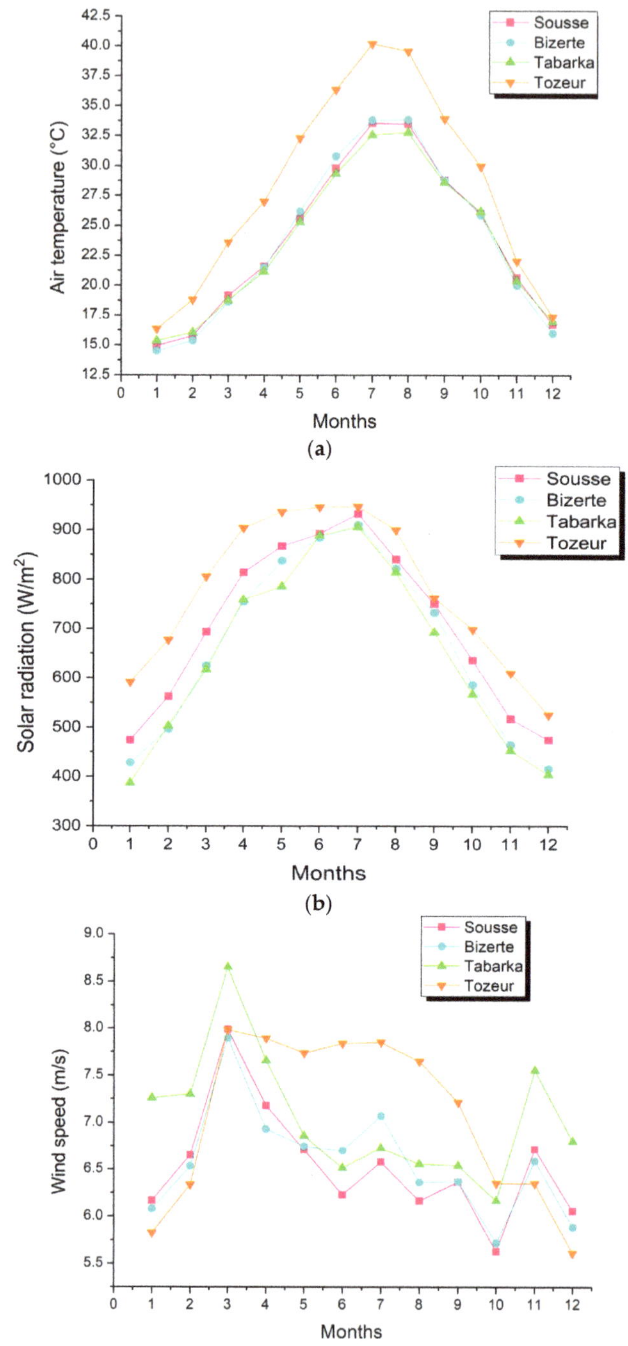

(a)

(b)

(c)

Figure 4. Climatic data (**a**) air temperature, (**b**) solar radiation, and (**c**) wind speed.

2.5. Wall and Roof Structure in the Building

In Tunisia, the brick wall is the conventional structure for residential building construction [37]. Here, the external building wall without PCM (control wall) is composed of three layers: plaster coating, brick, and cement (Figure 5a). On the building roof, four layers were considered: cement, concrete, reinforced concrete, and plaster coating (Figure 5b). Two different configurations were considered for PCM integration into the control wall: (i) PCM located in the outside of the wall (Case 1, Figure 5c) and (ii) PCM located in the inside of the wall (Case 2, Figure 5c). It should be noted that PCM is employed in the wall as an extra layer. The PCM layer thickness is varied between e = 10 mm and 30 mm for these configurations. The thermal performances of these materials, as described in Table 3, were also taken into account in the modeling process.

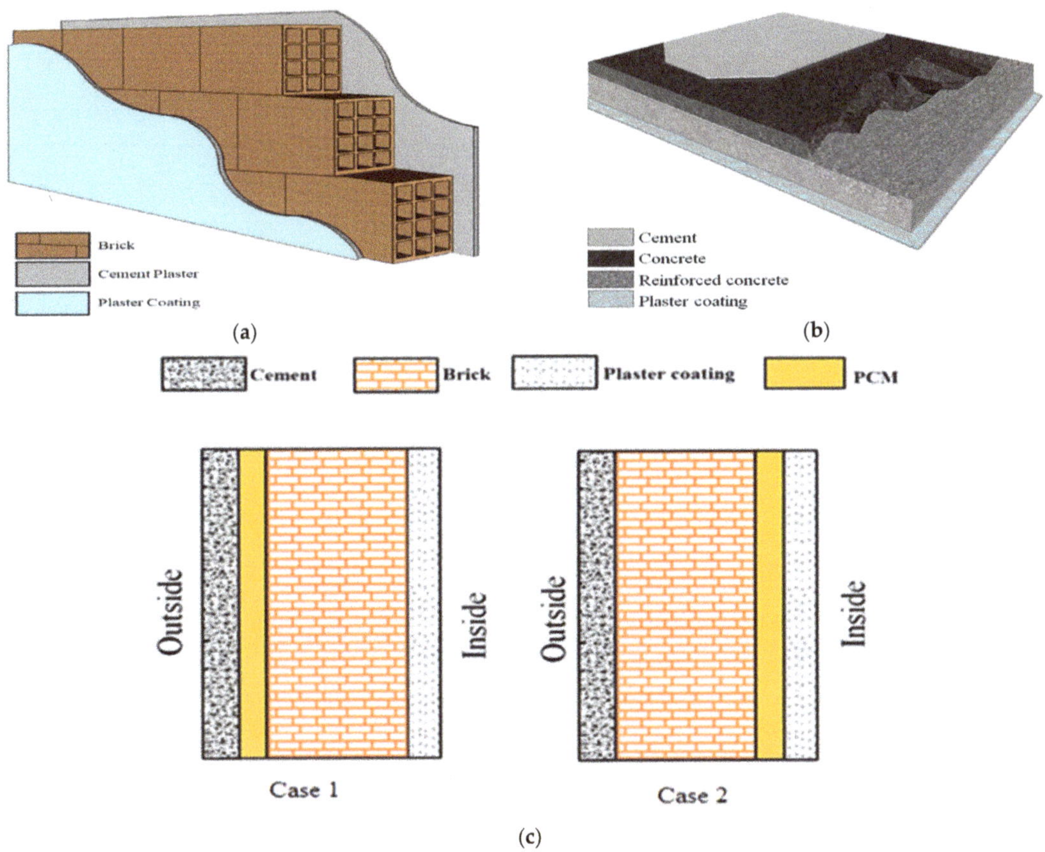

Figure 5. Schematic representation of (**a**) external wall, (**b**) roof, and (**c**) PCM integration.

To evaluate the effectiveness of the implementation of the PCM layer in the building's external envelope, two indicators are used:

(i) Energy savings: It gives the total decrease in energy consumption resulting from the integration of PCMs.

This can be deduced from Equations (4) and (5) for both heating and cooling energy savings, respectively.

$$E_{saving,heating} = E_{heating\ consumption,\ without\ PCM} - E_{heating\ consumption,\ with\ PCM} \quad (4)$$

$$E_{saving,cooling} = E_{cooling\ consumption,\ without\ PCM} - E_{cooling\ consumption,\ with\ PCM} \quad (5)$$

(ii) The annual percentage of heating and cooling energy consumption reduction due to the PCM integration, the rate R (in %), is given by the following equation:

$$R\ (\%) = \frac{E_{saving}}{E_{consumption\ ,without\ PCM}} * 100 \quad (6)$$

Table 3. Thermal properties of materials.

	k (W/mK)	*ρ* (kg/m³)	C_p (J/kgK)	*e* (m)
Brick	0.72	1920	840	0.20
Cement	1.4	2200	940	0.025
Plaster coating	1.2	1800	840	0.015
Concrete	1.4	2100	840	0.1
Reinforced concrete	2.3	2300	1000	0.15

3. Results and Discussion

3.1. *Effect of PCM Melting Temperature*

The effect of PCM melting temperature on the total energy saving was evaluated by testing different phase transition temperatures ranging from 18 °C to 29 °C. The thickness of the PCM layer was taken as 10 mm, and the PCM was located both near the interior as well as near the exterior surface of the south wall to also reveal the impact of the PCM location on the melting temperature. The obtained results related to the parameters above are given in Figure 6. From the figures, it can be deduced that the PCM21 (i.e., PCM with a 21 °C melting temperature) provided the highest energy saving for Bizerte, Tabarka, and Tozeur, whereas the PCM29 showed the best performance for Sousse. The highest energy-saving amounts were also attained by locating the PCM layer near the interior surface rather than near the exterior, regardless of the city except for the Sousse region. The same finding, that the PCM layer near the indoor outperforms that near the outdoor, was also pointed out in the literature.

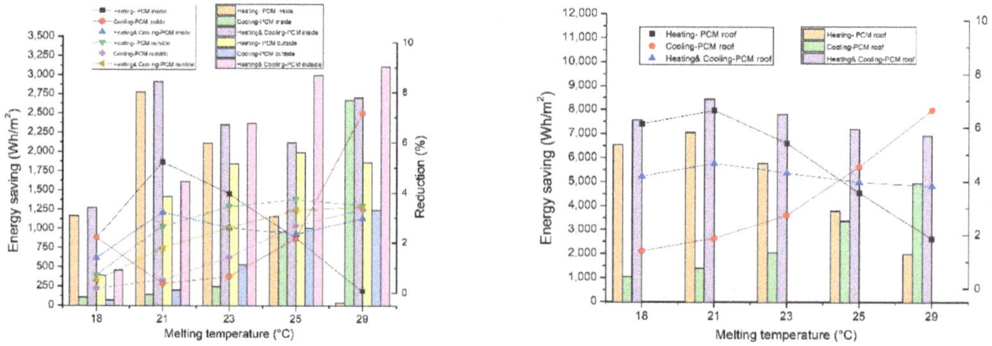

(**a**) Sousse

Figure 6. *Cont.*

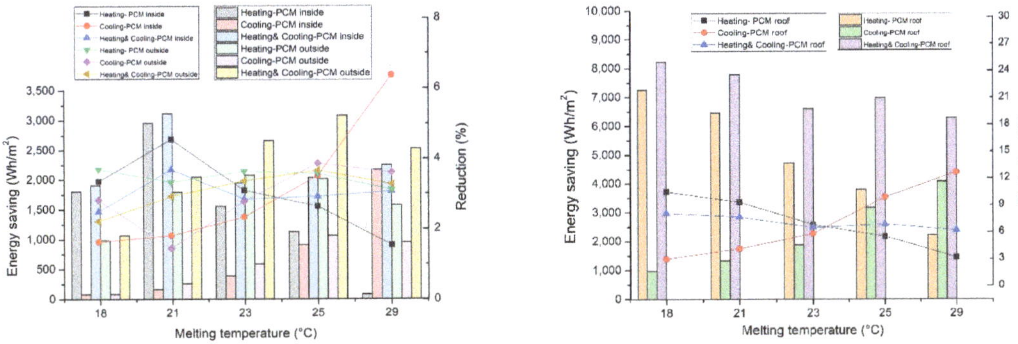

(**b**) Bizerte

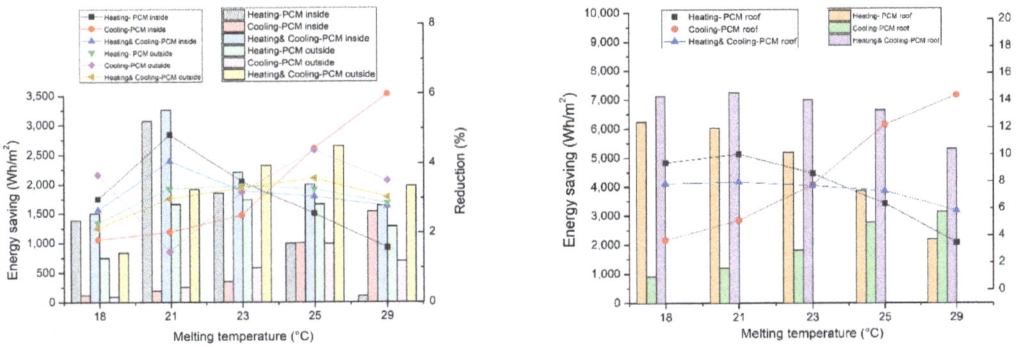

(**c**) Tabarka

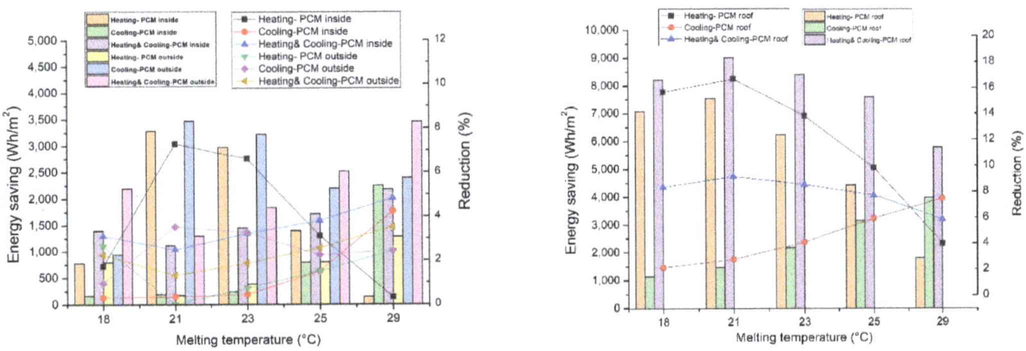

(**d**) Tozeur

Figure 6. Effect of PCM melting temperature on energy saving and energy reduction.

Excluding the worst-case scenario for all the considered cities (PCM18), the total energy savings ranged from 3116.0 Wh/m^2 (PCM21) to 1946.0 Wh/m^2 (PCM23) for Bizerte, from 3259.9 Wh/m^2 (PCM21) to 1639.9 Wh/m^2 (PCM29) for Tabarka, and from 3474.21 Wh/m^2 (PCM21) to 2145.6 Wh/m^2 (PCM25) for Tozeur with PCM located near the interior. As previously mentioned, the optimum melting temperature for PCM-integrated walls near the indoor environment was found to be 21 °C, except for the Sousse. A total energy saving of 3104.7 Wh/m^2 can be attained with PCM29 located near the exterior surface of the wall under Sousse climatic conditions, indicating that the melting temperature should be 29 °C. It was noted that the PCM location inside the wall affected the optimum melting temperature for the PCM. For instance, the optimum melting temperature for all cities was found to be 29 °C when the PCM was located near the interior surface of the wall for the cooling energy. Overall, the PCM location inside the wall and the climatic conditions of the region significantly affected the thermal performance of the PCM application. It can be deduced from the results that the optimum melting temperature was close to the setpoint of the room [38]. For example, the melting temperature was equal to 21 °C for the heating load in each city. On the contrary, the optimum melting temperature was equal to 29 °C for cooling loads in all cities. Moreover, it was shown that higher energy-saving amounts can be attained in high-amplitude regions. When the PCM layer is located near the exterior or interior roof surface, the optimum melting temperature is always equal to 21°C for all regions except for Bizerte (found at 18 °C), which is located in a relatively cold region. In summary, it can be concluded that when modeling an implementation of PCM building technology, one should consider geographical and climatic factors such as temperature, wind profile, and solar irradiance [39].

3.2. *Effect of Building Envelope Type and PCM Location*

Utilization of PCM in different building envelope types was studied by integrating PCM in the roof, external or internal wall, and wall-roof of the building. In the wall-roof configuration, PCM21 is located in the interior surface of the wall, except in the Sousse region, where PCM29 is placed on the exterior surface of the wall. Moreover, PCM location was found to be an important parameter in the literature; therefore, two different locations, namely near the interior and near the exterior, were tested. The overall effect of these variables on the energy saving in heating, cooling, and total energy demand was then analyzed using the optimum melting temperature of PCM for each city. The obtained results related to optimal envelope type, location, and configuration are given in Figure 7.

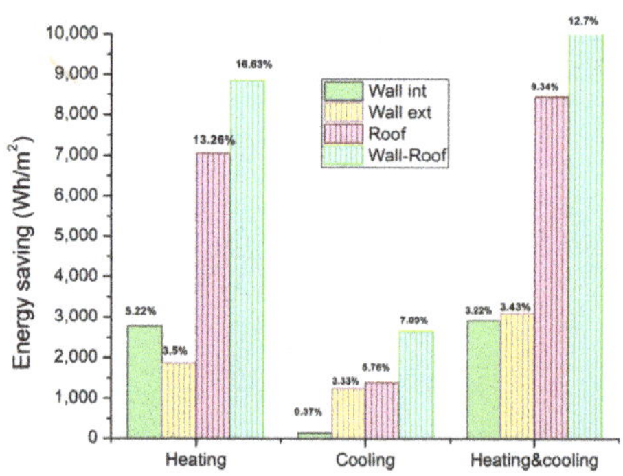

(**a**) Sousse

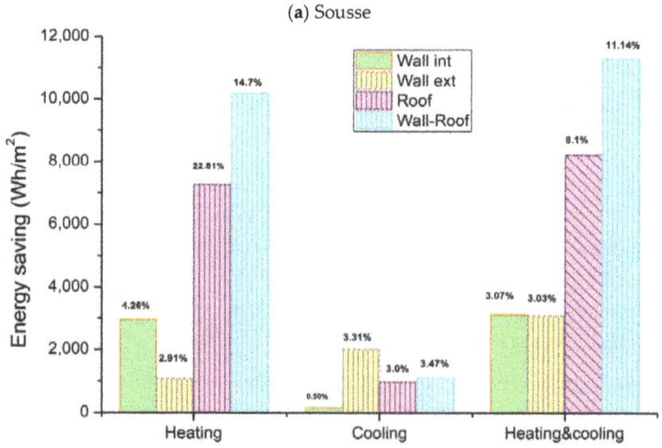

(**b**) Bizerte

Figure 7. *Cont.*

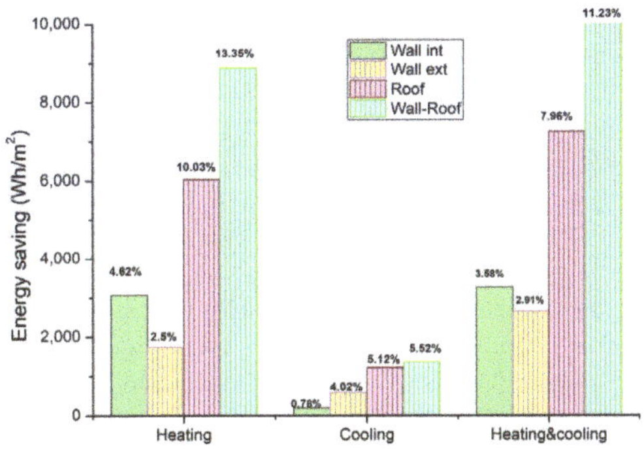

(**c**) Tabarka

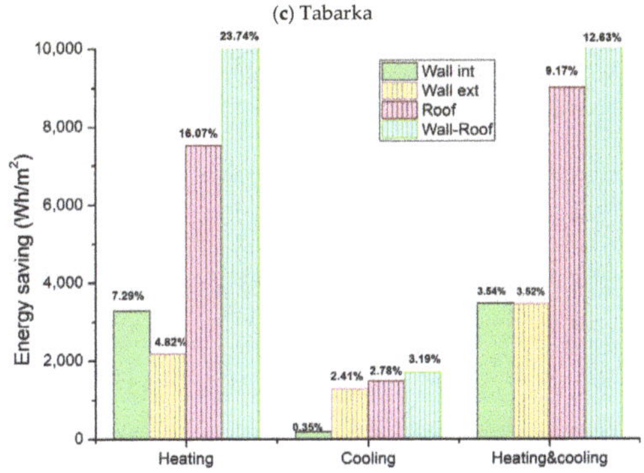

(**d**) Tozeur

Figure 7. Effect of envelope type and PCM location on the energy saving.

It can be seen from the given figure that the utilization of PCM provided significant energy savings both in heating and cooling applications for all of the considered cities, regardless of the PCM location (Wall int and Wall ext) as well as building envelope type (wall, roof, or wall-roof). The interior location for PCM21 was found to be more beneficial in terms of energy saving compared to the exterior location for the heating period. However, on the contrary, the exterior location saved more energy than the interior location for the cooling period. As compared to the traditional building envelopes, the reduction in the heating load reached 16.6%, 23.7%, 14.7%, and 13.6% in Sousse, Tozeur, Bizerte, and Tabarka, respectively, by implementing PCM21 in a wall-roof configuration. Although considerable reductions in energy consumption for cooling were achieved, these savings are marginal compared to the amount saved in heating energy. Using PCM in both the internal wall (with higher energy saving) and roof (wall-roof configuration) provided the highest energy saving amount irrespective of the city, i.e., climatic region. The least-efficient scenario was attained when the PCM was integrated into the exterior wall in the heating period and into the interior wall during the cooling period. The corresponding total energy reduction percentages for PCM integrated roofs are 9.34%, 9.2%, 8.1%, and 7.9% in Sousse, Tozeur, Bizerte, and Tabarka, respectively. In total, the PCM located near the interior (near the indoor) saved more energy than the PCM located near the exterior (near the outdoor) in all climatic regions except the Sousse region. This finding agrees with the findings reported in Refs. [40–42]. Consequently, approximately 12.4 kWh/m^2 energy savings were achieved for Tozeur (warm region) when the optimum configuration was applied, i.e., PCM integration in both wall and roof (wall-roof case).

3.3. *Effect of PCM Layer Thickness*

The impact of the PCM thickness on the energy-saving performance was investigated by analyzing different thicknesses, namely, 10 mm, 20 mm, 30 mm, and 40 mm. The optimized phase transition temperature of the PCM layer, i.e., the optimum PCM type, was used in the simulations. For heating, cooling, and heating and cooling, the PCM layer was tested in both types, near the interior surface of the wall and near the exterior surface of the roof. The results are given in Table 4. It is evident from the table that the total energy requirement (heating and cooling) is greater in the case of increasing the PCM layer thickness. This can be expected since a higher amount of latent heat can be stored due to the increased PCM amount [42]. Higher energy requirement reduction was achieved with a 40 mm PCM located in the roof and also in the wall. The heating and cooling energy needs were reduced by 26.7 kWh/m^2, 23.3 kWh/m^2, 21.6 kWh/m^2, and 30.8 kWh/m^2 with a 40 mm PCM located in the roof for the buildings in Sousse, Bizerte, Tabarka, and Tozeur, respectively. Following the PCM located in the roof, the best case was found for locating the PCM near the interior wall surface, regardless of the PCM layer thickness. These results also comply with the outcomes of the studies in the literature [38–44].

Table 4. Annual heating and cooling energy saving per 1 m^2 of three types of building envelopes and for five PCM layer thicknesses.

		e_1 = 10 mm	**e_2 = 20 mm**	**e_3 = 30 mm**	**e_4 = 40 mm**
Sousse		Heating (Wh/m^2)			
	PCM-wall	1862.87 (−3.50%)	3688.96 (−6.93%)	4834.89 (−9.09%)	5514.46 (−10.36%)
	PCM-roof	7056.94 (−13.26%)	13,703.54 (−25.76%)	19,543.38 (−36.74%)	21,476.14 (−40.37%)
		Cooling (Wh/m^2)			
	PCM-wall	1241.79 (−3.33%)	2105.36 (−5.65%)	2344.13 (−6.29%)	2330.63 (−6.26%)
	PCM-roof	1389.78 (−3.73%)	2689.36 (−7.22%)	3594.85 (−9.65%)	5257.56 (−14.12%)
		Heating and Cooling (Wh/m^2)			
	PCM-wall	3104.66 (−3.43%)	5794.32 (−6.40%)	7179.02 (−7.94%)	7845.09 (−8.67%)
	PCM-roof	8446.72 (−9.34%)	16,392.9 (−18.13%)	16,392.9 (−18.13%)	26,733.7 (−29.57%)
Bizerte		Heating (Wh/m^2)			
	PCM-wall	2953.44 (−4.26%)	4092.55 (−5.90%)	4565.82 (−6.59%)	4852.72 (−7.00%)
	PCM-roof	7253.05 (−10.47%)	12,206.94 (−17.62%)	16,656.06 (−24.04%)	18,763.45 (−27.08%)
		Cooling (Wh/m^2)			
	PCM-wall	162.6 (−0.50%)	299.2 (−0.93%)	387.45 (−1.20%)	425.04 (−1.32%)
	PCM-roof	963.84 (−3.00%)	2340.42 (−7.29%)	3102.68 (−9.67%)	4579.27 (−14.27%)
		Heating&Cooling (Wh/m^2)			
	PCM-wall	3116.04 (−3.07%)	4391.75 (−4.33%)	4953.27 (−4.88%)	5277.76 (−5.20%)
	PCM-roof	8216.89 (−8.10%)	14,547.36 (−14.35%)	19,758.74 (−19.49%)	23,342.72 (−23.03%)

Table 4. *Cont.*

		e_1 = 10 mm	e_2 = 20 mm	e_3 = 30 mm	e_4 = 40 mm
			Heating (Wh/m^2)		
Tabarka	PCM-wall	3067.38 (−4.62%)	3997.52 (−6.02%)	4721.15 (−7.11%)	5494.2 (−8.28%)
	PCM-roof	6029.15 (−9.09%)	11,606.52 (−17.50%)	15,731.54 (−23.72%)	17,727.22 (−26.73%)
			Cooling (Wh/m^2)		
	PCM-wall	192.55 (−0.78%)	336.41 (−1.37%)	444.68 (−1.81%)	515.54 (−2.10%)
	PCM-roof	1212.54 (−4.94%)	1001.02 (−4.07%)	734.94 (−2.99%)	958.68 (−3.9%)
			Heating and Cooling (Wh/m^2)		
	PCM-wall	3259.93 (−3.59%)	433.93 (−4.77%)	5165.83 (−5.68%)	6009.74 (−6.61%)
	PCM-roof	724,169 (−7.97%)	13,820.08 (−15.21%)	18,680.04 (−20.55%)	21,634.4 (−23.80%)
			Heating (Wh/m^2)		
Tozeur	PCM-wall	3285.02 (−7.29%)	4414.55 (−9.80%)	5320.52 (−11.81%)	5860.36 (−13.01%)
	PCM-roof	7525.08 (−16.70%)	15,295.52 (−33.95%)	20,391.24 (−45.26%)	25,419.04 (−56.42%)
			Cooling (Wh/m^2)		
	PCM-wall	189.19 (−0.36%)	355.44 (−0.67%)	516.83 (−0.97%)	649.39 (−1.22%)
	PCM-roof	1478.44 (−2.79%)	2810.79 (−5.30%)	4370.75 (−8.24%)	5407.76 (−10.19%)
			Heating and Cooling (Wh/m^2)		
	PCM-wall	3474.21 (−3.54%)	4769.99 (−4.86%)	5837.35 (−5.95%)	6509.75 (−6.64%)
	PCM-roof	9003.52 (−9.18%)	18,106.31 (−18.45%)	24,761.99 (−25.24%)	30,826.8 (−31.42%)

3.4. *Effect of Double-Layer PCMs*

The double-PCM layer model was considered to benefit from the thermal properties of two PCMs at different locations with different melting temperatures. The model was formed by two PCMs with the same thickness as the PCM single layer. In order to analyze the efficiency of a double-PCM layer and energy saving, two configurations of a double-PCM layer were proposed in Figure 8. In Case 1, the first PCM layer was placed on the exterior surface with a different melting temperature, and the second was located in the interior surface of the wall with a fixed melting temperature (the optimum melting temperature is 21 °C) in all the cities. In Case 2, the first PCM layer (PCM1) represents the optimum melting temperature, allowing for more heating energy saving (21 °C), and the second PCM layer (PCM2) provides the PCM gives more reduction in cooling energy consumption. Based on the previous simulations, PCM21 is used as the first layer of the double-layer system due to the corresponding higher heating energy saving, while PCM29 and PCM25, which perform better cooling energy saving in warm and cold regions, respectively, are applied to the second layer. In Case 2, two double-PCM layer systems were considered: PCM21-PCM29 and PCM21-PCM25. The obtained results for the considered cases are given in Table 5. Accordingly, in Case 1, the PCM21-PCM29 double layers represent the best energy saving in warm regions (Sousse and Tozeur). Whereas, it does not apply to cold regions. The same results are obtained in Case 2, where an improvement in cooling and heating energy saving is achieved for the PCM21-PCM29 double layer for Sousse and Tozeur as well.

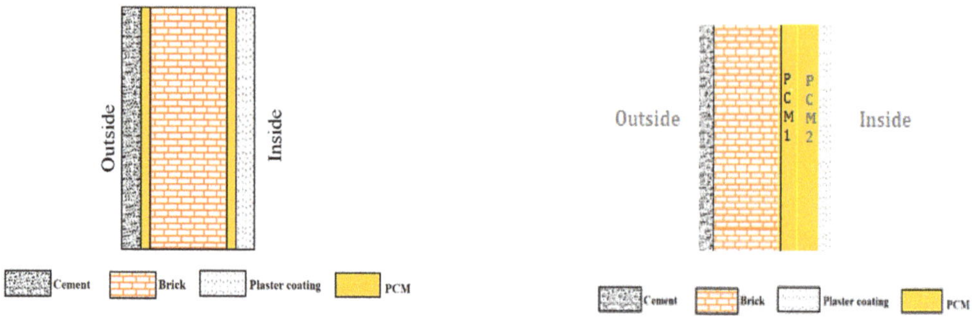

(a) (b)

Figure 8. Schematic representation of double layer of PCM location in wall (**a**) Case 1 and (**b**) Case 2.

Table 5. Effect of PCM double layers (10 mm) on energy saving for four regions.

		PCM	Heating/Cooling Energy Saving (Wh/m^2)		Heating and Cooling Energy Saving (Wh/m^2)	
			Heating (Wh/m^2)	Cooling (Wh/m^2)	Energy Saving (Wh/m^2)	Reduction (%)
Sousse	Single layer	PCM21	2776.46	140.77	2917.23	3.22
		PCM23	2107.87	246.35	2354.22	2.60
		PCM25	1161.38	959.39	2120.77	2.34
		PCM29	32.77	2668.14	2700.91	2.98
	Double layer (Case 1)	PCM29-PCM21	2512.36	554.99	3067.35	3.39
		PCM25-PCM21	2236	544.3	2780.3	3.07
		PCM23-PCM21	2157.28	329.89	2487.17	2.75
		PCM21-PCM21	2012.97	167.12	2180.09	2.41
	Double layer (Case 2)	PCM21-PCM29	1828.2	1982.51	3810.71	4.21
		PCM21-PCM25	2074.1	608.87	2682.97	2.96
Bizerte	Single layer	PCM21	2953.44	162.6	3116.04	3.07
		PCM23	1555.36	390.6	1945.96	1.92
		PCM25	1130.85	906.57	2037.42	2.01
		PCM29	82.01	2168.96	2250.97	2.22
	Double layer (Case 1)	PCM29-PCM21	2150.6	406.79	2557.39	2.52
		PCM25-PCM21	1980.6	583.17	2563.77	2.52
		PCM23-PCM21	1884.18	364.32	2248.5	2.21
		PCM21-PCM21	1724.75	213.3	1938.05	1.91
	Double layer (Case 2)	PCM21-PCM29	1385.87	1712.64	3098.51	3.06
		PCM21-PCM25	1640.34	672.65	2312.99	2.28
Tabarka	Single layer	PCM21	3067.38	192.55	3259.93	3.58
		PCM23	1851.81	353.44	2205.25	2.42
		PCM25	995.52	1005.07	2000.59	2.20
		PCM29	106.89	1533.01	1639.9	1.80
	Double layer (Case 1)	PCM29-PCM21	2357.6	345.36	2702.96	2.97
		PCM25-PCM21	2311.21	547.57	2858.78	3.14
		PCM23-PCM21	2247.06	379.76	2626.82	2.89
		PCM21-PCM21	2182.08	230.29	2412.37	2.65
	Double layer (Case 2)	PCM21-PCM29	1903.66	1280.05	3183.71	3.50
		PCM21-PCM25	2005.68	714.41	2720.09	2.99
Tozeur	Single layer	PCM21	3285.02	189.19	3474.21	3.54
		PCM23	2977.94	241.46	3219.4	3.28
		PCM25	1399.59	795.99	2195.58	2.23
		PCM29	144.85	2249.85	2394.7	2.44
	Double layer (Case 1)	PCM21-PCM29	2893	606.61	3499.61	3.56
		PCM21-PCM25	2429.33	484.71	2914.04	2.97
		PCM23-PCM21	2386.02	296.7	2682.72	2.73
		PCM21-PCM21	2335.7	187.68	2523.38	2.57
	Double layer (Case 2)	PCM21-PCM29	2244.52	1670.59	3915.11	3.99
		PCM21-PCM25	2713.79	493.31	3207.1	3.27

The results show that double-layer systems allow for a higher reduction in energy consumption than single-layer systems. This is in good agreement with several previous works [45–47]. In addition, annual heating and cooling energy savings and energy saving rates were calculated for single and double PCM layer systems (Table 5). For all PCMs used as a single layer (10 mm), it can be seen that the annual energy saving is 3.22% in Sousse and 3.54% in Tozeur. Whereas, in the case of double-layer systems, these values become 4.21% and 3.99% for Sousse and Tozeur, respectively. Moreover, the higher energy saving rate value was obtained when using the double-layer PCM29-PCM21 for warm regions and PCM25-PCM21 for cold regions. Hence, the proposed double-PCM layer systems integrated into buildings represent an effective way to improve energy efficiency.

3.5. CO_2 Emission Reduction in the Optimum Cases

As shown previously, the optimal conditions for minimizing the heating and cooling energy requirement vary from one climatic region (city) to another. Therefore, in this section, the energy conserved with these optimized cases for each respective region was calculated and the corresponding CO_2 emission reductions were revealed. The optimum melting temperature for each city was selected; the PCM was located at the best location, and the 40 mm PCM thickness was taken. It was seen that a significant reduction in heating, cooling, and total energy needs can be attained with the optimized integration of PCM. For instance, the total energy need reduction (heating and cooling) reached a value of 37.34%, 26.77%, 34.37%, and 41.61% for Sousse, Bizerte, Tabarka, and Tozeur, respectively (Figure 9). As a result of these reductions in the energy requirement of the building for heating and cooling applications, a considerable amount of CO_2 emission reductions were achieved, corresponding to 34.22%, 25.93%, 34.37%, and 38.74% for Sousse, Bizerte, Tabarka, and Tozeur, respectively (Figure 10). Consequently, the PCM integration with optimized parameters such as melting point, location, and thickness can significantly contribute to energy savings as well as a reduction in CO_2 emissions.

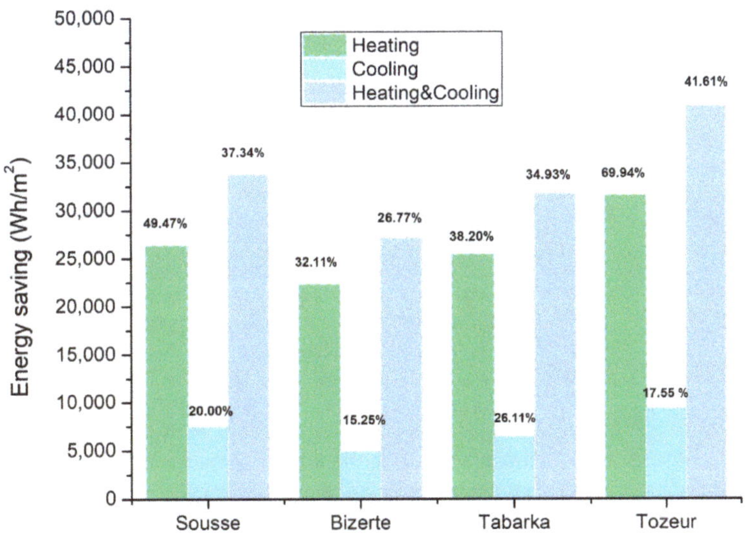

Figure 9. Energy saving in the optimum case.

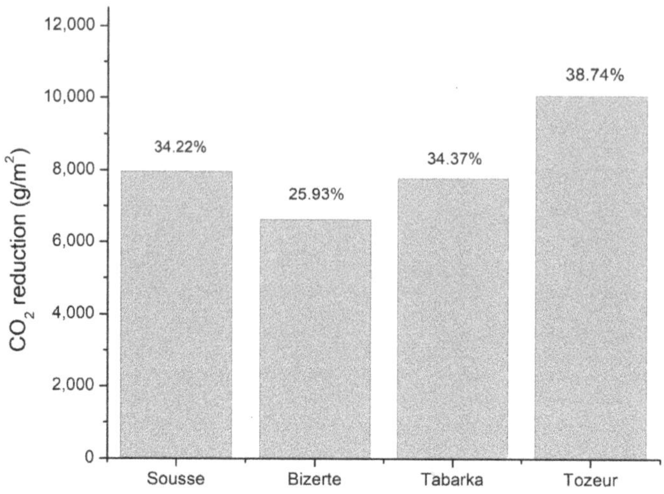

Figure 10. CO_2 emission reduction in the optimum case.

4. Conclusions

The annual energy savings of residential buildings incorporating PCM have been numerically examined in four different climatic regions of Tunisia by considering four cities located in those climates. The potential impacts of PCM melting temperature, location, and thickness were evaluated, along with PCM integration into different building envelopes (roof and external wall). The optimized configuration with respect to the melting point, location, thickness, and building envelope type was found, and the annual energy savings as well as CO_2 emission reductions were explored. Therefore, the following important results were reached in the study:

- Integrating PCM, either in the roof or wall, has a negligible influence on the cooling energy saving.
- The optimal location for the PCM implementation was found to be near the indoor, except for the Sousse region. The percentage energy saving for this optimal location and for the optimal PCM thickness of 40 mm was calculated as 41.61%, 37.34%, 34.93%, and 26.77% for Tozeur, Sousse, Tabarka, and Bizerte, respectively.
- Increasing the PCM thickness in the roof allows an increment in the total annual energy reduction, which varies from 7.97% to 31.42% depending on the climatic region.
- The PCM melting temperature has an important role in energy savings. The best melting temperature was 21 °C, the closest to the set point temperature, which is 21 °C in winter. However, in lower semi-arid regions (Sousse), the melting temperature, which allows the highest reduction in energy need, was 29 °C.
- The use of double-layer PCM with different melting temperatures at different locations represented an alternative for reducing energy consumption. The PCM with a low melting temperature (21 °C) favors heating energy savings, while PCM with a high melting point (29 °C) favors cooling energy savings. Besides, the double-layer systems composed of two different PCM represent a higher efficiency than a PCM single layer mainly in warm and arid regions (Sousse and Tozeur).
- Under optimal conditions of PCM integration in buildings, up to a 38.74% reduction in CO_2 emissions can be achieved in Tozeur.

Author Contributions: S.D.: Conceptualization, Methodology, Software, Investigation, Writing—Original Draft, Visualization, E.T.: Methodology, Software, Formal Analysis, Investigation, Writing—Original Draft, Visualization, O.K.: Formal Analysis, Writing—Original Draft, M.A.: Conceptualization, Formal Analysis, Writing—Review & Editing, Supervision, J.S.: Formal Analysis, Writing—Review & Editing. All authors have read and agreed to the published version of the manuscript.

Funding: This research received no external funding.

Data Availability Statement: Not applicable.

Conflicts of Interest: The authors declare no conflict of interest.

References

1. Al-Rashed, A.; Alnaqi, A.; Alsarraf, J. Usefulness of loading PCM into envelopes in arid climate based on Köppen–Geiger classification-Annual assessment of energy saving and GHG emission reduction. *J. Energy Storage* **2021**, *43*, 103152. [CrossRef]
2. Beltrán, R.D.; Martínez-Gomez, J. Analysis of phase change materials (PCM) for building wallboards based on the effect of environment. *J. Build. Eng.* **2019**, *24*, 100726. [CrossRef]
3. Li, Q.; Wang, Y.; Ma, L.; Arıcı, M.; Li, D.; Yıldız; Zhu, Y. Effect of sunspace and PCM louver combination on the energy saving of rural residences: Case study in a severe cold region of China. *Sustain. Energy Technol. Assess.* **2021**, *45*, 101126. [CrossRef]
4. Al-Yasiri, Q.; Szabó, M. Incorporation of phase change materials into building envelope for thermal comfort and energy saving: A comprehensive analysis. *J. Build. Eng.* **2021**, *36*, 102122. [CrossRef]
5. Farouk, N.; El-Rahman, M.A.; Sharifpur, M.; Guo, W. Assessment of CO_2 emissions associated with HVAC system in buildings equipped with phase change materials. *J. Build. Eng.* **2022**, *51*, 104236. [CrossRef]
6. Wang, X.; Li, W.; Luo, Z.; Wang, K.; Shah, S.P. Design, characteristic and application of phase change materials for sustainable and energy efficient buildings: A review. *Energy Build.* **2022**, *260*, 111923. [CrossRef]
7. Arumugam, C.; Shaik, S. Air-conditioning cost saving and CO2 emission reduction prospective of buildings designed with PCM integrated blocks and roofs. *Sustain. Energy Technol. Assess.* **2021**, *48*, 101657. [CrossRef]
8. Zavrl, E.; Stritih, U. Improved thermal energy storage for nearly zero energy buildings with PCM integration. *Sol. Energy* **2019**, *190*, 420–426.
9. Krarti, M. Multiple-Benefit Analysis of Scaling-Up Building Energy Efficiency Programs: The Case Study of Tunisia. *ASME J. Eng. Sustain. Build. Cities* **2020**, *1*, 1–35. [CrossRef]
10. Akeiber, H.; Nejat, P.; Majid, M.Z.A.; Wahid, M.A.; Jomehzadeh, F.; Famileh, I.Z.; Calautit, J.K.; Hughes, B.R.; Zaki, S.A. A review on phase change material (PCM) for sustainable passive cooling in building envelopes. *Renew. Sustain. Energy Rev.* **2016**, *60*, 1470–1497. [CrossRef]
11. Tunçbilek, E.; Arıcı, M.; Krajčík, M.; Li, Y.; Jurčević, M.; Nižetić, S. Impact of nano-enhanced phase change material on thermal performance of building envelope and energy consumption. *Int. J. Energy Res.* **2022**, *46*, 20249–20264. [CrossRef]
12. Surulivel, T.; Geetha, N.B.; Rajkumar, S. Parametric analysis of thermal behavior of the building with phase change materials for passive cooling. *Energy Sources Part A Recover. Util Env. Eff.* **2022**, *44*, 5627–5639.
13. Chelliah, A.; Saboor, S.; Ghosh, A.; Kontoleon, K.J. Thermal behaviour analysis and cost-saving opportunities of PCM-integrated terracotta brick buildings. *Adv. Civ. Eng.* **2021**, *2021*, 1–15. [CrossRef]
14. Li, D.; Yang, R.; Arıcı, M.; Wang, B.; Tunçbilek, E.; Wu, Y.; Liu, C.; Ma, Z.; Ma, Y. Incorporating phase change materials into glazing units for building applications: Current progress and challenges. *Appl. Therm. Eng.* **2022**, *210*, 118374. [CrossRef]
15. Zhang, S.; Hu, W.; Li, D.; Zhang, C.; Arıcı, M.; Yıldız; Zhang, X.; Ma, Y. Energy efficiency optimization of PCM and aerogel-filled multiple glazing windows. *Energy* **2021**, *222*, 119916. [CrossRef]
16. Lu, S.; Xu, B.; Tang, X. Experimental study on double pipe PCM floor heating system under different operation strategies. *Renew. Energy* **2020**, *145*, 1280–1291. [CrossRef]
17. González, B.; Prieto, M. Radiant heating floors with PCM bands for thermal energy storage: A numerical analysis. *Int. J. Therm. Sci.* **2021**, *162*, 106803. [CrossRef]
18. Meng, E.; Wang, J.; Yu, H.; Cai, R.; Chen, Y.; Zhou, B. Experimental study of the thermal protection performance of the high reflectivity-phase change material (PCM) roof in summer. *Build. Environ.* **2019**, *164*, 106381. [CrossRef]
19. Hu, J.; Yu, X. Adaptive building roof by coupling thermochromic material and phase change material: Energy performance under different climate conditions. *Constr. Build. Mater.* **2020**, *262*, 120481. [CrossRef]
20. Rathore PK, S.; Shukla, S.K. Potential of macroencapsulated PCM for thermal energy storage in buildings: A comprehensive review. *Constr. Build. Mater.* **2019**, *225*, 723–744. [CrossRef]
21. Hamidi, Y.; Aketouane, Z.; Malha, M.; Bruneau, D.; Bah, A.; Goiffon, R. Integrating PCM into hollow brick walls: Toward energy conservation in Mediterranean regions. *Energy Build.* **2021**, *248*, 111214. [CrossRef]
22. Kenzhekhanov, S.; Memon, S.A.; Adilkhanova, I. Quantitative evaluation of thermal performance and energy saving potential of the building integrated with PCM in a subarctic climate. *Energy* **2020**, *192*, 116607. [CrossRef]
23. Navarro, L.; de Gracia, A.; Castell, A.; Álvarez, S.; Cabeza, L.F. PCM incorporation in a concrete core slab as a thermal storage and supply system: Proof of concept. *Energy Build.* **2015**, *103*, 70–82. [CrossRef]

24. Royon, L.; Karim, L.; Bontemps, A. Thermal energy storage and release of a new component with PCM for integration in floors for thermal management of buildings. *Energy Build.* **2013**, *63*, 29–35. [CrossRef]
25. Hu, J.; Yu, X. Thermo and light-responsive building envelope: Energy analysis under different climate conditions. *Sol. Energy* **2019**, *193*, 866–877. [CrossRef]
26. Kumar, S.; Sheeja, R.; Jospher, A.J.; Krishnan, G.S.; Chandrasekar; AroulRaj, A. Energy-saving potential of a passive cooling system for thermal energy management of a residential building in Jaipur City, India. *Mater. Today: Proc.* **2020**, *43*, 1471–1477. [CrossRef]
27. Mechouet, A.; Oualim, E.M.; Mouhib, T. Effect of mechanical ventilation on the improvement of the thermal performance of PCM-incorporated double external walls: A numerical investigation under different climatic conditions in Morocco. *J. Energy Storage* **2021**, *38*, 102495. [CrossRef]
28. Hagenau, M.; Jradi, M. Dynamic modeling and performance evaluation of building envelope enhanced with phase change material under Danish conditions. *J. Energy Storage* **2020**, *30*, 101536. [CrossRef]
29. Fateh, A.; Borelli, D.; Weinläder, H.; Devia, F. Cardinal orientation and melting temperature effects for PCM-enhanced light-walls in different climates. *Sustain. Cities Soc.* **2019**, *51*, 101766. [CrossRef]
30. Lee, K.O.; Medina, M.A.; Sun, X. Development and verification of an EnergyPlus-based algorithm to predict heat transfer through building walls integrated with phase change materials. *J. Build. Phys.* **2016**, *40*, 77–95. [CrossRef]
31. Tabares-Velasco, P.C.; Christensen, C.; Bianchi, M. Verification and validation of EnergyPlus phase change material model for opaque wall assemblies. *Build. Environ.* **2012**, *54*, 186–196. [CrossRef]
32. Kuznik, F.; Virgone, J. Experimental assessment of a phase change material for wall building use. *Appl. Energy* **2009**, *86*, 2038–2046. [CrossRef]
33. Pedersen, C.O. Advanced zone simulation in EnergyPlus: Incorporation of variable properties and phase change material (PCM) capability. In Proceedings of the Building Simulation, Beijing, China, 15 September 2007; pp. 1341–1345.
34. Nghana, B.; Tariku, F. Phase change material's (PCM) impacts on the energy performance and thermal comfort of buildings in a mild climate. *Build. Environ.* **2016**, *99*, 221–238. [CrossRef]
35. Kottek, M.; Grieser, J.; Beck, C.; Rudolf, B.; Rubel, F. World map of the Köppen-Geiger climate classification updated. *Meteorol. Z.* **2006**, *15*, 259–263. [CrossRef]
36. National Institute of Meteorology of Tunisia. Available online: https://www.meteo.tn/en/ (accessed on 15 May 2022).
37. Available online: http://re.jrc.ec.europa.eu/pvg_tools/en/tools.html#HR (accessed on 15 May 2022).
38. Znouda, E.; Ghrab-Morcos, N.; Hadj-Alouane, A. Un algorithme génétique pour l'optimisation énergétique et économique des bâtiments méditerranéens. In Proceedings of the 6ème conférence Francophone de MOdélisation et SIMulation–MOSIM, Rabat, Morocco, 1 January 2006; Volume 6.
39. Saffari, M.; de Gracia, A.; Fernández, C.; Cabeza, L.F. Simulation-based optimization of PCM melting temperature to improve the energy performance in buildings. *Appl. Energy* **2017**, *202*, 420–434. [CrossRef]
40. Qu, Y.; Zhou, D.; Xue, F.; Cui, L. Multi-factor analysis on thermal comfort and energy saving potential for PCM-integrated buildings in summer. *Energy Build.* **2021**, *241*, 110966. [CrossRef]
41. Sovetova, M.; Memon, S.A.; Kim, J. Thermal performance and energy efficiency of building integrated with PCMs in hot desert climate region. *Sol. Energy* **2019**, *189*, 357–371. [CrossRef]
42. Zhang, Y.; Lin, K.; Jiang, Y.; Zhou, G. Thermal storage and nonlinear heat-transfer characteristics of PCM wallboard. *Energy Build.* **2008**, *40*, 1771–1779. [CrossRef]
43. Tunçbilek, E.; Arıcı, M.; Bouadila, S.; Wonorahardjo, S. Seasonal and annual performance analysis of PCM-integrated building brick under the climatic conditions of Marmara region. *J. Therm. Anal. Calorim.* **2020**, *141*, 613–624. [CrossRef]
44. Tong, X.; Xiong, X. A Parametric Investigation on Energy-Saving Effect of Solar Building Based on Double Phase Change Material Layer Wallboard. *Int. J. Photoenergy* **2018**, *2018*, 1829298. [CrossRef]
45. Cascone, Y.; Capozzoli, A.; Perino, M. Optimisation analysis of PCM-enhanced opaque building envelope components for the energy retrofitting of office buildings in Mediterranean climates. *Appl. Energy* **2018**, *211*, 929–953. [CrossRef]
46. Zhu, N.; Hu, P.; Xu, L. A simplified dynamic model of double layers shape-stabilized phase change materials wallboards. *Energy Build.* **2013**, *67*, 508–516. [CrossRef]
47. Louanate, A.; El Otmani, R.; Kandoussi, K.; Boutaous, M. Dynamic modeling and performance assessment of single and double phase change material layer–integrated buildings in Mediterranean climate zone. *J. Build. Phys.* **2021**, *44*, 461–478. [CrossRef]

buildings MDPI

Article

Potential Benefits of Thermal Insulation in Public Buildings: Case of a University Building

Reyhan Kaya and Semih Caglayan *

Department of Civil Engineering, Sakarya University of Applied Sciences, 54050 Sakarya, Turkey; y190007004@subu.edu.tr
* Correspondence: semihcaglayan@subu.edu.tr

Abstract: Global energy demand continues to rise due to advances in both developed and developing countries. Energy-efficient technologies and eco-friendly policies have been insufficient to counterbalance the increasing demand and, thus, the national strategies of many countries have been shaped by energy conservation considerations. Buildings are responsible for more than one third of the global final energy consumption and the energy use in buildings is expected to grow more than 40% in the next 20 years. Even though the energy-efficient retrofits and thermal insulation of the building envelope have been widely studied in academia, the case of existing public buildings has been largely neglected. To fill the gap, this study investigates the thermal insulation of existing public buildings and unveils its potential benefits. An administrative building of a public university has been the subject of financial analysis to observe the feasibility of insulation applications and to identify the most feasible insulation application. The results reveal that (i) the most feasible application depends considerably on the financial scenarios and (ii) the feasibility of insulation applications is greatly influenced by the building geometry. This study contributes to the literature by demonstrating the feasibility of energy retrofits in an administrative public building and proposing an alternative way to achieve national energy efficiency objectives.

Keywords: thermal insulation; public buildings; national energy policies; energy efficiency; optimum insulation application

check for updates

Citation: Kaya, R.; Caglayan, S. Potential Benefits of Thermal Insulation in Public Buildings: Case of a University Building. *Buildings* **2023**, *13*, 2586. https://doi.org/10.3390/buildings13102586

Academic Editors: Zhenjun Ma and Müslüm Arıcı

Received: 16 September 2023
Revised: 11 October 2023
Accepted: 12 October 2023
Published: 13 October 2023

1. Introduction

Rapid industrialization in recent decades has greatly accelerated the use of fossil fuels. Deriving the majority of energy consumption from nonrenewable energy sources leads to an undesired condition for the environment, emphasizing the need to reduce nonrenewable energy consumption on the global scale [1]. Even though the technological advancements and encouraging policies have gradually enhanced the efficiency of energy end-use services, the increasing demand for energy services has not been counterbalanced [2].

Achieving carbon neutrality by 2050 necessitates efficiency and a reduction in energy demand to ensure flexible selection of the available decarbonization options that avoid social and environmental side-effects [3]. Aside from benefitting the mitigation of climate change and national security of energy supply, energy savings owing to the conservation of energy can provide improvements in local pollution, productivity, competitiveness of companies, household energy expenditure, and health of building occupants [2].

Energy use in buildings and building construction sectors corresponds to more than one third of global final energy consumption and is responsible for nearly 40% of total (both direct and indirect) CO_2 emissions [4]. Furthermore, the energy demand of buildings is expected to grow by more than 40% in the next 20 years [5] due to urbanization and climate change, such as global warming and extreme weather events [6]. Therefore, providing energy-efficient buildings would be critical for the prevention of the increase in the energy demand.

The measures to achieve energy efficiency in buildings can be divided into two categories, namely soft and hard measures. The former implies education and behavior changes, while the latter contains investment in energy efficiency including equipment upgrades [7]. In spite of their high initial costs, the hard measures are especially effective for limiting energy consumption in buildings by means of well-proven solutions such as thermal insulation, the use of efficient glazing, the elimination of thermal bridges, and the installation of efficient heating/cooling generation and distribution systems [8].

The achievement of energy efficiency in new buildings has received a great deal of attention using a multidisciplinary approach throughout the life cycle including the pre-building, building, and post-building phases [9]. Equally important is the case of existing buildings where poor thermal properties lead to high energy demand [10]. In this direction, the trend toward re-engineering or retrofitting existing buildings has accelerated in recent years. Energy-efficient retrofits cover the improvement in the building envelope through building-integrated renewable energy technologies, climate control strategies, and insulation [11].

The renovations in buildings with structural vulnerability need to consider both the seismic and energy retrofitting concurrently. Especially in regions with seismic hazard, the interventions should address seismic and energy performance for buildings not designed to modern standards. The types of integrated retrofitting solutions proposed in the literature include (i) exoskeleton interventions, (ii) enhancements in envelope elements to achieve better energy/seismic performance, and (iii) replacements of envelope elements by higher-performance elements [12].

This study examines the potential benefits of energy-efficient retrofits, more specifically, the thermal insulation of existing public buildings. An administrative building in Sakarya University of Applied Sciences was subjected to investigation. The financial analysis of thermal insulation considered the cost of insulation application and potential savings owing to the reduction in annual energy requirements. Financial parameters were calculated to observe the feasibility of insulation applications on existing public buildings and to identify the optimum insulation application. Scenario analysis was conducted to pay regard to the probable deviations in inflation and interest rates.

This paper is organized as follows: Section 1 presents the motivation of this study and summarizes previously conducted studies on energy efficiency in buildings; Section 2 illustrates the flowchart of methodology and explains the steps in detail; Section 3 introduces and discusses the results; and Section 4 emphasizes main observations, clarifies the contribution to the literature, and proposes future studies.

1.1. Role of Energy Efficiency in the National Strategies

The oil crisis in the early 1970s brought about the emergence of conservation of energy and energy efficiency as the key pillars of national energy policies [13]. Countries from all over the world started to shape their national energy policies in compliance with these energy conservation considerations [14]. These policies have been especially effective in Europe, where the European Union has been the tower of strength. The Member States have been motivated to join the energy efficiency movement.

A critical part of the European Union climate and energy strategies is to decrease the energy requirement of buildings through the implementation of energy efficiency policies. The concept of energy efficiency first appeared in the European Union energy policy agenda in the 1970s and gradually gained importance with the increasing concern for global energy and climate priorities [15]. The Paris Agreement in December 2015 accelerated the attempts to mitigate the effects of global warming and climate change [16].

Energy efficiency has been proposed as a way to promote sustainability and competitiveness of the European economy. It is recognized as a cost-effective solution to concurrently enhance the security of supply and contribute to the energy and climate objectives. The European Union has set the target to achieve an energy efficiency of 32.5%

by 2030. The National Energy Efficiency Action Plans of the Member States involve radical energy efficiency measures to accomplish the national energy efficiency objectives [7].

The main piece of European Union legislation that imposes binding measures for the Member States to reach the objectives is the Energy Efficiency Directive (2012/27/EU) [17]. The Directive covers a number of binding measures including energy efficiency policies, the provisions on the setting of energy efficiency targets, and legal obligations to establish energy conservation schemes in Member States. It instructs the Member States to draft national energy efficiency action plans proposing a structural framework and an implementation methodology on energy efficiency [7].

The instructions of the directive have encouraged the Member States to develop energy efficiency policies. To illustrate, Italy promoted financial incentives to renovations and gave tax credits up to 110% of the intervention value [18]. Even though the binding measures are not applicable to the United Kingdom after leaving the European Union on the Brexit deal, it still follows through the international commitments. The United Kingdom aims to cut carbon emissions to combat climate change with the recent strategy "the Clean Growth Strategy: Leading the way to a low carbon future" [7].

To harmonize with the Energy Efficiency Directive, Turkish officials also implemented a policy, namely the National Energy Efficiency Action Plan (NEEAP). Within this context, the goals of the NEEAP were also included in the National Energy and Mining Policy issued by the Ministry of Energy and Natural Resources (MENR). The action plan, which was implemented in the period of 2017–2023, contained 55 actions defined under six categories, namely transport, industry and technology, buildings and services, energy, agriculture, and cross-cutting areas [19].

The primary energy consumption of Türkiye was 147.2 Mtoe in 2020 and is expected to reach up to 205.3 Mtoe by 2035. The shares of fossil resources and renewable energy sources in primary energy consumption in 2020 were 83.3% and 16.7%, respectively. The final energy consumption, which was 105.5 Mtoe in 2020, is expected to move up to 148.5 Mtoe by 2035. As of 2020, residential buildings were responsible for 24.5% of the total final energy consumption [20].

With the help of measures taken in the 2000–2020 period, Türkiye could reduce the energy intensity by 25%, which was still less than the 28%–36% reduction achieved by developed countries, namely France and Germany. The objective was indicated as the reduction in the energy intensity by 35.3% in the 2020–2035 period. It was also stated that meeting the objective would require a major transformation in all sectors and a systematic approach unlike that which had been previously followed [20].

1.2. *Previous Studies on Energy Efficiency in Buildings*

Energy efficiency has been an attractive topic in academia for decades. Researchers have conducted numerous studies to promote energy efficiency, especially in buildings, where energy use accounts for a considerable part of the global primary energy consumption. The topic has been mainly investigated through the following aspects: evaluation of the accuracy and optimality of national standards, prediction of building energy consumption, review and classification of energy efficiency measures, and analysis of the impact of energy efficiency measures.

A number of studies have questioned the accuracy and optimality of national standards. Caglayan et al. [21] examined the optimality of the limits stated in the Turkish national standard for thermal insulation requirements. Hussein et al. [22] assessed the benefits of an updated building energy code. They focused on the heat transfer coefficient of the building envelope to reduce the future energy demand. He et al. [23] analyzed the impact of upgrading the ASHRAE 90.1–2016 to 2019 in sixteen climate zones in the United States. Wang et al. [24] calculated the difference between the actual energy use and regulated energy consumption by design standards for residential buildings in China.

Multiple studies have attempted to predict the energy consumption of buildings. Runge and Zmeureanu [25] reviewed studies that had utilized artificial neural networks to forecast building energy use and demand. Le et al. [26] forecasted the heating load of buildings' energy efficiency by developing four artificial intelligence techniques including the combination of the artificial neural network with artificial bee colony optimization, particle swarm optimization, the imperialist competitive algorithm, and the genetic algorithm. Pham et al. [27] utilized machine learning algorithms to predict the short-term energy consumption in an hourly resolution in several buildings.

Certain studies have reviewed and classified the energy efficiency measures utilized in the literature. Belussi et al. [28] summarized the state of the art of zero-energy building performances and related technical solutions. Lidelöw et al. [29] performed a literature review of the energy efficiency measures for heritage buildings. Farzaneh et al. [30] reviewed the application of artificial intelligence technologies in smart buildings to decrease energy consumption through better control, improved reliability, and automation. Nair et al. [31] reviewed the energy efficiency retrofit measures and revealed the technical challenges and possibilities.

Several studies have focused on the impact of energy efficiency measures on building energy efficiency. Serale et al. [32] highlighted the application of model predictive control in improving energy efficiency in buildings. Bughio et al. [33] investigated the influence of passive energy efficiency measures on the cooling energy demand. Chippagiri et al. [34] tested the effects of sustainable prefabricated wall technology on energy consumption. The peak cooling load was reduced by six times. Meena et al. [35] assessed the potential of utilizing solar energy for water heating. Albatayneh et al. [36] investigated the shading effect of solar photovoltaic rooftop panels on the roof surface.

Researchers have paid great attention to the thermal insulation of the building envelope, but relatively few studies have specifically addressed the case of existing public buildings. On the government side, officials have frequently encouraged private home owners to pursue energy efficiency. The issue seems to have been neglected for public buildings that account for a significant portion of the total building stock. Based on the fact that meeting the national energy objectives requires a systematic approach unlike that previously followed, this study considers promoting the energy efficiency of public buildings as an alternative path to decrease the total energy consumption and to realize the national strategy in energy efficiency.

2. Research Methodology

2.1. The Building Profile

The profile of the aforementioned building and its floor plan are shown in Figure 1. It is a four-story university building located next to the Sapanca lake in the city of Sakarya, Türkiye. The building currently belongs to Sakarya University of Applied Sciences and is labeled as the T2 building. The building comprises various units including the office of the rectorate, dean's office, registrar's office, directorate of information technologies, department of civil engineering, and conference hall. The building has a total area of 4486 m^2 and a gross volume of 16,140 m^3. The areas of the windows are 167 m^2, 134 m^2, 7.2 m^2, and 10 m^2 in the south, north, east, and west directions, respectively.

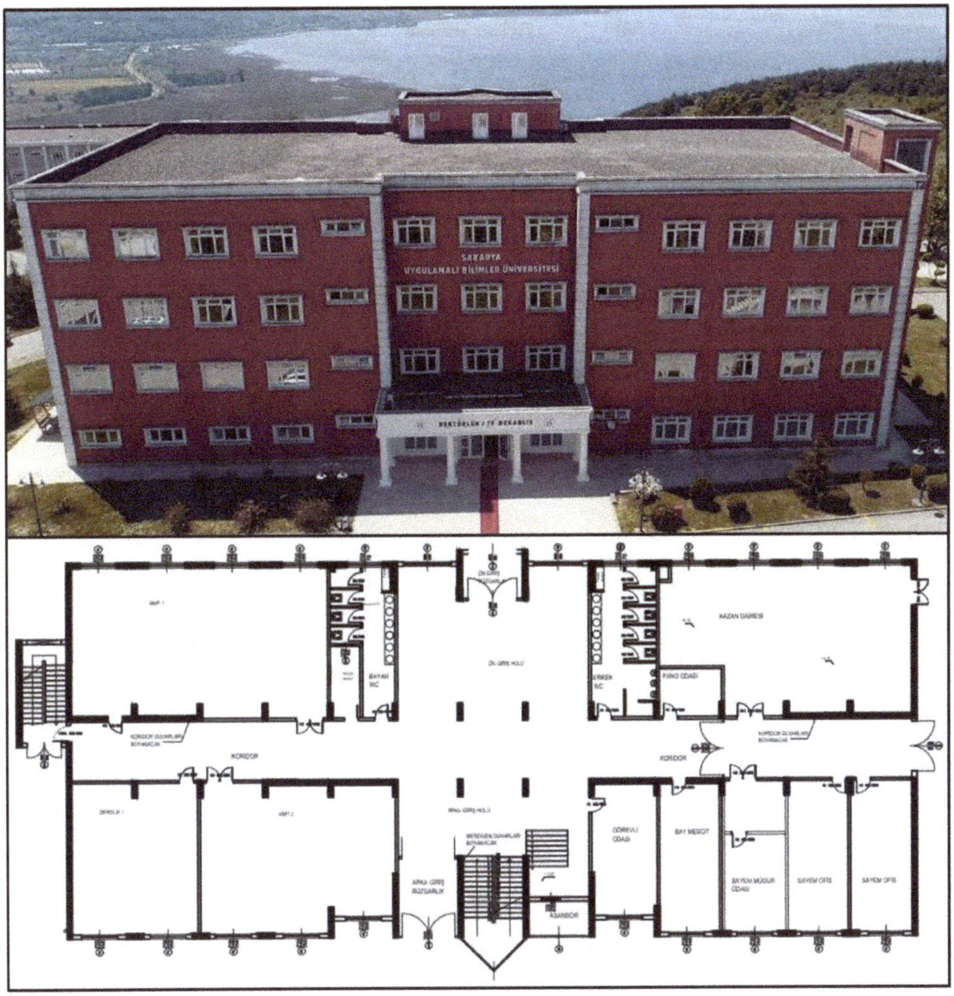

Figure 1. The profile and floor plan of T2 building.

The construction of the building was completed on 1 March 2011. The reinforced concrete building was designed and constructed in accordance with the Turkish Earthquake Code (TEC) 2007 [37]. The history of code revisions shows that the earthquakes in Turkey and the code revisions occur at similar times. The code came into force after the Great Marmara Earthquake on 17 August 1999. The earthquake caused a financial cost of USD 1.1–4.5 billion and a loss of approximately 25,000 lives [38]. Sakarya was categorized in the first seismic zone in TEC 2007. The building was designed according to an effective ground acceleration coefficient of 0.40 and building importance factor of 1.4. The building has had no structural damage. Therefore, this study focuses solely on the energy retrofitting.

2.2. *The Flowchart of Research Methodology*

The flowchart of the research methodology is illustrated in Figure 2. It is composed of a total of four phases, including the (i) thermal insulation options, (ii) annual energy requirement, (iii) life cycle costing analysis, and (iv) alternative design evaluation. In the first phase, thermal insulation options were determined. Insulation application included insulation of the exterior walls and insulation of the ceiling. It was assumed that the exterior walls would be insulated with expanded polystyrene (EPS). The lack of an inclined roof (a non-heated attic with sloping roof pitches) made extruded polystyrene (XPS) the only applicable insulation material to be applied on the ceiling. The costs of the insulation applications were determined by taking offers from companies for each insulation thickness (from 0 to 20 cm). The annual energy requirement of the building was calculated in the second phase. Space heating was calculated for the uninsulated and insulated cases to observe the potential saving. The calculations were based on the national standard, Turkish Standard (TS) 825 [39], which mainly considered the building geometry and climate.

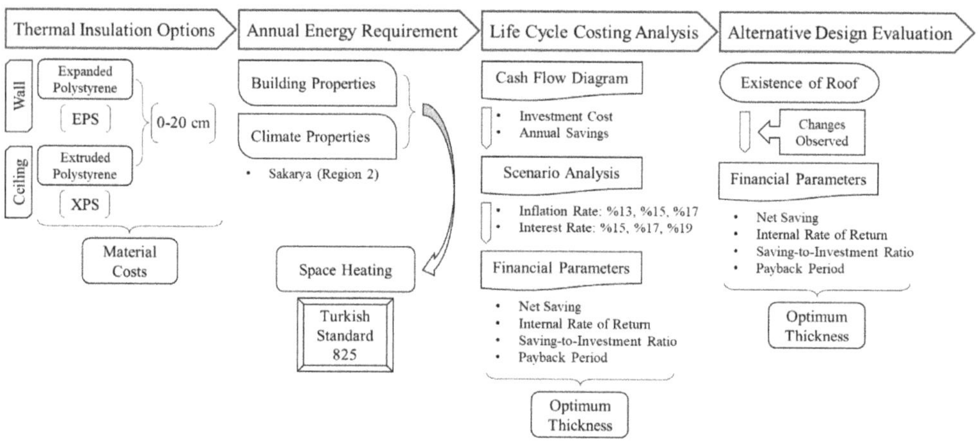

Figure 2. Flowchart of research methodology.

The third phase involved the life cycle costing analysis and detection of the optimum insulation thickness. A cash flow diagram was generated for each insulation alternative. The cash flow diagram covered the cost of investment and annual savings obtained in the following years. Financial parameters were determined for different scenarios of inflation and interest rates. The insulation alternative resulting in the greatest net saving was regarded as the optimum alternative. In the fourth phase, focus was placed on discovering the changes in results if the building had an inclined roof. The presence of a non-heated attic with sloping roof pitches would enable the application of alternative insulation materials such as stone wool on the ceiling. This could consequently lead to the achievement of greater financial parameters and different optimum thicknesses.

2.3. *Determination of Thermal Insulation Options*

The most common way of applying thermal insulation on existing buildings is to insulate the exterior walls and ceiling [21]. EPS is the most preferred insulation material for the exterior walls. Nevertheless, the situation is quite different for the ceiling. The preferred insulation material largely depends on the presence of the inclined roof. In case of the inclined roof, it would be possible to apply a variety of materials and the insulation material would be simply spread over the ceiling. The lower cost of stone wool makes it the most appropriate material for inclined roofs. The lack of the inclined roof restricts the types of applicable materials because the insulation material is applied on the interior side of the ceiling. XPS is the most commonly used material in this case. A cross-section of the insulated building envelope is presented in Figure 3. The area of both the ceiling and basement is 1040 m^2. The areas of the infilled and reinforced concrete walls are 1190 m^2 and 876 m^2, respectively.

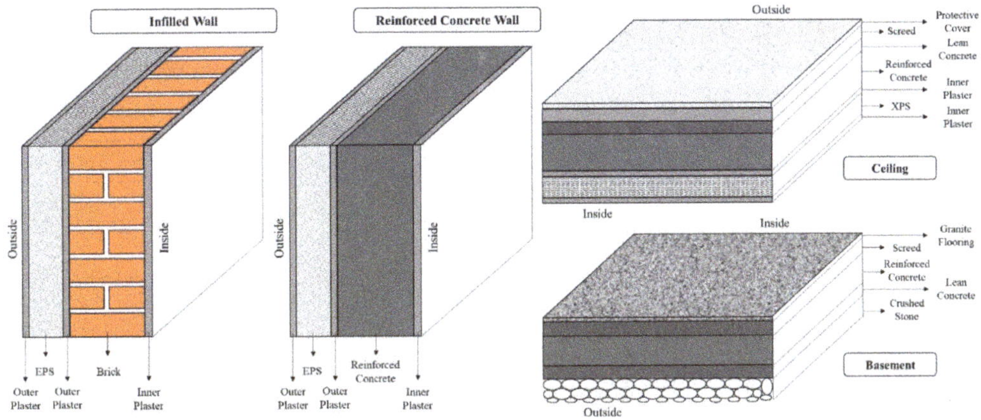

Figure 3. Cross-section of the building envelope.

2.4. Calculation of Annual Energy Requirement

The annual energy requirement was calculated by using the national standard TS 825 "thermal insulation requirements for buildings" published by the Turkish Standards Institute [39]. The standard mainly considers the building geometry and climate properties. Cities in Türkiye are categorized into four climate regions including region 1, region 2, region 3, and region 4. Sakarya belongs to region 2 in this category, where region 1 represents the warmest and region 4 covers the coldest cities. The yearly heating degree-days of Sakarya was calculated as 2154 for a base temperature of 19.5 °C [40].

The annual heating energy consumption (Q_{year}) is equal to the sum of monthly heating energy consumptions (Q_m).

$$Q_{year} = \sum Q_m \tag{1}$$

$$Q_m = [H * (\theta_{in} - \theta_{out}) - \eta * (\varphi_{in} + \varphi_s)] * t \tag{2}$$

The specific heat loss (H) of the building equals to the sum of the heat losses caused by conduction and convection (H_{tr}) and ventilation (H_{ven}).

$$H = H_{tr} + H_{ven} \tag{3}$$

H_{tr} is obtained as follows:

$$H_{tr} = \sum A * U = U_{ew} * A_{ew} + U_{gl} * A_{gl} + U_{ed} * A_{ed} + U_{ce} * A_{ce} + (0.5) * U_{fl} * A_{fl} \tag{4}$$

A and U represent the area and heat transfer coefficient, respectively. In the case of an inclined roof, $U_{ce} * A_{ce}$ is multiplied by 0.8.

According to the national standard, the heat loss due to thermal bridges is calculated separately. In this study, it was assumed that necessary precautions had been taken to prevent the occurrence of thermal bridges.

H_{ven} is calculated as follows:

$$H_{ven} = (0.264) * n_a * V_{gross} \tag{5}$$

V_{gross} is the gross building volume and n_a is the air changing volume. n_a was taken as 0.8 for natural ventilation.

The monthly average heat gain (φ_{in}) is equal to

$$\varphi_{in} \leq 5 * A_n \tag{6}$$

A_n is the building usage area.

$$A_n = (0.32) * V_{gross} \quad (7)$$

The monthly average solar energy gain (φ_s) is equal to

$$\varphi_{s,j} = \sum_k r_j * G_j * I_{j,k} * A_{gl,k} \quad (8)$$

r is the monthly average shading factor of the transparent surfaces, $A_{gl,k}$ is the total glazing area in direction k, and G is the solar energy permeation factor of the transparent elements. r was considered as 0.8 for detached buildings. The monthly average solar radiation intensities ($I_{j,k}$) are given in Table 1 [39].

Table 1. Monthly average solar radiation intensities (W/m^2).

Solar Radiation	Months											
	January	February	March	April	May	June	July	August	September	October	November	December
I_{south}	72	84	87	90	92	95	93	93	89	82	67	64
I_{north}	26	37	52	66	79	83	81	73	57	40	27	22
$I_{east/west}$	43	57	77	90	114	122	118	106	81	59	41	37

Solar energy permeation factor (G) is equal to

$$G_j = F_w * g_\perp \quad (9)$$

F_w is the correction factor for windows and $g_\perp$ is the solar energy permeation factor measured under laboratory conditions for the rays striking the surface vertically. F_w was assumed as 0.8 and $g_\perp$ was considered as 0.75 for colorless glass.

The monthly average usage factor of heat gain (η) is equal to

$$\eta = 1 - e^{(-1/GLR)} \quad (10)$$

GLR is the gain/loss ratio and is equal to

$$GLR = \frac{(\varphi_{in} + \varphi_s)}{H * (\theta_{in} - \theta_{out})} \quad (11)$$

The GLR formula in Equation (11) is inserted in Equation (10) and η becomes

$$\eta = 1 - e^{\left(\frac{H*(\theta_{out} - \theta_{in})}{(\varphi_{in} + \varphi_s)}\right)} \quad (12)$$

The monthly average indoor temperature (θ_{in}) is assumed as 20 °C in the national standard. The monthly average outdoor temperatures (θ_{out}) are presented in Table 2. Having a GLR value equal to or greater than 2.5 implies that no heat loss occurs in the corresponding month.

2.5. Limitations of the National Standard

The national standard requires that when the whole or independent parts of existing buildings are subjected to substantial repair or amendment, the resulting heat transfer coefficients of the exterior wall (U_{ew}), ceiling (U_{ce}), basement (U_{bs}), and glazing (U_{gl}) should be equal to or smaller than the limiting values indicated in the standard (Table 3). As the insulation is implemented to the exterior wall and ceiling, the resulting heat transfer coefficients (U_{ew} and U_{ce}) need to be less than 0.60 and 0.40 W/m^{2}K, respectively.

Table 2. Monthly average outdoor temperatures (°C).

Month	Region 1	Region 2	Region 3	Region 4
January	8.4	2.9	−0.3	−5.4
February	9.0	4.4	0.1	−4.7
March	11.6	7.3	4.1	0.3
April	15.8	12.8	10.1	7.9
May	21.2	18.0	14.4	12.8
June	26.3	22.5	18.5	17.3
July	28.7	24.9	21.7	21.4
August	27.6	24.3	21.2	21.1
September	23.5	19.9	17.2	16.5
October	18.5	14.1	11.6	10.3
November	13.0	8.5	5.6	3.1
December	9.3	3.8	1.3	−2.8

Table 3. Limiting heat transfer coefficients for existing buildings (W/m^{2}K) [39].

Region	U_{ew}	U_{ce}	U_{bs}	U_{gl}
Region 1	0.70	0.45	0.70	2.40
Region 2	0.60	0.40	0.60	2.40
Region 3	0.50	0.30	0.45	2.40
Region 4	0.40	0.25	0.40	2.40

The heat transfer coefficients of the insulated building are summarized in Table 4. The coefficients that satisfy the standard limits are colored in gray. The EPS insulation applied to the exterior walls needs to satisfy the requirements of the standard for both the infilled and reinforced concrete (RC) walls. Thus, the minimum applicable insulation thickness of EPS on exterior walls is 5 cm. A minimum XPS thickness of 8 cm can satisfy the requirements for the ceiling. This results in 17 insulation alternatives for the exterior walls (none or 5–15 cm) and 14 insulation alternatives for the ceiling (none or 8–20 cm), the combination of which leads to a total of 238 different insulation applications.

Table 4. Heat transfer coefficients of insulation applications.

U_{ew} Coefficient (W/m^{2}K)			U_{ce} Coefficient (W/m^{2}K)	
Thickness (Thickness of EPS insulation on the exterior wall (cm))	Infilled Wall	RC Wall	Thickness (Thickness of XPS insulation on the ceiling (cm))	Ceiling
0	1.670	3.239	0	2.479
1	1.102	1.620	1	1.451
2	0.838	1.107	2	1.026
3	0.676	0.841	3	0.793
4	0.576	0.678	4	0.647
5	0.488	0.568	5	0.546
6	0.428	0.489	6	0.472
7	0.381	0.429	7	0.416
8	0.344	0.382	8	0.372
9	0.313	0.344	9	0.336
10	0.287	0.314	10	0.307
11	0.266	0.288	11	0.282
12	0.247	0.266	12	0.261
13	0.231	0.247	13	0.243
14	0.216	0.231	14	0.227
15	0.204	0.217	15	0.213
16	0.193	0.204	16	0.201
17	0.183	0.193	17	0.190
18	0.173	0.183	18	0.180
19	0.165	0.174	19	0.171
20	0.158	0.165	20	0.163

2.6. Life Cycle Costing Analysis

The financial benefits of 238 different insulation applications were determined based on the life cycle costing analysis. From the financial perspective, insulation application implies an initial cost and annual savings in the following years. The initial cost is the cost of implementing insulation, which can be subdivided into the material cost, auxiliary item cost, and application cost. The initial cost was determined by taking offers from construction companies. Annual savings occur due to the decrease in the annual energy requirement. The annual saving is equal to the difference between the annual energy requirement of the uninsulated and insulated building.

A cash flow diagram was created for each insulation application. The diagram considered a period of 20 years in line with the assumptions made in the literature [21]. Financial parameters such as the net savings (NS), internal rate of return (IRR), savings-to-investment ratio (SIR), and payback period (PBP) were calculated for each insulation application to discover the potential benefits of insulation applications and determine the optimum alternative. Scenario analysis was conducted to observe the effect of changes in inflation and interest rates on the financial parameters and optimum insulation alternative.

NS measures the cost effectiveness of the benefits to be achieved from the investments. It is obtained by subtracting the present value of investment costs from the present value of the savings.

$$NS = \sum_{t=0}^{N} \frac{S_t}{(1+i)^t} - \frac{Inv_t}{(1+i)^t} \tag{13}$$

where S is the saving, Inv is the investment, i is the interest rate, t is the time, and N is the period of the study, which was assumed as 20 years.

IRR represents the annual rate of return to be earned on a project. It is equal to the discount rate that makes the net present value of a project zero.

$$0 = \sum_{t=0}^{N} \frac{S_t}{(1+IRR)^t} - \frac{Inv_t}{(1+IRR)^t} \tag{14}$$

SIR is the ratio of the net present value of the savings to the net present value of the investment. It should be greater than 1.0 to be considered as an alternative and regarded as cost-effective.

$$SIR = \frac{\sum_{t=0}^{N} \frac{S_t}{(1+i)^t}}{\sum_{t=0}^{N} \frac{Inv_t}{(1+i)^t}} \tag{15}$$

PBP shows the minimum time satisfying the condition that cash inflows offset the investment costs.

2.7. Evaluation of Alternative Design

The university building under investigation had no inclined roof and, thus, the insulation implementation on the ceiling was restricted to XPS insulation on the interior side. In an attempt to observe the effect of the building geometry (presence of the inclined roof) on the results, the process was repeated with the assumption that the building had an inclined roof. The presence of the non-heated attic with sloping roof pitches could enable the application of alternative insulation materials such as stone wool on the exterior side, which would lead to a significant decrease in the initial cost. The probable changes in the financial parameters and optimum insulation alternative were noted.

3. Research Results and Discussion

3.1. Annual Energy Requirement and Saving

The annual energy requirement and saving are summarized in Table 5 and the annual energy requirement is graphically presented in Figure 4. Each cell in the table is composed of the values where the upper represents the annual energy requirement and the lower stands for the annual energy saving. The annual energy requirement of the uninsulated building (the current form) is 615,056 kWh/year. Insulating the building according to the minimum thicknesses that satisfy the standard limitations decreases the annual energy requirement to 259,360 kWh/year, providing a saving of 57.8%. Increasing the insulation thicknesses can increase the saving amount up to 66.4%. Nevertheless, it should be noted that the increasing insulation thickness also results in a greater initial cost and, thus, does not guarantee better financial results.

Table 5. Annual energy requirement/saving (kWh/year).

	Exterior Wall Insulation Thickness (cm)																
Ceiling Insulation Thickness (cm)	0	5	6	7	8	9	10	11	12	13	14	15	16	17	18	19	20
0	615,056 0	387,987 227,069	379,657 235,399	373,268 241,788	368,210 246,846	364,107 250,950	360,710 254,346	357,852 257,205	355,413 259,643	353,308 261,748	351,473 263,583	349,858 265,198	348,427 266,630	347,149 267,907	346,001 269,055	344,965 270,092	344,024 271,032
8	481,319 133,737	259,360 355,696	251,509 363,547	245,501 369,555	240,754 374,303	236,908 378,149	233,728 381,328	231,056 384,000	228,778 386,278	226,814 388,243	225,102 389,954	223,597 391,459	222,264 392,793	221,074 393,982	220,006 395,050	219,042 396,014	218,168 398,909
9	479,072 135,984	257,282 357,775	249,441 365,615	243,440 371,616	238,700 376,357	234,859 380,197	231,684 383,372	229,016 386,041	226,742 388,315	224,780 390,276	223,071 391,985	221,568 393,488	220,237 394,819	219,049 396,007	217,983 397,073	217,021 398,036	216,148 400,572
10	477,220 137,836	255,569 359,487	247,737 367,319	241,743 373,313	237,008 378,049	233,172 381,885	230,001 385,056	227,335 387,721	225,064 389,992	223,105 391,951	221,398 393,658	219,898 395,159	218,568 396,488	217,382 397,674	216,317 398,739	215,356 399,700	214,484 401,966
11	475,667 139,389	254,135 360,922	246,309 368,747	240,321 374,736	235,590 379,466	231,758 383,299	228,590 386,466	225,928 389,129	223,659 391,398	221,702 393,355	219,997 395,060	218,498 396,559	217,170 397,887	215,985 399,071	214,921 400,135	213,961 401,095	213,090 403,151
12	474,346 140,710	252,915 362,142	245,095 369,961	239,112 375,945	234,385 380,672	230,556 384,501	227,391 387,666	224,731 390,326	222,464 392,593	220,509 394,548	218,805 396,251	217,308 397,749	215,981 399,075	214,797 400,259	213,735 401,321	212,776 402,281	211,906 404,170
13	473,208 141,848	251,865 363,192	244,050 371,006	238,071 376,985	233,347 381,709	229,521 385,535	226,359 388,698	223,701 391,355	221,436 393,621	219,482 395,574	217,780 397,276	216,284 398,772	214,958 400,098	213,776 401,281	212,714 402,342	211,756 403,301	210,886 405,056
14	472,218 142,838	250,952 364,105	243,142 371,915	237,166 377,890	232,445 382,611	228,622 386,435	225,461 389,595	222,805 392,251	220,542 394,515	218,590 396,467	216,889 398,167	215,394 399,663	214,069 400,987	212,887 402,169	211,826 403,230	210,869 404,187	210,000 405,834
15	471,349 143,707	250,150 364,906	242,344 372,712	236,372 378,684	231,654 383,402	227,832 387,224	224,674 390,382	222,019 393,037	219,757 395,299	217,806 397,250	216,107 398,950	214,612 400,444	213,289 401,768	212,108 402,949	211,048 404,009	210,091 404,966	209,223 406,522
16	470,580 144,476	249,441 365,615	241,639 373,417	235,669 379,387	230,954 384,103	227,134 387,922	223,977 391,079	221,324 393,732	219,063 395,993	217,114 397,943	215,415 399,642	213,921 401,135	212,598 402,458	211,418 403,638	210,358 404,698	209,402 405,654	208,535 407,135
17	469,895 145,162	248,809 366,247	241,010 374,046	235,043 380,013	230,330 384,727	226,512 388,544	223,357 391,700	220,705 394,351	218,445 396,611	216,496 398,560	214,798 400,258	213,306 401,751	211,983 403,073	210,804 404,253	209,745 405,312	208,789 406,268	207,922 407,684
18	469,280 145,777	248,243 366,813	240,447 374,610	234,482 380,574	229,770 385,286	225,954 389,102	222,800 392,256	220,150 394,907	217,891 397,166	215,943 399,114	214,246 400,811	212,754 402,303	211,432 403,624	210,253 404,804	209,194 405,862	208,239 406,817	207,372 408,179
19	468,725 146,331	247,732 367,324	239,939 375,118	233,976 381,080	229,266 385,790	225,451 389,605	222,298 392,758	219,649 395,407	217,391 397,665	215,444 399,613	213,747 401,309	212,256 402,800	210,935 404,122	209,756 405,300	208,698 406,358	207,743 407,313	206,877 408,179
20	468,222 146,834	247,269 367,787	239,478 375,578	233,517 381,539	228,809 386,247	224,995 390,061	221,844 393,213	219,195 395,861	216,938 398,118	214,992 400,065	213,296 401,761	211,805 403,251	210,484 404,572	209,306 405,750	208,249 406,808	207,294 407,762	206,428 408,628

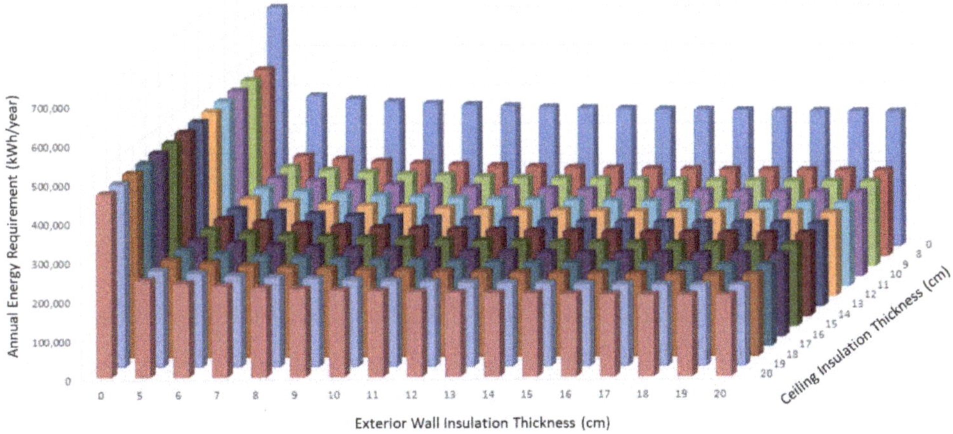

Figure 4. Graphical presentation of annual energy requirement.

3.2. Cost of Insulation

The insulation cost was determined by receiving offers from construction firms (Table 6). The cost of an insulation application includes the cost of material, auxiliary items, and application. It was initially assumed that EPS insulation would be applied on the exterior walls and XPS insulation would be applied on the interior side of the ceiling. In order to evaluate the changes in case of the inclined roof, the cost of stone wool application on the exterior side of the ceiling was also obtained. As the stone wool is simply spread over the ceiling, the cost only includes the cost of the material. It does not necessarily require auxiliary items and the application cost becomes negligible. It was observed that the presence or lack of the inclined roof caused a considerable difference in the cost of insulation application on the ceiling and consequently in the initial cost.

Table 6. Cost of insulation applications (USD/m^2).

Thickness (cm)	Expanded Polystyrene (EPS)				Extruded Polystyrene (XPS)				Stone Wool
	Mat.	Aux.	App.	Total	Mat.	Aux.	App.	Total	
1	0.68	2.57	3.90	7.15	1.17	2.57	3.90	7.64	0.43
2	1.35	2.57	3.90	7.82	2.34	2.57	3.90	8.81	0.86
3	2.02	2.57	3.90	8.49	3.51	2.57	3.90	9.98	1.29
4	2.70	2.57	3.90	9.17	4.69	2.57	3.90	11.16	1.72
5	3.38	2.63	3.90	9.91	5.85	2.63	3.90	12.38	2.15
6	4.05	2.63	3.90	10.58	7.02	2.63	3.90	13.55	2.58
7	4.72	2.67	3.95	11.34	8.20	2.67	3.95	14.82	3.01
8	5.40	2.67	4.00	12.07	9.36	2.67	4.00	16.03	3.44
9	6.07	2.72	4.00	12.79	10.54	2.72	4.00	17.26	3.87
10	6.74	2.72	4.00	13.46	11.70	2.72	4.00	18.42	4.30
11	7.43	2.92	4.05	14.40	12.87	2.92	4.05	19.84	4.73
12	8.10	2.92	4.05	15.07	14.05	2.92	4.05	21.02	5.16
13	8.77	2.95	4.05	15.77	15.21	2.95	4.05	22.21	5.59
14	9.44	2.95	4.15	16.54	16.39	2.95	4.15	23.49	6.02
15	10.12	3.03	4.15	17.30	17.56	3.03	4.15	24.74	6.45
16	10.79	3.03	4.15	17.97	18.73	3.03	4.15	25.91	6.88
17	11.46	3.06	4.25	18.77	19.90	3.06	4.25	27.21	7.31
18	12.13	3.12	4.25	19.50	21.07	3.12	4.25	28.44	7.74
19	12.82	3.15	4.35	20.32	22.24	3.15	4.35	29.74	8.17
20	13.49	3.15	4.35	20.99	23.41	3.15	4.35	30.91	8.60

3.3. Cash Flow Diagrams

A cash flow diagram was generated for each insulation application and Table 7 illustrates the diagrams of certain applications for an inflation and interest rate of 15% and 17%, respectively. The annual savings were calculated based on the assumption that the cost of natural gas was 0.026 USD/kWh in the base year and would increase in harmony with the inflation rate. Financial parameters indicated the financial feasibility of insulation applications in existing public buildings. The cases showed that NS values were clearly positive, IRR values were mostly greater than 30%, SIR values were greater than 2, and PBP values were less than 7 years.

NS values of all insulation applications are summarized in Table 8 and Figure 5 for an inflation and interest rate of 15% and 17%, respectively. The insulation alternative with the greatest NS value is regarded as the optimum alternative. It is observed that insulating the building with minimum insulation thicknesses satisfying the standard limits could provide a saving of USD 112,838. However, greater savings could be attained by increasing the thicknesses of insulation materials. The optimum insulation application is identified as the application of 9 cm EPS on the exterior walls and 8 cm XPS on the ceiling, which results in a saving of USD 116,344.

Table 7. Cash flow diagram of certain insulation applications (USD).

Insulation	Year							NS (USD)	IRR (%)	SIR (-)	PBP (Years)
	0	1	2	5	13	17	20				
Wall: none Ceiling: 8 cm	−16,672	3867	4447	6763	20,688	36,184	55,031	39,718	37.55	3.38	5–6
Wall: 5 cm Ceiling: 8 cm	−37,140	10,284	11,827	17,987	55,024	96,237	146,364	112,838	42.30	4.04	4–5
Wall: 5 cm Ceiling: 20 cm	−52,612	10,634	12,229	18,599	56,894	99,508	151,339	102,464	34.30	2.95	6–7
Wall: 6 cm Ceiling: none	−21,854	6806	7827	11,904	36,414	63,689	96,863	77,401	45.88	4.54	3–4
Wall: 8 cm Ceiling: 15 cm	−50,664	11,085	12,748	19,388	59,310	103,733	157,764	110,996	36.13	3.19	5–6
Wall: 9 cm Ceiling: 8 cm	−43,101	10,933	12,573	19,123	58,497	102,311	155,603	116,344	39.86	3.70	4–5
Wall: 9 cm Ceiling: 20 cm	−58,573	11,278	12,970	19,725	60,340	105,534	160,504	105,895	33.24	2.81	6–7
Wall: 12 cm Ceiling: 15 cm	−56,852	11,429	13,144	19,990	61,150	106,951	162,660	109,824	34.19	2.93	6–7
Wall: 18 cm Ceiling: 20 cm	−72,438	11,762	13,526	20,572	62,930	110,065	167,395	99,091	29.79	2.37	7–8
Wall: 19 cm Ceiling: 14 cm	−66,400	11,686	13,439	20,439	62,525	109,356	166,317	104,023	31.37	2.57	6–7

Table 8. Net saving of insulation alternatives (USD).

Ceiling Insulation Thickness (cm)	Exterior Wall Insulation Thickness (cm)															
	0	5	6	7	8	9	10	11	12	13	14	15	16	17	18	19	20
0		75,275	77,401	78,523	79,143	79,384	79,429	78,709	78,351	77,790	76,971	76,078	75,295	74,179	73,153	71,912	70,922
8	39,718	112,838	114,762	115,723	116,212	116,344	116,299	115,500	115,074	114,454	113,583	112,643	111,819	110,666	109,606	108,335	108,169
9	39,389	112,438	114,358	115,316	115,802	115,932	115,884	115,084	114,657	114,035	113,163	112,222	111,397	110,243	109,183	107,911	107,594
10	38,957	111,947	113,863	114,819	115,303	115,431	115,382	114,580	114,151	113,529	112,656	111,714	110,889	109,734	108,672	107,400	106,969
11	38,139	111,080	112,993	113,946	114,428	114,554	114,504	113,701	113,271	112,648	111,774	110,832	110,005	108,850	107,788	106,516	105,996
12	37,472	110,370	112,281	113,231	113,712	113,837	113,785	112,981	112,551	111,927	111,052	110,109	109,283	108,127	107,065	105,791	105,202
13	36,708	109,569	111,477	112,426	112,906	113,029	112,977	112,172	111,741	111,116	110,241	109,297	108,470	107,314	106,251	104,978	104,332
14	35,797	108,626	110,532	111,480	111,958	112,081	112,027	111,221	110,789	110,164	109,288	108,345	107,517	106,360	105,297	104,023	103,331
15	34,868	107,668	109,573	110,519	110,996	111,118	111,063	110,257	109,824	109,199	108,322	107,378	106,550	105,393	104,330	103,056	102,326
16	33,968	106,743	108,646	109,591	110,067	110,188	110,133	109,326	108,893	108,267	107,390	106,445	105,617	104,460	103,396	102,122	101,360
17	32,909	105,661	107,563	108,507	108,982	109,102	109,047	108,239	107,806	107,179	106,302	105,357	104,529	103,371	102,307	101,033	100,244
18	31,894	104,625	106,526	107,469	107,943	108,063	108,006	107,198	106,764	106,137	105,260	104,315	103,486	102,328	101,264	99,990	99,178
19	30,768	103,481	105,381	106,323	106,796	106,915	106,858	106,050	105,616	104,989	104,111	103,166	102,336	101,178	100,114	98,839	97,818
20	29,768	102,464	104,362	105,304	105,777	105,895	105,838	105,029	104,594	103,967	103,089	102,143	101,314	100,156	99,091	97,816	96,795

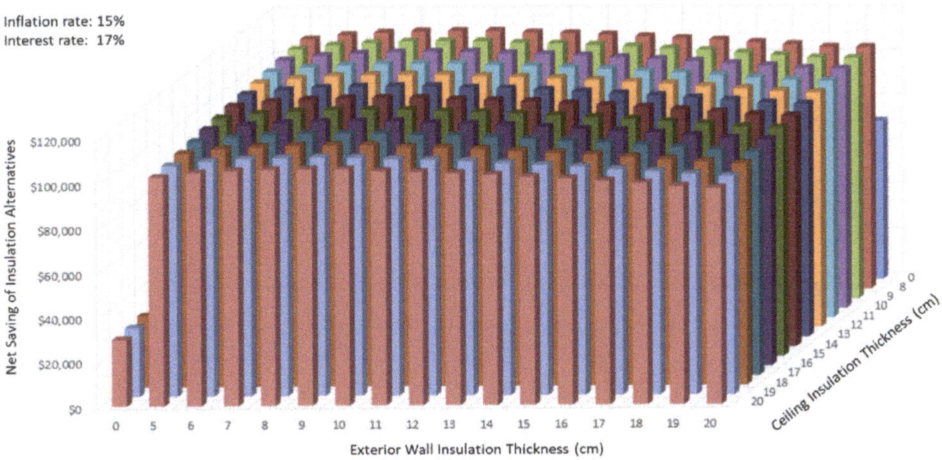

Figure 5. Graphical presentation of net saving.

3.4. Scenario Analysis and Optimum Insulation Thickness

Generation of the cash flow diagrams for each insulation application was repeated for varying interest and inflation rates. The analysis included a total of nine different scenarios, where the interest and inflation rates varied between 15–19% and 13–17%, respectively. The optimum insulation application was noted for each scenario and the financial results of these optimum thicknesses are presented in Table 9. The optimum EPS thickness on the exterior walls changed between 7 and 10 cm, while the optimum XPS thickness on the ceiling in all scenarios was 8 cm, which corresponded to the minimum insulation thickness satisfying the standard limits. Financial parameters of optimum insulation applications demonstrated the profitability of these public investments. The NS values ranged between USD 74,707 and 182,503, IRR values were more than two times the interest rates, SIR values were mostly greater than 3 and went up to 5, and the PBP values were only 4–5 years.

Table 9. Scenario analysis of financial parameters.

No	Interest Rate	Inflation Rate	Optimum Thickness		Financial Parameters			
			Wall	Ceiling	NS (USD)	IRR	SIR	PBP
1	%15	%13	9 cm	8 cm	118,680	37.89%	3.75	4–5 years
2	%15	%15	10 cm	8 cm	147,259	39.24%	4.31	4–5 years
3	%15	%17	10 cm	8 cm	182,503	41.21%	5.10	4–5 years
4	%17	%13	8 cm	8 cm	94,015	38.57%	3.26	4–5 years
5	%17	%15	9 cm	8 cm	116,344	39.86%	3.70	4–5 years
6	%17	%17	10 cm	8 cm	143,982	41.21%	4.24	4–5 years
7	%19	%13	7 cm	8 cm	74,707	39.24%	2.86	4–5 years
8	%19	%15	8 cm	8 cm	92,401	40.54%	3.22	4–5 years
9	%19	%17	9 cm	8 cm	114,074	41.83%	3.65	4–5 years

3.5. Case of the Inclined Roof

Investigation of the financial feasibility and the optimization process were repeated for the case of the inclined roof, and changes in the financial parameters and optimum insulation thicknesses were observed (Table 10). The optimum thickness of EPS insulation remained the same as expected. However, the notably lower cost of stone wool application increased the optimum insulation thickness applied on the ceiling to 10–14 cm. Financial parameters were also influenced by the reflections of this situation on the cash flow diagram. NS values had an increase of approximately 20%, IRR values became greater than 50%, SIR values had an increase of more than 50%, and PBP became less than 4 years.

Table 10. Optimum insulation thickness in case of the inclined roof.

No	Interest Rate	Inflation Rate	Optimum Thickness		Financial Parameters			
			Wall	Ceiling	NS (USD)	IRR	SIR	PBP
1	%15	%13	9 cm	11 cm	142,007	50.25%	5.53	3–4 years
2	%15	%15	10 cm	12 cm	172,673	50.51%	6.20	3–4 years
3	%15	%17	10 cm	14 cm	210,586	51.68%	7.18	3–4 years
4	%17	%13	8 cm	10 cm	115,587	52.24%	4.93	3–4 years
5	%17	%15	9 cm	11 cm	139,504	52.24%	5.45	3–4 years
6	%17	%17	10 cm	12 cm	169,154	52.49%	6.10	3–4 years
7	%19	%13	7 cm	10 cm	94,948	53.90%	4.40	3–4 years
8	%19	%15	8 cm	10 cm	113,861	54.23%	4.87	3–4 years
9	%19	%17	9 cm	11 cm	137,072	54.22%	5.37	3–4 years

4. Conclusions

This study discovered the potential benefits of insulation application in existing public buildings. Insulation applications satisfying the standard limits were considered as the alternatives and the optimum alternative was determined through the life cycle costing analysis with different scenarios of inflation and interest rates. Changes in the optimization process were observed for the case of an alternative building geometry. The findings show that

- The optimum insulation application depends considerably on the scenario of the inflation and interest rates;
- Benefits of the insulation application are greatly influenced by the building geometry, more specifically, the presence of the inclined roof.

Cash flow diagrams generated under different economic scenarios demonstrated the profitability of thermal insulation in public buildings. The diagrams of optimum applications produced NS values greater than zero, IRR values greater than the interest rates, SIR values greater than one, and payback period less than five years. The financial results (and thus the optimum insulation alternatives) varied with the scenario of the inflation and interest rates. It was observed that the presence of the inclined roof could provide even better financial results as it enabled the use of alternative cost-effective insulation materials on the ceiling.

Governments encourage their citizens to invest in energy conservation instruments in an attempt to actualize the national energy strategies focusing on decreasing the energy requirement and greenhouse gas emissions. These attempts are not perfectly effective, as convincing home owners for such investments requires raising awareness of the life cycle costing concept where the benefits are obtained in the course of time. This study demonstrates the profitability of thermal insulation in existing public buildings, which can be achieved with the governments' own initiatives. Investing in the thermal insulation of public buildings should be considered as an alternative path for governments to achieve the objectives of their national energy strategies.

This study examined the financial feasibility of thermal insulation for an administrative university building. The results cannot be generalized for residential buildings or other types of public buildings such as hospitals. This is mainly because the calculation of the annual energy requirement changes according to the building functionality. To illustrate, the monthly average indoor temperature (θ_{in}) is taken as 19 °C, 20 °C, and 22 °C for residential buildings, education buildings, and hospitals, respectively. The changing energy calculation method in the national standard necessitates repetition of the analysis for the other building types. Yet, the analysis provided with this demonstrates the potential benefits of energy efficiency measures in public buildings.

The illustrated optimization process can be repeated for existing public buildings in other cities or countries to observe the changes in potential benefits. The results may change with respect to the building geometry, building functionality, standard limitations in the corresponding country, availability and cost of insulation materials, and climate properties. The identified optimum insulation alternative and financial benefits can be compared to investigate potential differences across cities/countries and the reasons behind these differences can be discussed. Country-specific suggestions can be provided to promote energy efficiency and enhance the return on investment in corresponding countries.

Author Contributions: Conceptualization, S.C.; Methodology, R.K.; Validation, R.K.; Formal analysis, R.K.; Writing—review & editing, S.C.; Supervision, S.C. All authors have read and agreed to the published version of the manuscript.

Funding: This research received no external funding.

Data Availability Statement: Data should be made available on request.

Conflicts of Interest: The authors declare no conflict of interest.

Nomenclature

A	area (m^2)
A_n	building usage area (m^2)
F_w	correction factor for windows
G	solar energy permeation factor of the transparent elements
$g_{\perp}$	solar energy permeation factor measured under laboratory conditions
H	specific heat loss of the building (W/K)
I	monthly average solar radiation intensity (W/m^2)
i	interest rate (%)
Inv	investment (USD)
IRR	internal rate of return (%)
n_a	air changing ratio
N	period of the study (year)
NS	net savings (USD)
PBP	payback period (year)
Q_m	monthly heating energy consumption (kWh/month)
Q_{year}	annual heating energy consumption (kWh/year)
r	monthly average shading factor of the transparent surfaces
S	saving
SIR	savings-to-investment ratio
t	time
U	heat transfer coefficient ((W/m^2)/K)
V	volume (m^3)
η	average usage factor of heat gain
θ	temperature (°C)
φ	average heat gain (W)
Subscripts	
ce	ceiling
ed	exterior door
ew	exterior wall
fl	floor
gl	glazing
in	inside
j	month
k	direction
out	outside
s	solar
tr	transfer
ven	ventilation

References

1. Zune, M.; Pantua, C.A.J.; Rodrigues, L.; Gillott, M. A review of traditional multistage roofs design and performance in vernacular buildings in Myanmar. *Sustain. Cities Soc.* **2020**, *60*, 102240. [CrossRef]
2. Bertoldi, P. Policies for energy conservation and sufficiency: Review of existing policies and recommendations for new and effective policies in OECD countries. *Energy Build.* **2022**, *264*, 112075. [CrossRef]
3. Buonomano, A.; Barone, G.; Forzano, C. Advanced energy technologies, methods, and policies to support the sustainable development of energy, water and environment systems. *Energy Rep.* **2022**, *8*, 4844–4853. [CrossRef]
4. International Energy Agency. *Tracking Buildings 2020*; International Energy Agency: Paris, France, 2020.
5. Mariano-Hernández, D.; Hernández-Callejo, L.; Zorita-Lamadrid, A.; Duque-Pérez, O.; García, F.S. A review of strategies for building energy management system: Model predictive control, demand side management, optimization, and fault detect & diagnosis. *J. Build. Eng.* **2021**, *33*, 101692.
6. Lu, C.; Li, S.; Lu, Z. Building energy prediction using artificial neural networks: A literature survey. *Energy Build.* **2022**, *262*, 111718. [CrossRef]
7. Malinauskaite, J.; Jouhara, H.; Ahmad, L.; Milani, M.; Montorsi, L.; Venturelli, M. Energy efficiency in industry: EU and national policies in Italy and the UK. *Energy* **2019**, *172*, 255–269. [CrossRef]
8. Aslani, A.; Bakhtiar, A.; Akbarzadeh, M.H. Energy-efficiency technologies in the building envelope: Life cycle and adaptation assessment. *J. Build. Eng.* **2019**, *21*, 55–63. [CrossRef]

9. Najjar, M.; Figueiredo, K.; Hammad, A.W.; Haddad, A. Integrated optimization with building information modeling and life cycle assessment for generating energy efficient buildings. *Appl. Energy* **2019**, *250*, 1366–1382. [CrossRef]
10. Kalmár, F. Summer operative temperatures in free running existing buildings with high glazed ratio of the facades. *J. Build. Eng.* **2016**, *6*, 236–242. [CrossRef]
11. Sharma, S.K.; Mohapatra, S.; Sharma, R.C.; Alturjman, S.; Altrjman, C.; Mostarda, L.; Stephan, T. Retrofitting existing buildings to improve energy performance. *Sustainability* **2022**, *14*, 666. [CrossRef]
12. Pohoryles, D.; Bournas, D.; Da Porto, F.; Caprino, A.; Santarsiero, G.; Triantafillou, T. Integrated seismic and energy retrofitting of existing buildings: A state-of-the-art review. *J. Build. Eng.* **2022**, *61*, 105274. [CrossRef]
13. Grossman, P.Z. Energy shocks, crises and the policy process: A review of theory and application. *Energy Policy* **2015**, *77*, 56–69. [CrossRef]
14. Caglayan, S.; Ozorhon, B.; Kurnaz, L. Nationwide mapping of optimum wall insulation thicknesses: A stochastic approach. *J. Therm. Sci. Technol.* **2022**, *42*, 169–202. [CrossRef]
15. Economidou, M.; Todeschi, V.; Bertoldi, P.; D'Agostino, D.; Zangheri, P.; Castellazzi, L. Review of 50 years of EU energy efficiency policies for buildings. *Energy Build.* **2020**, *225*, 110322. [CrossRef]
16. Kinley, R. Climate change after Paris: From turning point to transformation. *Clim. Policy* **2017**, *17*, 9–15. [CrossRef]
17. European Parliament and Council. Directive 2012/27/EU of the European Parliament and of the Council of 25 October 2012 on energy efficiency, amending Directives 2009/125/EC and 2010/30/EU and repealing Directives 2004/8/EC and 2006/32/EC. *Off. J. Eur. Union* **2012**, *14112012*, 1–56.
18. Presidente Della Repubblica Italiana. *'Decree 34/2020. Misure Urgenti in Materia di Salute, Sostegno al Lavoro e All'economia, Nonché di Politiche Sociali Connesse All'emergenza Epidemiologica da COVID-19 Gazzetta Ufficiale Della Repubblica Italiana n. 21 del 19.05.2020'*; Gazzetta Ufficiale Della Repubblica Italiana: Rome, Italy, 2020.
19. MENR. *The National Energy Efficiency Action Plan*; Republic of Türkiye Ministry of Energy and Natural Resources: Ankara, Turkey, 2018.
20. MENR. *Türkiye National Energy Plan*; Republic of Türkiye Ministry of Energy and Natural Resources: Ankara, Turkey, 2022.
21. Caglayan, S.; Ozorhon, B.; Ozcan, G.; Yigit, S. A life cycle costing approach to determine the optimum insulation thickness of existing buildings. *J. Therm. Sci. Technol.* **2020**, *40*, 1–14.
22. Hussein, M.H.; Monna, S.; Abdallah, R.; Juaidi, A.; Albatayneh, A. Improving the thermal performance of building envelopes: An approach to enhancing the building energy efficiency code. *Sustainability* **2022**, *14*, 16264. [CrossRef]
23. He, Y.; Chen, Y.; Chen, Z.; Deng, Z.; Yuan, Y. Impacts of occupant behavior on building energy consumption and energy savings analysis of upgrading ASHRAE 90.1 energy efficiency standards. *Buildings* **2022**, *12*, 1108. [CrossRef]
24. Wang, X.; Yuan, J.; You, K.; Ma, X.; Li, Z. Using real building energy use data to explain the energy performance gap of energy-efficient residential buildings: A case study from the hot summer and cold winter zone in China. *Sustainability* **2023**, *15*, 1575. [CrossRef]
25. Runge, J.; Zmeureanu, R. Forecasting energy use in buildings using artificial neural networks: A review. *Energies* **2019**, *12*, 3254. [CrossRef]
26. Le, L.T.; Nguyen, H.; Dou, J.; Zhou, J. A comparative study of PSO-ANN, GA-ANN, ICA-ANN, and ABC-ANN in estimating the heating load of buildings' energy efficiency for smart city planning. *Appl. Sci.* **2019**, *9*, 2630. [CrossRef]
27. Pham, A.-D.; Ngo, N.-T.; Truong, T.T.H.; Huynh, N.-T.; Truong, N.-S. Predicting energy consumption in multiple buildings using machine learning for improving energy efficiency and sustainability. *J. Clean. Prod.* **2020**, *260*, 121082. [CrossRef]
28. Belussi, L.; Barozzi, B.; Bellazzi, A.; Danza, L.; Devitofrancesco, A.; Fanciulli, C.; Ghellere, M.; Guazzi, G.; Meroni, I.; Salamone, F.; et al. A review of performance of zero energy buildings and energy efficiency solutions. *J. Build. Eng.* **2019**, *25*, 100772. [CrossRef]
29. Lidelöw, S.; Örn, T.; Luciani, A.; Rizzo, A. Energy-efficiency measures for heritage buildings: A literature review. *Sustain. Cities Soc.* **2019**, *45*, 231–242. [CrossRef]
30. Farzaneh, H.; Malehmirchegini, L.; Bejan, A.; Afolabi, T.; Mulumba, A.; Daka, P.P. Artificial intelligence evolution in smart buildings for energy efficiency. *Appl. Sci.* **2021**, *11*, 763. [CrossRef]
31. Nair, G.; Verde, L.; Olofsson, T. A review on technical challenges and possibilities on energy efficient retrofit measures in heritage buildings. *Energies* **2022**, *15*, 7472. [CrossRef]
32. Serale, G.; Fiorentini, M.; Capozzoli, A.; Bernardini, D.; Bemporad, A. Model predictive control (MPC) for enhancing building and HVAC system energy efficiency: Problem formulation, applications and opportunities. *Energies* **2018**, *11*, 631. [CrossRef]
33. Bughio, M.; Khan, M.S.; Mahar, W.A.; Schuetze, T. Impact of passive energy efficiency measures on cooling energy demand in an architectural campus building in Karachi, Pakistan. *Sustainability* **2021**, *13*, 7251. [CrossRef]
34. Chippagiri, R.; Gavali, H.R.; Ralegaonkar, R.V.; Riley, M.; Shaw, A.; Bras, A. Application of sustainable prefabricated wall technology for energy efficient social housing. *Sustainability* **2021**, *13*, 1195. [CrossRef]
35. Meena, C.S.; Prajapati, A.N.; Kumar, A.; Kumar, M. Utilization of solar energy for water heating application to improve building energy efficiency: An experimental study. *Buildings* **2022**, *12*, 2166. [CrossRef]
36. Albatayneh, A.; Albadaineh, R.; Juaidi, A.; Abdallah, R.; Zabalo, A.; Manzano-Agugliaro, F. Enhancing the energy efficiency of buildings by shading with PV panels in semi-arid climate zone. *Sustainability* **2022**, *14*, 17040. [CrossRef]
37. TEC. *Turkish Earthquake Code: Regulations on Structures Constructed in Disaster Regions;* Ministry of Public Works and Settlement: Ankara, Turkiye, 2007.

38. Arslan, M.H. An evaluation of effective design parameters on earthquake performance of RC buildings using neural networks. *Eng. Struct.* **2010**, *32*, 1888–1898. [CrossRef]
39. *TS 825*; Thermal İnsulation Requirements for Buildings. Turkish Standard Institution: Ankara, Turkiye, 2008.
40. Dombaycı, A. Degree-days maps of Turkey for various base temperatures. *Energy* **2009**, *34*, 1807–1812. [CrossRef]

buildings

MDPI

Article

Exergy Analysis of Transcritical CO_2 Air-Source Heat Pump with Honeycomb Gas Cooler

Yujia Zhang [1], Zhihua Wang [1,*], Pengfei Zhang [1], Xin Jiang [1], Fenghao Wang [1], Chao Huan [2] and Zhenjun Ma [3]

[1] School of Human Settlements and Civil Engineering, Xi'an Jiaotong University, Xi'an 710049, China
[2] College of Energy Engineering, Xi'an University of Science and Technology, Xi'an 710054, China
[3] Sustainable Buildings Research Centre (SBRC), University of Wollongong, Wollongong, NSW 2522, Australia
* Correspondence: wangzh065@xjtu.edu.cn; Tel.: +86-29-8896-5117; Fax: +86-29-8339-5100

Abstract: In order to build an efficient and energy-saving CO_2 heat pump system and to improve the heat transfer efficiency of the gas cooler, a novel honeycomb gas cooler with a compact structure, high heat transfer efficiency, and high pressure-bearing capacity was proposed in our previous work. To clarify the components in the system that need further optimization and to improve its performance, an exergy analysis of a transcritical CO_2 air-source heat pump system with the novel honeycomb gas cooler is studied in this paper. Based on the second law of thermodynamics, the exergy model of each component in the heat pump system is established, and the irreversible loss of each component is analyzed. In addition, the degree of energy loss of the honeycomb gas cooler is clarified, and the possibility and direction of system optimization are pointed out. The results show that the exergy efficiency of the system is 35.33% under nominal operating conditions, and there is a lot of room for improvement in its energy utilization. The three components with the largest exergy destruction percentage are the compressor, throttle valve, and evaporator in the order of 36.13%, 22.90%, and 19.51%, respectively. These components with high exergy destruction percentages are the main reasons for the large irreversible losses of the system.

Keywords: biomimetic honeycomb gas cooler; discharge pressure control; exergy analysis; irreversible loss; transcritical CO_2 heat pump

Citation: Zhang, Y.; Wang, Z.; Zhang, P.; Jiang, X.; Wang, F.; Huan, C.; Ma, Z. Exergy Analysis of Transcritical CO_2 Air-Source Heat Pump with Honeycomb Gas Cooler. *Buildings* **2023**, *13*, 2147. https://doi.org/10.3390/buildings13092147

Academic Editor: Yuehong Su

Received: 3 July 2023
Revised: 20 August 2023
Accepted: 22 August 2023
Published: 24 August 2023

1. Introduction

Energy is not only the basic driving force of social development and an important support for the modern economy but is also a leading factor in the progress of human civilization. With accelerated industrialization, energy crisis and environmental problems are becoming more and more prominent, which have become an important limitation to economic development. Therefore, exploring and optimizing the efficient utilization of energy technology has become a hot spot for research in various countries [1]. Nowadays, abundant nonfossil energy resources, especially renewable energy resources, have become an important part of energy resource endowment [2]. The development potential of wind, solar, and air energy remains huge. Promoting energy transformation and increasing renewable energy development and utilization is conducive to providing abundant kinetic energy to promote green and low-carbon development [3].

One of the applications of air energy is the air-source heat pump (Carnot cycle), which can realize the use of evaporators to absorb low-grade heat energy from the air, which is converted into high-grade heat energy through the compressor compression and condenser condensation processes, and which is then applied to different heat-using situations. Its thermal efficiency is greater than one and can reach more than three when used for the production of low-grade hot water [4]. Natural refrigerants that are not harmful to nature have become a very promising choice for heat pumps. Carbon dioxide

(CO_2) is considered an ideal refrigerant for heat pump water heaters because of its superior environmental friendliness; excellent thermodynamic properties; and safety properties, such as nonflammability, nontoxicity, and chemical stability [5–10].

In order to improve and evaluate the energy efficiency of the transcritical CO_2 heat pump, the method of exergy analysis has become the focus and topicality of current research [11–13]. The exergy analysis method is based on the second law of thermodynamics; the method cannot only clarify the energy conversion efficiency and the degree of effective energy utilization and analyze whether the degree of energy utilization is reasonable, but it can also clarify the size and location distribution of the irreversible losses in the system and propose ways to improve and predict the effect of optimized energy utilization [14,15]. The exergy analysis method has been applied by many scholars in different system studies due to its superiority in analyzing the effectiveness of energy utilization [16–18].

Liu et al. [19] divided the total exergy destruction of heat exchanger components into dissipative exergy destruction and thermal exergy destruction, and the ratio of total exergy destruction to thermal exergy destruction was defined as the temperature matching degree. On this basis, a method for evaluating the effect on the thermal performance of heat pump systems in complex heat transfer processes at nonlinear sliding temperatures was proposed. Maddah et al. [20] investigated the heating and cooling modes of a CO_2 heat pump cycle at different ambient temperatures and gas cooler CO_2 outlet temperatures. The COP and exergy efficiencies were (2.762, 0.293) and (2.395, 0.292) at the ambient temperature of 0 °C and the gas cooler CO_2 outlet temperatures of 30 °C and 35 °C, respectively. Wang et al. [21] performed an energy analysis, exergy analysis, and exergy economic analysis on a transcritical CO_2 heat pump with/without an internal heat exchanger (IHX). They indicated that increasing the efficiency of the IHX was beneficial for improving the system performance factor, the irreversible losses of the throttle valve were reduced by 40.36~50.73%, and the total system irreversible losses were reduced by 2.74~8.24%. Ghazizade et al. [22] analyzed in depth the effect of the expander and the IHX on the exergy efficiency. They showed that the use of an IHX in a CO_2 transcritical cycle with an expansion valve always leads to an increase in the system COP and exergy efficiencies. However, the use of an IHX in a CO_2 transcritical cycle with an expander has a smaller effect on the COP and exergy efficiencies. Wang et al. [23] studied a CO_2/R32 refrigerant blend for a transcritical CO_2 heat pump and determined that the optimum concentration of the CO_2/R32 blend was 0.9/0.1. The increase in the COP and exergy efficiencies was more significant in cold climates. Zhang et al. [24] proposed a novel dual heat-source compression/injection transcritical CO_2 heat pump system and compared its performance with a conventional one. It was found that the maximum exergy destruction of the new system at 20 Hz~30 Hz occurred in the gas cooler, while the maximum exergy destruction at 35 Hz was in the compressor.

To solve the problems of the large irreversible losses of the gas cooler and the poor control of the optimal discharge pressure of the system, which still exist in the practical application of the transcritical CO_2 heat pump, and to improve its performance, the authors' team [25–27] proposed a honeycomb heat exchanger based on the biomimetic principle that can effectively take advantage of the microchannel heat exchanger with a compact structure, high heat transfer efficiency, and high pressure-bearing capacity. In previous studies, the flow and heat transfer characteristics of supercritical CO_2 in a honeycomb heat exchanger were mainly investigated and compared with a printed circuit heat exchanger. The results showed that the heat transfer coefficient of the biomimetic honeycomb fractal heat exchanger was improved by 144.6% compared to the printed circuit heat exchanger.

Meanwhile, this novel honeycomb gas cooler is applied to a transcritical CO_2 air-source heat pump system, and in order to clarify the components in the system that need further optimization and improvement, it is necessary to conduct an exergy analysis study. Based on the second law of thermodynamics, this paper establishes the exergy model of each component in the heat pump system, analyzes the magnitude of the irreversible losses of each component, clarifies the degree of energy loss of the honeycomb gas cooler, summarizes the effects of an ambient temperature and of inlet and outlet water temperature changes on the exergy destruction of the components, and points out the possibility and direction of system optimization. The research results of this paper can promote the popularization and application of the transcritical CO_2 heat pump and can help to reduce the energy consumption of buildings to save energy and reduce carbon emissions.

2. Model of CO_2 Heat Pump System

2.1. Introduction of Biomimetic Honeycomb Gas Cooler

To improve the heat transfer efficiency of the gas cooler and the performance of the heat pump, the authors' team proposed a honeycomb gas cooler. Figure 1 shows the cross-sectional schematic of the honeycomb gas cooler's different honeycomb layers. Figure 2 shows the physical model structure of the honeycomb gas cooler with three layers as an example. The physical characteristics and mathematical model of this type of gas cooler are shown in the literature [24] and have the following advantages:

1. Each hexagonal unit of this honeycomb structure uses the walls of its surrounding units as its own heat transfer ribs, increasing the heat transfer area, and the monolithic honeycomb plays an intensive effect, enhancing the heat transfer capacity of individual units and significantly improving the heat transfer effect of the whole heat exchanger.
2. The structure makes the hot and cold fluids in the heat exchange in reverse cross-cut not only to facilitate heat transfer but also to have a self-scouring effect to prevent fluid scaling in the heat exchanger wall.
3. The honeycomb unit can make the cross-sectional flow velocity in the direction of fluid flow converge while driving the cross-sectional temperature gradient to tend to be uniform with the ideal field synergy effect, improving the efficiency of heat transfer.
4. The honeycomb structure has strong pressure-bearing capacity and has the characteristics of periodic topological distribution. Honeycomb can absorb the energy of load action through the bending deformation of the cell wall with excellent stiffness, strength, and energy absorption performance.

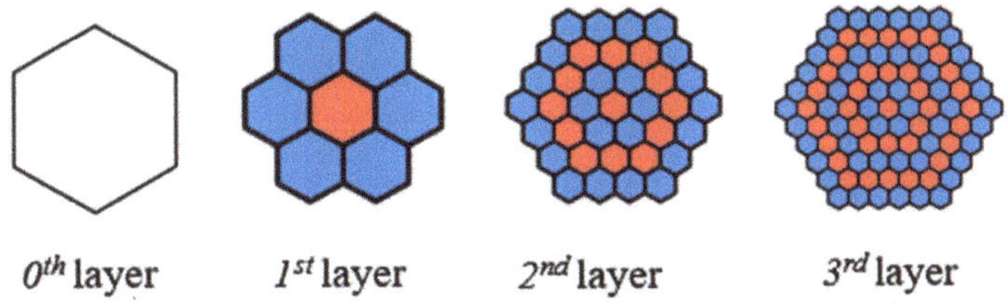

Figure 1. The cross-sectional schematic of the honeycomb gas cooler's different honeycomb layers [27].

2.2. System Model

As shown in Figure 3, the transcritical CO_2 air-source heat pump cycle studied in this paper consisted of five main components: compressor, honeycomb gas cooler, evaporator, expansion valve, and IHX. The transcritical CO_2 air-source heat pump cycle with IHX consisted of the following five processes:

1. Compression process $1'$–2: low-temperature and low-pressure CO_2 was compressed to high-temperature and high-pressure state; in this process, due to the compressor's unequal entropy compression, the actual outlet temperature of the compressor was greater than the temperature of equal entropy compression.
2. Cooling process 2–3: CO_2 was cooled from high-temperature and high-pressure state to low-temperature and high-pressure state in the gas cooler.

3. Internal heat exchange processes 3–3′ and 1–1′: the CO_2 from the gas cooler and the CO_2 from the evaporator heat exchanged and were used to reduce the CO_2 temperature at the throttle valve inlet and to increase the CO_2 temperature at compressor suction.
4. Throttling process 3′–4: CO_2 in a high-pressure state moved through the throttle valve, throttling into low-pressure two-phase state.
5. Evaporation process 4–1: the two-phase CO_2 evaporated in the evaporator and became saturated or superheated.

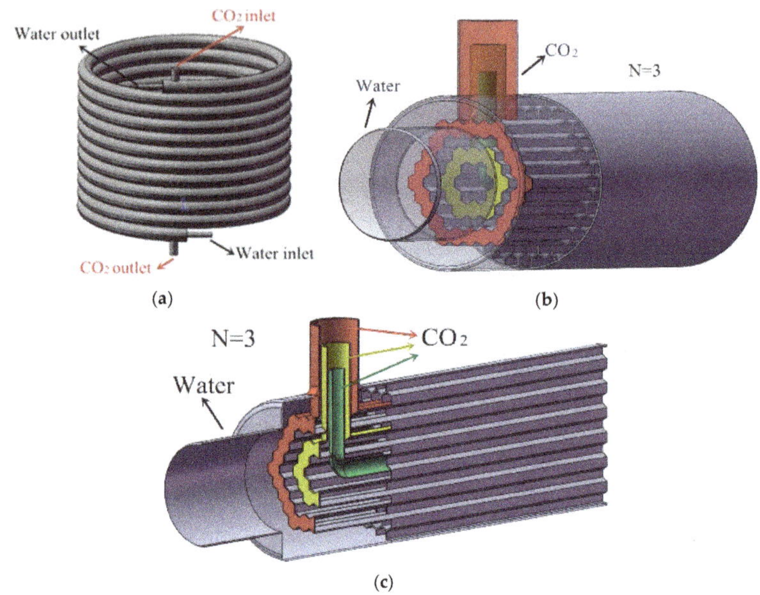

Figure 2. Physical model structure of the 3-layers honeycomb gas cooler: (**a**) external view, (**b**) perspective view, and (**c**) cutaway view [27].

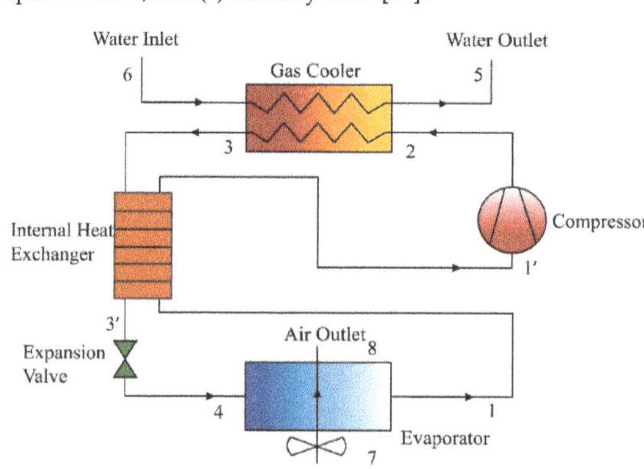

Figure 3. Transcritical CO_2 heat pump cycle with IHX.

The mathematical models of all components are described in turn below.

Compressor:

$$m_r = \rho_{1'} V_{com} \quad (1)$$

where m_r is mass flow rate of CO_2 in $kg{\cdot}s^{-1}$, $\rho_{1'}$ is density at the inlet of the compressor in $kg{\cdot}m^{-3}$, and V_{com} is the actual compressor gas delivery volume in $m^3{\cdot}r^{-1}$.

$$h_2 = h_{1'} + \frac{h_{2,is} - h_{1'}}{\eta_{is}} \tag{2}$$

where $h_{1'}$ is compressor inlet enthalpy in $J{\cdot}kg^{-1}$, h_2 is actual enthalpy of compressor outlet in $J{\cdot}kg^{-1}$, $h_{2,is}$ is isentropic compression with respect to the compressor outlet enthalpy in $J{\cdot}kg^{-1}$, and η_{is} is isentropic efficiency.

$$\mathrm{a} = 1 \tag{3}$$

where W is compressor power consumption in kW.

Honeycomb gas cooler [27]:

$$Q_j = m_{w,j}c_{p,w}(t_{wout,j} - t_{win,j}) \tag{4}$$

$$Q_j = m_{r,j}[h_{rin,j} - h_{rout,j}] \tag{5}$$

$$LMTD_j = \frac{(t_{rin,j} - t_{wout,j}) - (t_{rout,j} - t_{win,j})}{\ln(\frac{t_{rin,j} - t_{wout,j}}{t_{rout,j} - t_{win,j}})} \tag{6}$$

$$Q_j = \frac{LMTD_j}{\frac{1}{h_{r,j}A_{r,j}} + \frac{\delta}{\lambda_{wall}A_{r,j}} + \frac{1}{h_{w,j}A_{w,j}}} \tag{7}$$

where Q is the heat transfer capacity in W, m is the mass flow rate in kg/s, $c_{p,w}$ is the specific heat capacity of water in J/kg·K, t is the temperature in K, $h(t,p)$ is the enthalpy in J/kg, h is the heat transfer coefficient in $W/(m^2{\cdot}K)$, $A_{w,j}$ is the heat exchange area in $(6 + 24(n^2 - n))aL$ m^2, δ is the wall thickness in m, and λ_{wall} is the thermal conductivity of the wall. In this study, the material of the tube is copper with 398 $W/(m^2{\cdot}K)$.

Evaporator:

$$Q_{e,r} = m_r(h_{r,in} - h_{r,out}) = \alpha_r A_i(T_{wall} - \frac{T_{r,in} + T_{r,out}}{2}) \tag{8}$$

where $Q_{e,r}$ is heat transfer capacity of refrigerant side in W, $h_{r,in}$ is inlet enthalpy of evaporator microelement section in $J{\cdot}kg^{-1}$, α_r is refrigerant-side heat transfer coefficient in $W{\cdot}(m^2{\cdot}K)^{-1}$, $h_{r,out}$ is evaporator microelement section outlet enthalpy in $J{\cdot}kg^{-1}$, A_i is evaporator microelement section in-tube heat transfer area in m^2, T_{wall} is evaporator wall temperature in °C, $T_{r,in}$ is inlet temperature of evaporator microelement section in °C, and $T_{r,out}$ is outlet temperature of evaporator microelement section in °C.

$$Q_{e,a} = m_a(h_{a,in} - h_{a,out}) = \xi\alpha_a A_o(\frac{T_{a,in} + T_{a,out}}{2} - T_{wall}) \tag{9}$$

where $Q_{e,a}$ is heat transfer capacity of air side in W, m_a is air mass flow rate in $kg{\cdot}s^{-1}$, $h_{a,in}$ is air inlet enthalpy of the evaporator microelement section in $J{\cdot}kg^{-1}$, α_a is air-side heat transfer coefficient in $W{\cdot}(m^2{\cdot}K)^{-1}$, $h_{a,out}$ is air outlet enthalpy of evaporator microelement section in $J{\cdot}kg^{-1}$, A_o is heat transfer area outside the tube of evaporator microelement section in m^2, and ξ is coefficient of moisture analysis.

$$Q_{e,a} = Q_{e,r} \tag{10}$$

Expansion valve:

$$h_{3'} = h_4 \tag{11}$$

where $h_{3'}$ is enthalpy of expansion valve inlet in J·kg^{-1} and where h_4 is enthalpy of expansion valve outlet in J·kg^{-1}.

IHX:

$$m_r(h_3 - h_{3'}) = m_r(h_{1'} - h_1) \tag{12}$$

where h_3 is the return heater high-pressure-side inlet enthalpy in J·kg^{-1}, h_1 is the return heater low-pressure-side inlet enthalpy in J·kg^{-1}, $h_{3'}$ is the return heater high-pressure-side outlet enthalpy in J·kg^{-1}, and $h_{1'}$ is the return heater low-pressure-side outlet enthalpy in J·kg^{-1}.

Based on the mathematical model of each main component, this paper established a simulation model of the supercritical cycle of the CO_2 heat pump. The final results were obtained through iterative calculation of each component so that each parameter met the convergence accuracy requirements. In order to facilitate the computational analysis, the following assumptions were made:

1. The CO_2 heat pump water heater was in steady-state operation.
2. Neglected the heat exchange between the system piping, compressor, etc., and the external environment.
3. Ignored the pressure drop in the elbow and the piping connected between the components.
4. The inlet flow rate of each component should be equal to the outlet flow rate of the previous component. The calculation flow of this procedure is shown in Figure 4.

2.3. Model Validation

Because the structure and heat transfer characteristics of the tube-in-tube gas cooler were similar to those of the 1st-layer honeycomb gas cooler, in order to reduce the occurrence of unsafe accidents and waste of resources, the reliability of the simulation model was initially verified in the research stage of this paper by using a CO_2 air-source heat pump test bench with a tube-in-tube gas cooler.

The schematic diagram of the test bench and the testing procedure are detailed in the literature [28]. Photographs of the test bench are shown in Figure 5, and detailed parameters of the system components are shown in Table 1. The specific experimental conditions are shown in Table 2.

Table 1. Table of structural parameters of each component of the experimental prototype.

Components	Type	Structural Parameters
Compressor	Rotor compressor	Motor frequency: 50 Hz Power: 2.19 kW Theoretical volume: 1.44 m^3/h
Gas cooler	Coaxial tube-in-tube type of heat exchanger	Inner tube: 22.2 mm diameter, 1.2 mm wall thickness, made of stainless steel Inner tube: 19 mm diameter, 0.9 mm wall thickness, made of copper Length of heat exchange section: 18.57 m
Evaporator	Finned-tube heat exchanger	Tube diameter: 7.94 mm Tube wall thickness: 0.7 mm Tube material: copper Fin thickness: 0.11 mm Fin material: aluminum sleeve Fin spacing: 2.0 mm Number of tubes in a single row: 30 Number and length of loops: 5/14.4 m
IHX	Coaxial casing type of heat exchanger	Outer tube: 22 mm diameter, made of copper Inner tube: tube diameter of 12.7 mm, material is copper Length of heat exchange section: 1.2 m
Expansion valve	CO_2 heat pump cross-critical cycle special electronic expansion valve	Valve orifice diameter: Ø 1.4 mm Maximum pressure limit: 14 MPa

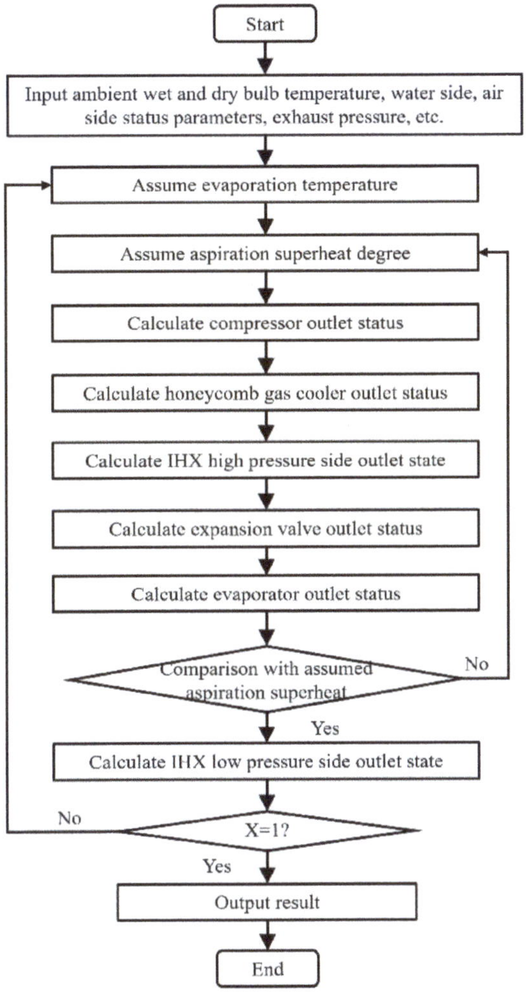

Figure 4. CO_2 heat pump cycle calculation flow chart.

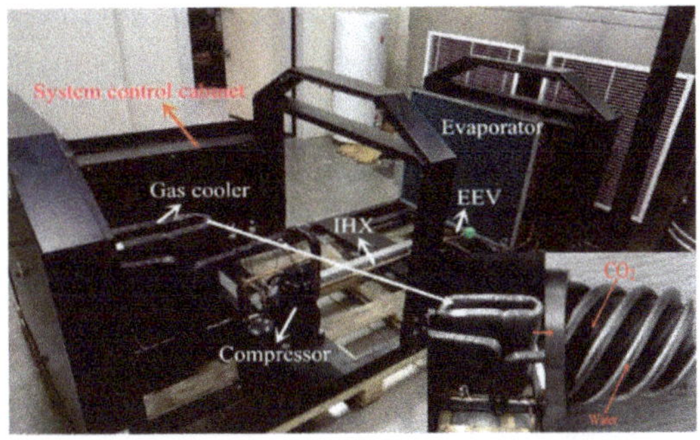

Figure 5. Photo of the experimental prototype [28].

Table 2. Table of working condition parameters.

Working Condition	Case 1	Case 2	Case 3	Case 4	Case 5
Ambient temperature/°C	−10	−5	0	5	10
Water flow/$m^3 \cdot h^{-1}$			0.2		
Inlet water temperature/°C			35		

As shown in Figures 6 and 7, the simulation results of heat pump water supply temperature and system performance COP compared with experimental values verified the correctness and reliability of the model, whose maximum errors were less than 3% and 10%, respectively.

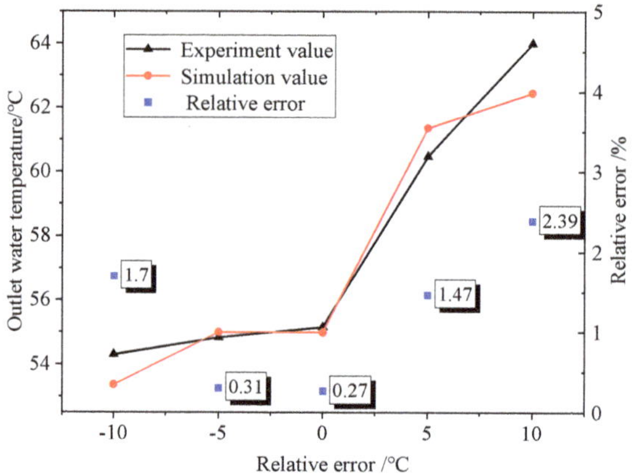

Figure 6. CO_2 heat pump water supply temperature comparison.

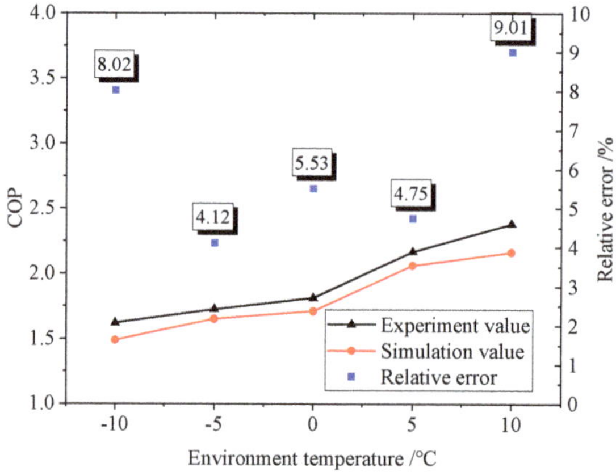

Figure 7. CO_2 heat pump COP comparison.

3. Exergy Analysis Model

3.1. Theoretical Basis

When a system changes reversibly from an arbitrary state to a state in equilibrium with a given environment, a portion of the energy can theoretically be converted indefinitely to any other form, called exergy and also called effective energy. The other part of the energy that cannot be used or lost is called anergy. During the thermal process, the sum of both remains constant. Energy can only be converted from the high quality of exergy to the low quality of anergy. The exergy efficiency of a system gives an indication of the extent to which the system utilizes energy and reflects the rationality of the system structure and efficiency.

The exergy analysis method is based on the second law of thermodynamics, and it can be used to determine the location and magnitude of the irreversible losses in a system and can be used as a basis for making recommendations for system improvements. In heat pump systems, the temperature difference between the hot and cold fluids in the heat exchanger and the irreversibility of the compression and throttling processes can cause the system to suffer from exergy destruction. In this paper, based on the established simulation model of the CO_2 heat pump, the parameters of various state points of each component in a stable operation are calculated. It also calculates the system and components' exergy efficiency and exergy destruction according to the methodology of the exergy analysis, thus pointing out the potential for the improvement of the system and the priority of the improved components. This work is beneficial for improving the system performance and expanding the application fields.

3.2. Exergy Modeling

Exergy can be divided into chemical exergy, kinetic exergy, latent exergy, and physical exergy. In the CO_2 heat pump system described in this paper, no chemical reaction occurs, and the system is a one-dimensional stable system; therefore the chemical exergy, kinetic exergy, and latent exergy can be neglected, and only the physical exergy should be considered in the exergy analysis of the system.

The specific physical exergy e^{PH} is the physical exergy per unit mass, and it contains the thermal exergy e^T and mechanical exergy e^M; the former is related to temperature, and the latter is caused by pressure as shown in Equation (13).

$$\begin{aligned} e_j &= e_j^{PH} = e_j^T + e_j^M \\ &= [(h_j - h_{j,x}) - T_0(s_j - s_{j,x})]_{P=\text{const}} \\ &+[(h_{j,x} - h_{j,0}) - T_0(s_{j,x} - s_{j,0})]_{T=\text{const}} \end{aligned} \tag{13}$$

where the subscript j indicates the state point, the subscript 0 indicates the selected reference ambient state, and the subscript x refers to the condition at the reference ambient temperature T_0 and the pressure P_j at the state point j.

The evaluation indexes of the system exergy analysis mainly include the exergy destruction E_D, exergy efficiency ε, exergy destruction percentage y, and exergy destruction coefficient Ω.

In the thermodynamic process, the fuel exergy E_F is the input energy of a component, and the product exergy E_P refers to the part of the input energy that is utilized to produce the due thermodynamic effect. The difference between the two is the exergy destruction as shown in Equation (14). For a transcritical CO_2 heat pump, the fuel exergy of the system is the energy consumed by the compressor, i.e., the compressor power consumption, as in Equation (15). The product exergy of the system is the product energy of the gas cooler as in Equation (16), and the system exergy destruction is the sum of the exergy destruction of all the components as in Equation (17). Since all the thermodynamic processes in nature are irreversible and necessarily involve energy loss and dissipation, the exergy destruction is constantly greater than zero.

$$E_D = E_F - E_P \tag{14}$$

$$E_{F,\text{sys}} = E_{F,\text{comp}} = W_t \tag{15}$$

$$E_{P,sys} = E_{P,gc} \tag{16}$$

$$E_{D,sys} = \sum E_{D,k} \tag{17}$$

The exergy efficiency ε is the ratio of the components' product exergy to their fuel exergy, which can indicate the extent of the components' utilization of the input energy and can provide guidance for targeted component optimization and modification. The exergy efficiency can be calculated by Equation (18).

$$\varepsilon_k = \frac{E_{P,k}}{E_{F,k}} \tag{18}$$

The exergy destruction percentage y is the proportion of the exergy destruction of a component to the total exergy destruction of the system, and the sum of the exergy destruction percentages of all the components is one. This parameter can clarify the contribution of the exergy destruction of each component in the system to the total exergy destruction of the system and can provide a basis for the optimization order of each component. The exergy destruction percentage can be calculated by Equation (19):

$$y_k = \frac{E_{D,k}}{E_{D,sys}} \tag{19}$$

The exergy destruction coefficient Ω is the ratio of the exergy destruction of a component to the total fuel exergy of the system, which reveals the degree of impact of the exergy destruction of each component in the system. The exergy destruction coefficient is defined as shown in Equation (20):

$$\Omega_k = \frac{E_{D,k}}{E_{F,sys}} \tag{20}$$

Based on the above theoretical basis for exergy modeling, Table 3 collates the equations for the fuel exergy, product exergy, and exergy destruction of each component in the CO_2 heat pump and of the system.

Table 3. Calculation formula of the fuel exergy, product exergy, and exergy destruction of each component in the CO_2 heat pump system and of the system.

Component	Fuel Exergy	Product Exergy	Exergy Destruction
Compressor	$\dot{E}_{F,comp} = W_{comp}$	$E_{P,comp} = m_{1'}(e_2 - e_{1'})$	$E_{D,comp} = E_{F,comp} - E_{P,comp}$
Gas cooler	$E_{F,gc} = m_2(e_2 - e_3)$	$E_{P,gc} = m_5(e_6 - e_5)$	$E_{D,gc} = E_{F,gc} - E_{P,gc}$
Evaporator	$E_{F,eva} = m_4(e_4 - e_1)$	$E_{P,eva} = m_7(e_8 - e_7)$	$E_{D,eva} = E_{F,eva} - E_{P,eva}$
IHX	$E_{F,ihx} = m_3(e_3 - e_{3'})$	$E_{P,ihx} = m_1(e_{1'} - e_1)$	$E_{D,ihx} = E_{F,ihx} - E_{P,ihx}$
EEV	$E_{F,eev} = m_{3'}(e_{3'}^M - e_4^M + e_{3'}^T)$	$E_{P,eev} = m_{3'}e_4^T$	$E_{D,eev} = E_{F,eev} - E_{P,eev}$
System	$E_{F,sys} = E_{F,comp}$	$E_{P,sys} = E_{P,gc}$	$E_{D,sys} = \sum E_D$

4. Results and Discussion

4.1. Exergy Destruction and Exergy Efficiency of the System and Each Component

In this section, the exergy destruction and exergy efficiency are calculated for each component and the whole system based on Equations (13)–(20) under nominal operating conditions; i.e., the inlet and outlet water temperatures are 17 ± 1.0 °C and 65 ± 2.0 °C, and the air dry- and wet-bulb temperatures are 16 ± 1.0 °C and 12 ± 1.0 °C, respectively. The exergy analysis results are presented in Table 4, which shows the distribution and magnitude of the irreversible losses in each component of the system.

Table 4. Exergy analysis results of CO_2 heat pump system under nominal working conditions.

Component	$\dot{E}_{F,k}$/kW	$\dot{E}_{P,k}$/kW	$\dot{E}_{D,k}$/kW	$\dot{E}_{L}$/kW	ε_k/%	Ω_k/%	y_k
Compressor	1.840	1.414	0.426	-	76.85	23.15	36.13
Gas cooler	0.860	0.650	0.210	-	75.58	11.41	17.81
Evaporator	1.050	0.820	0.230	-	78.10	12.50	19.51
IHX	0.053	0.010	0.043	-	18.87	2.34	3.65
EEV	0.350	0.080	0.270	-	22.86	14.67	22.90
System	1.840	0.650	1.179	0.011	35.33	64.08	100.00

From Figure 8, it can be seen that the exergy efficiency of the system is 35.33% under nominal working conditions, which means that the heat pump system still has a large potential for optimization in terms of energy utilization. The three components with the largest exergy destruction percentage are the compressor, throttle valve, and evaporator, whose exergy destruction percentages are 36.13%, 22.90%, and 19.51%, respectively. These high-exergy-destruction-percentage components are the main reason for the large irreversible losses of the system, so optimizing and improving their performance is the key direction for the optimization and upgrading of the system. The throttle valve shows a low exergy efficiency, which is only 22.86%. The reason for this phenomenon is to throttle the CO_2 gas at the outlet of the gas cooler to a two-phase state with equal enthalpy, and in this process, it generates the conversion of multiple states of the refrigerant, which makes the availability of energy decrease. Therefore, to improve the system performance and energy efficiency, the throttle valve must be optimized. Common measures include structural optimization or replacing the throttle valve with an expander, injector, etc. The exergy efficiency of the honeycomb gas cooler is 75.58%, which is relatively high, meaning that the honeycomb gas cooler is able to meet the demand for hot water preparation for design conditions with a high energy efficiency. The exergy destruction of the honeycomb gas cooler is relatively small, which is 17.81%. The exergy efficiency of the IHX is extremely low at 18.87%, but at the same time, the exergy destruction of the IHX is the smallest among the system components, accounting for only 3.65%.

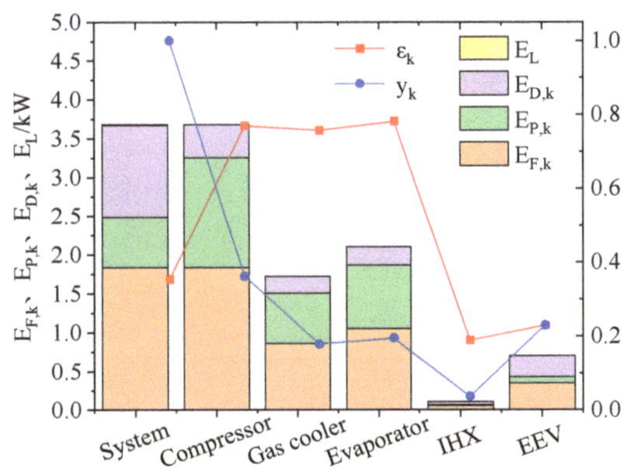

Figure 8. Exergy analysis results of systems and components under nominal operating conditions.

4.2. *Effect of Different Forms of Gas Coolers on Exergy Destruction and Exergy Efficiency of Each Component*

In order to clarify the difference in energy utilization between the honeycomb type of gas cooler and the tube-in-tube type of gas cooler, the thermogravimetric ratio of the two forms of gas coolers is kept consistent in this section. A comparative analysis is conducted under four typical operating conditions throughout the year (Table 5) [29], and the results are shown in Figure 9. It can be seen that the COP of the heat pump with the honeycomb gas cooler is significantly higher than that of the heat pump with the tube-in-tube gas cooler under the same conditions. The difference in the COP is even greater in summer conditions, and the COP of the heat pump with the honeycomb gas cooler is about twice as high as that of the tube-in-tube type. As the typical operating conditions change from summer to a low temperature, the exergy efficiency of the gas cooler increases; this is because the inlet water temperature is lower under low-temperature operating conditions, the heat transfer temperature difference is smaller, and the exergy destruction is lower. The system exergy efficiency decreases because the ambient temperature decreases and because the compressor power consumption increases; i.e., the denominator of the system exergy efficiency increases. Its growth rate is greater than the growth rate of the numerator (the gas cooler product exergy), so the system exergy efficiency is decreasing. In addition, the exergy efficiency of the honeycomb gas cooler is 20.51%, 14.34%, 9.76%, and 7.20% higher under the four operating conditions, and the system exergy efficiency is 52.78%, 55.19%, 53.68%, and 43.89% higher, respectively, than that of the basic cycle. This means that the honeycomb gas cooler has a higher energy efficiency.

Table 5. Four typical operating conditions throughout the year.

Condition	Water Inlet Temperature/°C	Water Outlet Temperature/°C	Air Dry-Bulb Temperature/°C	Air Wet-Bulb Temperature/°C
Summer	24 ± 1.0	65 ± 2	25 ± 1.0	21 ± 1.0
Nominal	17 ± 1.0	65 ± 2	16 ± 1.0	12 ± 1.0
Winter	9 ± 1.0	65 ± 2	7 ± 1.0	6 ± 1.0
Low temperature	5 ± 1.0	65 ± 2	-12 ± 1.0	-14 ± 1.0

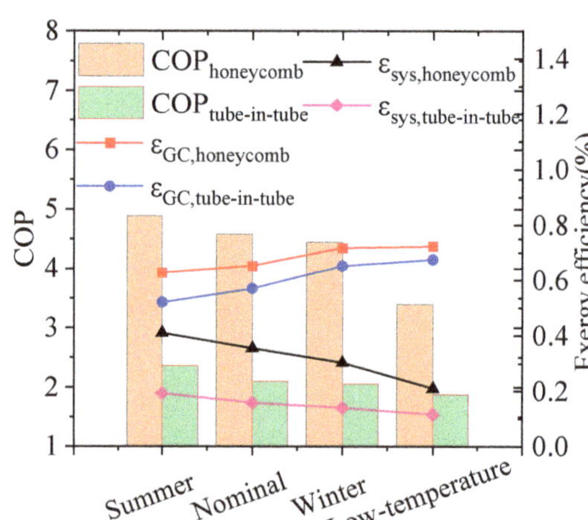

Figure 9. Exergy comparison results of honeycomb gas cooler and tube-in-tube gas cooler under typical operating conditions.

4.3. *Effect of Different Working Conditions on Exergy Destruction and Exergy Efficiency of Each Component*

In this section, the exergy efficiency and exergy destruction of the CO_2 heat pump system with or without the IHX are investigated under the different ambient temperature conditions. Namely, the inlet and outlet water temperatures are 17/65 °C, respectively. As shown in Figure 10, the exergy efficiency of the system increases as the ambient temperature increases, and the exergy efficiency of the system with the IHX is higher than that without the IHX under each operating condition. When the ambient temperature rises from −15 °C to 25 °C, the exergy efficiency of the system with the IHX increases from 19.16% to 39.88%, increasing by 108.09%; the exergy efficiency of the system without the IHX increases from 18.87% to 30.06%, increasing by 59.29%. The exergy efficiency of the honeycomb gas cooler decreases with an increasing ambient temperature; however, the rate of the decrease is small, and the exergy efficiency of the gas cooler with the IHX is higher than that without the IHX under each operating condition. When the ambient temperature increases from −15 °C to 25 °C, the exergy efficiency of the gas cooler with the IHX decreases from 72.7% to 67.95%, decreasing by 6.53%, while the exergy efficiency of the gas cooler without it decreases from 72.47% to 66.26%, decreasing by 8.58%.

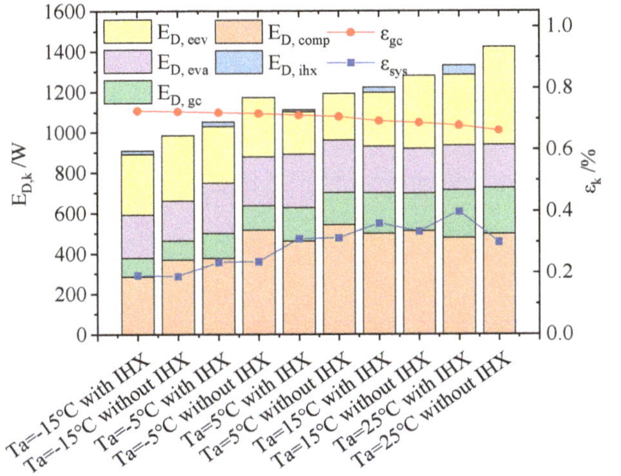

Figure 10. Exergy analysis results of each component at different ambient temperatures.

In terms of energy loss, with an increase in the ambient temperature, the system exergy destruction gradually increases, and the heat pump system exergy destruction with the IHX is smaller than the system exergy destruction without the IHX. When the ambient temperature rises from −15 °C to 25 °C, the system exergy destruction with the IHX increases from 909 W to 1331 W, increasing by 46.49%; the exergy destruction of the system without the IHX increases from 983 W to 1473 W, increasing by 49.74%. Meanwhile, the exergy destruction of the honeycomb gas cooler gradually increases with a rising ambient temperature, so the transcritical CO_2 heat pump is superior in a low-temperature environment.

As shown in Figure 11, the exergy efficiency and exergy destruction of the CO_2 heat pump system with or without the IHX at different inlet water temperatures are analyzed. The outlet water temperature is 65 °C, and the ambient temperature is 16 °C. The exergy efficiency of the system decreases as the inlet temperature increases; the exergy efficiency of the system with the IHX is higher than the exergy efficiency of the system without the IHX under each operating condition, but the difference is small. When the inlet water temperature increases from 5 °C to 45 °C, the exergy efficiency of the system with the IHX decreases from 38.95% to 32.45%, decreasing by 16.69%; the exergy efficiency of the system without the IHX decreases from 38.17% to 32.11%, decreasing by 15.88%. The exergy efficiency of the honeycomb gas cooler decreases as the inlet water temperature increases, and the exergy efficiency of the gas cooler with the IHX is higher than the exergy efficiency of the gas cooler without the IHX under each operating condition. When the inlet water temperature increases from 5 °C to 45 °C, the exergy efficiency of the gas cooler with the IHX decreases from 77.98% to 60.75%, decreasing by 22.10%; the exergy efficiency of the gas cooler without the IHX decreases from 77.28% to 60.13%, decreasing by 22.18%.

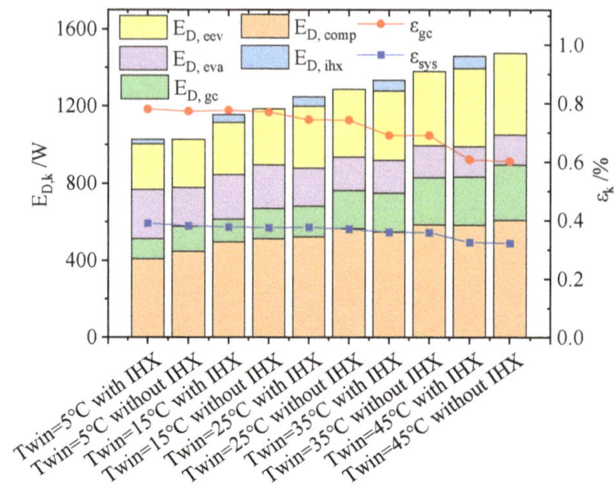

Figure 11. Exergy analysis results of each component at different inlet water temperatures.

In terms of energy loss, with an increase in the inlet water temperature, the system exergy destruction gradually increases, and when the inlet water temperature increases from 5 °C to 45 °C, the system exergy destruction with the IHX increases from 1028 W to 1458 W, increasing by 41.85%; the system exergy destruction without the IHX increases from 1027 W to 1474 W, increasing by 43.50%. Under the same inlet water temperature, the exergy destruction of the heat pump system with the IHX is smaller than that without the IHX. Taking the inlet water temperature of 5 °C as an example, the exergy destruction of the system with the IHX is about 0.07% lower than that of the system without the IHX. Taking the inlet water temperature of 45 °C as an example, the exergy destruction of the system with the IHX is about 1.08% lower than that of the system without the IHX. Meanwhile, the exergy destruction of the compressor increases gradually with an increase in the inlet water temperature, the exergy destruction of the honeycomb gas cooler increases gradually with an increase in the inlet water temperature, and the exergy destruction of the throttle valve increases gradually with an increase in the inlet water temperature.

As shown in Figure 12, the exergy efficiency and exergy destruction of the CO_2 heat pump system with or without the IHX at different outlet water temperatures are analyzed. The inlet water temperature is 9 °C, and the ambient temperature is 16 °C. The exergy efficiency of the system increases with an increase in the outlet water temperature, and the exergy efficiency of the system with the IHX is equal to or slightly higher than the exergy efficiency of the heat pump system without the IHX under each operating condition. When the outlet water temperature increases from 50 °C to 90 °C, the exergy efficiency of the system with the IHX increases from 31.65% to 41.18%, increasing by 30.12%; the exergy efficiency of the system without the IHX increases from 31.65% to 39.80%, increasing by 25.77%. The exergy efficiency of the honeycomb gas cooler increases with an increase in the outlet water temperature, and when the outlet water temperature increases from 50 °C to 90 °C, the exergy efficiency of the gas cooler with the IHX increases from 60.73% to 77.93%, increasing by 28.32%; the exergy efficiency of the gas cooler without the IHX increases from 60.73% to 76.33%, increasing by 25.68%.

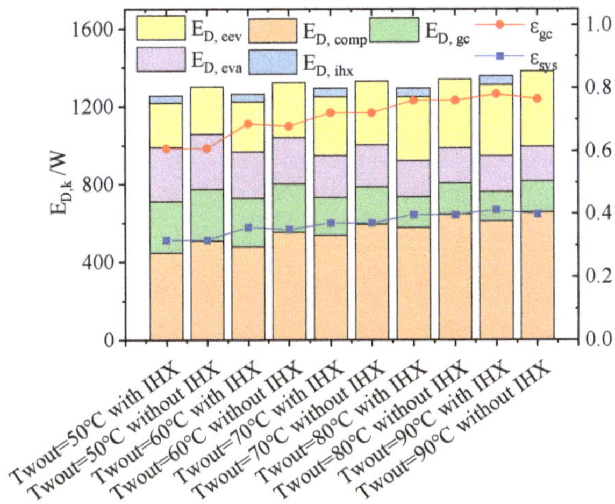

Figure 12. Exergy analysis results of each component at different outlet water temperatures.

In terms of energy loss, with an increase in the outlet water temperature, the system exergy destruction gradually increases, and when the outlet water temperature rises from 50 °C to 90 °C, the system exergy destruction with the IHX increases from 1254 W to 1355 W, increasing by 8.05%; the system exergy destruction without the IHX increases from 1300 W to 1381 W, increasing by 6.19%. Under the same outlet water temperature conditions, the exergy destruction of the heat pump system with the IHX is less than the exergy destruction of the system without the IHX, and taking the outlet water temperature of 50 °C as an example, the exergy destruction of the system with the IHX is about 3.58% lower than the exergy destruction of the system without the IHX. Taking the outlet water temperature of 90 °C as an example, the exergy destruction of the system with the IHX is about 1.90% lower than the exergy destruction of the system without the IHX. Meanwhile, the exergy destruction of the compressor gradually increases with an increase in the outlet water temperature, the exergy destruction of the honeycomb gas cooler gradually decreases with an increase in the outlet water temperature, and the exergy destruction of the throttle valve gradually increases with an increase in the inlet water temperature.

4.4. System Improvement and Optimization Suggestions

In summary, the exergy destruction and exergy destruction percentage of the compressor are the largest among the system components, which is mainly due to the nonisentropic compression caused by the friction and other factors in the compressor. In order to reduce the exergy destruction of the compressor, consideration can be given to improving the structural design, selecting higher-quality lubricating oil to reduce friction, reducing the heat dissipation loss, and improving the isentropic compression efficiency of the compressor.

Secondly, the next largest component is the exergy destruction of the throttle valve, whose main function is to throttle the CO_2 gas at the outlet of the gas cooler to a two-phase state with equal enthalpy, and in this process, it produces the conversion of multiple states of the refrigerant, which makes the availability of energy decrease. Therefore, the throttle valve must be optimized to improve the system performance and energy efficiency. Common measures include structural optimization or replacing the throttle valve with an expander, injector, etc.

The exergy destruction ranking of each component under different working conditions is either third or fourth for the evaporator and the honeycomb gas cooler. For the evaporator, this part of exergy destruction is caused by the temperature difference between the hot and cold fluid inside the heat exchanger. The larger the temperature difference is, the larger the irreversible loss and the lower the exergy efficiency of the heat exchanger. The exergy efficiency can be improved by increasing the heat transfer area and reducing the temperature difference. However, a larger heat transfer area makes the heat exchanger cost more expensive, so thermodynamic knowledge should be considered to estimate its cost. For the honeycomb gas cooler, it is known from the literature [24] that its heat transfer area is large and that the heat transfer performance is good, and the difference between the CO_2 outlet temperature of the gas cooler and the inlet water temperature can reach within 5 °C. It is suggested that the exergy efficiency can be further improved by reducing the inlet water temperature or increasing the outlet water temperature.

The exergy efficiency of the IHX is low; however, the percentage of exergy destruction is small, and the improved gain is not very large. The application of the IHX under different operating conditions reduces the exergy destruction and increases the exergy efficiency of the CO_2 heat pump system, so the application of the IHX has positive significance for the energy utilization of the CO_2 heat pump system.

5. Conclusions

In this work, an exergy analysis is performed for a CO_2 system operation integrated with a honeycomb gas cooler and is compared with a heat pump system with a tube-in-tube gas cooler. Also, the optimization potential of the system is pointed out. The main conclusions are as follows:

1. The system exergy efficiency is 35.33% under nominal operating conditions, which has a large potential for improvement in energy utilization. The three components with the largest exergy destruction percentage are the compressor, throttle valve, and evaporator in the order of 36.13%, 22.90%, and 19.51%, respectively.
2. Under the same conditions, the COP of the heat pump with the honeycomb gas cooler is significantly higher than that of the heat pump with the tube-in-tube gas cooler. The exergy efficiency of the honeycomb gas cooler is 20.51%, 14.34%, 9.76%, and 7.20% higher than that of the gas cooler type under the four typical operating conditions. The system exergy efficiencies are 52.78%, 55.19%, 53.68%, and 43.89% higher, respectively. This implies that the honeycomb gas cooler has higher energy utilization.
3. Suggestions for the improvement and optimization of the key components are made: The compressor is the key object of system improvement, and the exergy destruction of the compressor can be reduced by improving the structural design, using higher-quality lubricating oil, etc. Next is the throttle valve; common measures include structural optimization or using an expander or injector instead. The exergy destruction of the evaporator is mainly caused by the heat exchange temperature difference in the heat exchanger; the exergy efficiency can be improved by increasing the heat exchange area and reducing the heat exchange temperature difference.

Author Contributions: Conceptualization, Z.W. and F.W.; methodology, Z.W. and P.Z.; software, Y.Z.; validation, Y.Z.; formal analysis, X.J. and P.Z.; investigation, Y.Z.; resources, Y.Z.; data curation, Y.Z.; writing—original draft preparation, Y.Z.; writing—review and editing, Z.M.; visualization, Z.W.; supervision, Z.W. and F.W.; project administration, C.H.; funding acquisition, F.W. All authors have read and agreed to the published version of the manuscript.

Funding: This research was funded by the National Natural Science Foundation of China (No. 52076158 and 52004207), the Xi'an Municipal Science and Technology Foundation (No. 21XJZZ0038), the Natural Science Basic Research Plan in the Shaanxi Province of China (No. 2023-JC-YB-456), and the Research and Development Program of the Shaanxi Province (No. 2023-YBGY-283).

Data Availability Statement: Not applicable.

Conflicts of Interest: The authors declare no conflict of interest.

Nomenclature

A	Heat exchange area, m^2
c_p	Specific heat capacity, J/kg·K
E_D	Exergy destruction, kW
E_F	Fuel exergy, kW
E_P	Product exergy, kW
P	Pressure of CO_2, MPa
Q	Heat transfer capacity, kW
V	Gas delivery volume of the compressor, m^3/r
W	Compressor power consumption, kW
h	Enthalpy, kJ/ kg
m	Mass flow rate, kg/s
n	Number of layers in the honeycomb CO_2 gas cooler
t	Temperature of fluid, °C
y	Exergy destruction percentage, %
Ω	Exergy destruction coefficient, %
α	Heat transfer coefficient, $W/(m^2 \cdot K)$
δ	Wall thickness, m
ε	Exergy efficiency, %
η	Efficiency, %
λ	Thermal conductivity, W/(m·K)
ρ	Fluid density, kg/m^3

Abbreviations

EEV	Electronic expansion valve
IHX	Internal heat exchanger

Subscript

a	Air
com	Compressor
eev	Electronic expansion valve
eva	Evaporator
gc	Gas cooler
ihx	Internal heat exchanger
k	Component
f	Fluid
in	Inlet
is	Isentropic
j	Microelement
out	Outlet
r	Refrigerant
sys	System
w	Water
wall	Tube wall

References

1. Liu, S.C.; Zheng, L.; Dai, B.M.; Zhong, Z.; Li, H.; Song, M.; Sun, Z. Energetic, economic and environmental analysis of air source transcritical CO_2 heat pump system for residential heating in China. *Appl. Therm. Eng.* **2019**, *148*, 1425–1439. [CrossRef]
2. Zhang, Y.X.; Wang, S.N.; Shao, W.; Hao, J. Feasible distributed energy supply options for household energy use in China from a carbon neutral perspective. *Int. J. Environ. Res. Public Health* **2021**, *18*, 12992. [CrossRef]
3. Barta, R.B.; Groll, E.A.; Ziviani, D. Review of stationary and transport CO_2 refrigeration and air conditioning technologies. *Appl. Therm. Eng.* **2021**, *185*, 116422. [CrossRef]
4. Nawaz, K.; Shen, B.; Elatar, A.; Baxter, V.; Abdelaziz, O. Performance optimization of CO_2 heat pump water heater. *Int. J. Refrig.* **2018**, *85*, 213–228. [CrossRef]

5. Wang, J.; Belusko, M.; Semsarilar, H.; Evans, M.; Liu, M.; Bruno, F. An optimisation study on a real-world transcritical CO_2 heat pump system with a flash gas bypass. *Energy Convers. Manag.* **2022**, *251*, 114995. [CrossRef]
6. Cao, F.; Ye, Z.L.; Wang, Y.K. Experimental investigation on the influence of internal heat exchanger in a transcritical CO_2 heat pump water heater. *Appl. Therm. Eng.* **2020**, *168*, 114855. [CrossRef]
7. Wang, J.; Belusko, M.; Evans, M.; Liu, M.; Zhao, C.; Bruno, F. A comprehensive review and analysis on CO_2 heat pump water heaters. *Energy Convers. Manag. X* **2022**, *15*, 100277. [CrossRef]
8. Cui, C.; Zong, S.; Song, Y.L.; Yin, X.; Cao, F. Experimental investigation of the extreme seeking control on a transcritical CO_2 heat pump water heater. *Int. J. Refrig.* **2022**, *133*, 111–122. [CrossRef]
9. Wang, Y.K.; Ye, Z.L.; Song, Y.L.; Yin, X.; Cao, F. Energy, exergy, economic and environmental analysis of refrigerant charge in air source transcritical carbon dioxide heat pump water heater. *Energy Convers. Manag.* **2020**, *223*, 113209. [CrossRef]
10. Wang, Z.H.; Li, G.C.; Wang, F.H. Performance analysis and operation optimization of air-to-water CO_2 heat pump with phase change thermal storage. *Energy Build.* **2020**, *209*, 109738. [CrossRef]
11. Dai, B.; Liu, C.; Liu, S.; Wang, D.; Wang, Q.; Zou, T.; Zhou, X. Life cycle techno-enviro-economic assessment of dual-temperature evaporation transcritical CO_2 high-temperature heat pump systems for industrial waste heat recovery. *Appl. Therm. Eng.* **2023**, *219*, 119570. [CrossRef]
12. Illán-Gómez, F.; Sena-Cuevas, V.F.; García-Cascales, J.R.; Velasco, F.J. Analysis of the optimal gas cooler pressure of a CO_2 heat pump with gas bypass for hot water generation. *Appl. Therm. Eng.* **2021**, *182*, 116110. [CrossRef]
13. Peng, X.; Wang, D.B.; Wang, G.H.; Yang, Y.; Xiang, S. Numerical investigation on the heating performance of a transcritical CO_2 vapor-injection heat pump system. *Appl. Therm. Eng.* **2020**, *166*, 114656. [CrossRef]
14. Wu, Z.X.; Zhang, Y.F.; Sheng, Y. Energy, exergy, economic(3E) analysis and multi-objective optimization of a novel dual functional integration system. *Energy Convers. Manag.* **2019**, *199*, 111962. [CrossRef]
15. Sun, Z.L.; Wang, Q.F.; Xie, Z.Y.; Liu, S.; Su, D.; Cui, Q. Energy and exergy analysis of low GWP refrigerants in cascade refrigeration system. *Energy* **2019**, *170*, 1170–1180. [CrossRef]
16. Patil, O.S.; Shet, S.A.; Jadhao, M.; Agrawal, N. Energetic and exergetic studies of modified CO_2 transcritical refrigeration cycles. *Int. J. Low Carbon Technol.* **2021**, *16*, 171–180. [CrossRef]
17. Wu, J.X.; Sun, S.J.; Song, Q.L.; Sun, D.; Wang, D.; Li, J. Energy, exergy, exergoeconomic and environmental (4E) analysis of cascade heat pump, recuperative heat pump and carbon dioxide heat pump with different temperature lifts. *Renew. Energy* **2023**, *207*, 407–421. [CrossRef]
18. Li, Y.M.; Wang, C.C. Investigation of the performance of a transcritical CO_2 heat pump system subject to heated water conditions: Perspective from the second law. *Appl. Therm. Eng.* **2021**, *193*, 116999. [CrossRef]
19. Liu, R.J.; Gao, F.L.; Liang, K.F.; Wang, L.; Wang, M.; Mi, G.; Li, Y. Thermodynamic Evaluation of Transcritical CO_2 Heat Pump Considering Temperature Matching under the Constraint of Heat Transfer Pinch Point. *J. Therm. Sci.* **2021**, *30*, 869–879. [CrossRef]
20. Maddah, S.; Safaei, M.R. Determination of the optimal discharge pressure of the transcritical CO_2 heat pump cycles for heating and cooling performances based on new correlation. *J. Therm. Anal. Calorim. Int. Forum Therm. Stud.* **2021**, *145*, 1537–1546. [CrossRef]
21. Wang, Y.K.; Ye, Z.L.; Yin, X.; Song, Y.; Cao, F. Energy, exergy and exergoeconomic evaluation of the air source transcritical CO_2 heat pump with internal heat exchanger for space heating. *Int. J. Refrig.* **2021**, *130*, 14–26. [CrossRef]
22. Ghazizade-Ahsaee, H.; Ameri, M. Effects of using expander and internal heat exchanger on carbon dioxide direct-expansion geothermal heat pump. *Appl. Therm. Eng.* **2018**, *136*, 389–407. [CrossRef]
23. Wang, Y.; He, Y.; Song, Y.; Yin, X.; Cao, F.; Wang, X. Energy and exergy analysis of the air source transcritical CO_2 heat pump water heater using CO_2-based mixture as working fluid. *Energies* **2021**, *14*, 4470. [CrossRef]
24. Zhang, Y.X.; Wei, X.L.; Qin, X. Experimental study on energy, exergy, and exergoeconomic analyses of a novel compression/ejector transcritical CO_2 heat pump system with dual heat sources. *Energy Convers. Manag.* **2022**, *271*, 116343. [CrossRef]
25. Wang, Z.H.; Li, G.C.; Zhang, Y.J.; Wang, F.; Jiang, X.; Ma, Z. Flow and heat transfer investigation of supercritical carbon dioxide in a novel biomimetic honeycomb fractal gas cooler of transcritical CO_2 heat pumps. *Therm. Sci. Eng. Prog.* **2023**, *37*, 101533. [CrossRef]
26. Li, G.C.; Wang, Z.H.; Wang, F.H.; Zhang, Y. Numerical investigation on the performance characteristics of a novel biomimetic honeycomb fractal gas cooler of transcritical CO_2 heat pump. *J. Build. Eng.* **2022**, *59*, 105091. [CrossRef]
27. Zhang, Y.J.; Wang, Z.H.; Jiang, X.; Xu, K.; Ma, Z.; Dou, W. Heat transfer and performance enhancement investigation of biomimetic honeycomb gas coolers in transcritical CO_2 heat pumps. *Appl. Therm. Eng.* **2023**, *230*, 120645. [CrossRef]
28. Wang, Z.H.; Wang, F.H.; Li, G.C.; Song, M.; Ma, Z.; Ren, H.; Li, K. Experimental investigation on thermal characteristics of transcritical CO_2 heat pump unit combined with thermal energy storage for residential heating. *Appl. Therm. Eng.* **2020**, *165*, 114505. [CrossRef]
29. Ma, Y.T.; Wang, P.; Zhang, Q.C.; Li, M.X.; Cai, S.Q. Development status of CO_2 heat pump water heaters. *Refrig. Air Cond.* **2018**, *18*, 67–71. (In Chinese)

buildings

MDPI

Article

Recycling Textile Waste to Enhance Building Thermal Insulation and Reduce Carbon Emissions: Experimentation and Model-Based Dynamic Assessment

Rabeb Ayed [1,2,*], Salwa Bouadila [2], Safa Skouri [2], Laura Boquera [3], Luisa F. Cabeza [3] and Mariem Lazaar [2]

[1] Faculté des Sciences de Tunis, Université de Tunis El Manar, Tunis 2092, Tunisia
[2] Centre de Recherches et des Technologies de l'Energie, Technopole de Borj-Cédria, BP: 95, Hamam Lif 2050, Tunisia
[3] GREiA Research Group, Universitat de Lleida, Pere de Cabrera s/n, 25001 Lleida, Spain
* Correspondence: ayedrabeb09@gmail.com

Abstract: By enhancing the thermal properties of cement-based building materials, energy consumption and carbon dioxide (CO_2) emissions related to space conditioning in buildings can be alleviated. This study aims to present cement-based composites reinforced by textile fibers for application in building and construction. Several lightweight coating mortars were produced by partially replacing the sand in the mix with different percentages of textile waste. Mechanical and thermal characterizations of the reinforced cementitious composites were performed. The results showed that the thermal conductivity of cementitious compounds decreased as the proportion of reinforcing material in the mixture increased. In terms of mechanical properties, the textile slightly reduced the compressive strength of cementitious mortar, while it improved the flexural strength. A numerical study was then performed to derive the actual impact of these reinforced materials on the thermal behavior of a building element using COMSOL Multiphysics. Numerous configurations of walls coated with different mortar mixtures were studied. The results showed that coating both sides of a building wall with 20 mm of textile-reinforced mortar reduced the internal temperature by 1.5 °C. Thus, the application of these thermally improved mortars as coating mortars appears to be a relevant solution to enhance the thermal performance of buildings.

Keywords: textile fiber waste; reinforced cementitious mortar; thermal insulation; heat transfer analysis

check for updates

Citation: Ayed, R.; Bouadila, S.; Skouri, S.; Boquera, L.; Cabeza, L.F.; Lazaar, M. Recycling Textile Waste to Enhance Building Thermal Insulation and Reduce Carbon Emissions: Experimentation and Model-Based Dynamic Assessment. *Buildings* **2023**, *13*, 535. https://doi.org/10.3390/buildings13020535

Academic Editors: Zhenjun Ma and Müslüm Arıcı

Received: 14 January 2023
Revised: 7 February 2023
Accepted: 13 February 2023
Published: 15 February 2023

1. Introduction

Due to population growth and improving living standards in general, the consumption of textiles has increased worldwide, which has led to an increase in textile production. The textile industry presents a consumption of clothing and textiles of 7 kg per year per capita, and more than 49 million tons of products are produced every year [1]. Meanwhile, the textile industry is one of the most worrying supply chains, having catastrophic social and environmental impacts worldwide [2]. In 2020, textile consumption in Europe had, on average, the fourth-highest impact on the environment and climate change [3]. Moreover, increases in textile global production result in the creation of a high amount of textile waste. Accordingly, the implementation of a circular model in the textile sector, taking into account the recycling of textile waste, is essential to ensure sustainability and mitigate the environmental impacts of this sector. There are two types of textile waste: reusable waste (production offcuts, reel offcuts, etc.) and recyclable waste (filament, etc.). The last category, in turn, is divided into two main categories: one of high value and the other of low value. It is estimated that 1/4 of textile waste consists of pure fibers: cotton, synthetics, and others. In Tunisia, there were more than 3000 companies in the textile and clothing industry in 2019 [4]. The total textile waste in Tunisia was about 31.1 kilotons, divided into 6.3 kilotons of reusable waste and 24.8 kilotons of recyclable waste. The latter, in turn, is divided into 9.8 kilotons in high value and 15.0 kilotons of low value waste [3].

The textile industry is not the only one with disastrous social and environmental impacts. The building sector has many drawbacks in terms of energy consumption and carbon dioxide (CO_2) emissions [5]. According to the International Energy Agency (IEA), the residential and industrial building sectors represent more than a third of global final energy consumption and nearly 40% of total direct and indirect carbon dioxide (CO_2) emissions [6]. Most residential and industrial buildings require high-energy consumption to ensure acceptable thermal comfort for their occupants. Therefore, the implementation of an energy-saving program in buildings by improving the thermal performance of construction elements has become a necessity, with the application of thermal insulation being a promising way to reduce energy losses [7–9]. The commonly used technique to improve the insulation capacity of walls is to embed the insulation layer in the erected foundation wall, which increases the thickness of the wall and delays the construction time [10]. To overcome these design disadvantages, several researchers are focusing on the integration of insulators in building materials such as plaster [11], concrete [12,13], mortar [14–16], and construction bricks [17–19]. Fibers are among the most widely studied insulators for use in cementitious materials and have proven to be an excellent solution for mitigating heat loss. In fact, several studies have been carried out in this field of research to study new cement-based composites reinforced with fibers such as wool fiber [20], Acaï [21], palm fiber [22,23], agricultural fiber wastes [24], coconut fiber [25], rice straw fibers [26], recycled brass fibers [27], and basalt fiber [28].

The past decades have witnessed a growing interest in textile-reinforced mortar (TRM), and it has become one of the most important fiber-reinforced materials due to its promising textile properties. Based on a review of the literature, the use of textile fiber as a reinforcing material has only been explored in a limited number of applications. Generally, textiles have mainly been incorporated into cement-based materials to improve their mechanical strength, ductility, toughness, and durability [29–38]. Sadrolodabaee et al. [39] investigated the mechanical and durability properties of cement composites reinforced with two types of textile waste, either a fraction of short fibers or non-woven fabric. Flexural strength, toughness, stiffness, and drying shrinkage developed composites were evaluated. A composite reinforced with six layers of non-woven fabric showed optimal performance with a flexural strength of 15.5 MPa and a toughness of 9.7 kJ/m^2. Gulinelli et al. [40] studied the performance of various wall configurations strengthened with textile-reinforced mortar. The structural performance was evaluated experimentally and numerically by conducting diagonal compression tests. The findings indicated that, compared to a simple wall, the reinforcement systems lead to a substantial improvement in both load-bearing capacity and rigidity. Abbas et al. [41] investigated the potential of using 16 different textile fabrics for structural applications. A series of tests, including microscopic analysis, mass per unit area, and tensile strength, were carried out on specimens made of textile-reinforced mortar to determine the most suitable fabric. The results showed that the mass per unit area of the tested fabrics varied between 117 and 1145 g/m^2. It was noticed that the tensile strength was greater in the warp direction compared to the weft direction, owing to the increased number of yarns in the warp direction. Moreover, it was found that plain weave fabrics had higher strength compared to twill weave fabrics. Plain weave was deemed to be the most suitable among the 16 selected fabrics for use in textile-reinforced mortar applications, due to its adequate spacing and alternating arrangement of yarns, leading to a stronger bond with the matrix and higher tensile strength. Regarding thermal performance studies, few authors have experimentally investigated the integration of textile fibers in cementitious mortars to enhance building thermal insulation [42]. Oliveira et al. [1], for example, experimentally investigated the effect of fabric shavings on the thermal and mechanical properties of mortars. The results showed that the mechanical strength of the fabric-yarns-reinforced mortar was lower than that of the reference mortar. Despite this reduction, the minimum values set by the standards for compressive and flexural strengths were met. Furthermore, when the textile-reinforced mortar was thermally tested at 60 °C, its internal surface temperature was about 12 °C lower than that of the reference. This property is extremely important for questions of durability, as it protects structures and walls from expansion and contraction movements. Briga et al. [43] also investigated the potential of using textile fibers in building construction applications. The thermal characterization of the developed composites was performed by analyzing the heat flows, internal surface temperatures, heat transfer coefficients, and infrared thermal imaging. The authors affirmed that increasing the proportion of textile fibers in the mix resulted in higher thermal stability. Moreover, the developed composite showed promising results compared to traditional building materials.

Only a few studies have focused on the incorporation of textile fibers as thermal reinforcing in cementitious mortars. Since the results were promising, a more concise understanding and exploration of the overall thermal performance of these composites is required. Thus, this study aims to evaluate the potential of incorporating recycled industrial textile waste into building components to enhance their thermal properties and reduce energy consumption in buildings. Several lightweight coating mortars were produced by partially replacing the sand in the mix with different percentages of textile waste as reinforcement material. The mechanical and thermo-physical characterizations of these composites were carried out. A numerical investigation was then performed to examine the thermal efficiency of the developed materials using COMSOL Multiphysics 5.5 software. Numerous configurations of walls coated with different coating mortar mixtures were studied. The study suggests using textile-reinforced cement-based materials as coating mortars to enhance the thermal performance of buildings and reduce energy consumption and CO_2 emissions.

2. Textile Reinforced Mortar: Preparing and Testing Methods

2.1. Samples Preparation

Fiber-reinforced mortars were prepared by replacing the sand of cement mortar with volume fractions of 0% (PM), 10% (M10), 20% (M20), 30% (M30), and 40% (M40) of textile. The cement used in this study was Portland cement in accordance with the terminology of the European standard EN-197-1 [44] and was supplied by Cementos Molins Industrial S.A. A natural sand AF-R-0/2-S was used as a fine aggregate. The textile fiber added as a reinforcing material is a waste generated at the end of the textile spinning process. Table 1 summarizes the thermal properties of the fibers used. Before mixing the mortar, the fiber was dispersed by injecting compressed air. Mortar mixes were prepared using a cement-to-sand volume ratio of 1:4 and the mix design is shown in Table 2 and Figure 1. The mortar mixing procedure was carried out according to the European standard EN 1015-2 [45]. Finally, the mixed slurry was poured into prismatic molds with dimensions of 40 × 40 × 160 mm (Figure 2). The blend was cast in two stages, each layer set being first tapped with a steel bar and then vibrated until the fresh mortar was completely leveled. Three samples of each mixture were made. The samples were left in the molds for 24 h at a laboratory temperature of 20 °C. After demolding, some of the specimens were cured in water for 7 days and others for 28 days.

Table 1. Thermal properties of textile-reinforced mortar components.

Material	Thermal Conductivity [W/m·K]	Thermal Diffusivity [mm^2/s]	Volumetric Heat Capacity [MJ/m^3 K]
Cement	0.140	0.201	0.694
Sand	0.335	0.278	0.278
Textile fibers	0.082	0.418	0.196

Table 2. Mixture designs for tested mortar samples, mass for 1 m^3.

Materials	PM	M10	M20	M30	M40
Cement	215	215	215	215	215
Sand	1540	1386	1232	1078	924
Textile	0	5.26	10.52	15.78	21.04

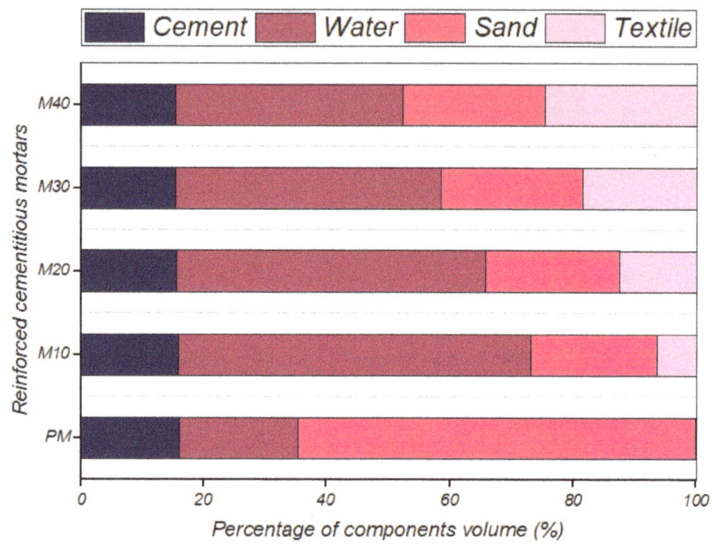

Figure 1. Mix design of the textile-reinforced cementitious mortars.

Figure 2. Textile-reinforced mortars (**a**) in molds, (**b**) during the curing period, and (**c**) in the hardened state.

2.2. Workability and Density Measurement

Both the flow behavior and the bulk density were tested in the fresh state. The fluidity of the cementitious mortar, which is expressed by the workability, was evaluated immediately after the mixing process using a flow table test (Figure 3a) in accordance with the standard EN 1015-3 [46]. The fresh bulk density was quantified by the European standard EN 1015-6 [47]. In the hardened state, the dry bulk density of the reinforced mortars was studied in accordance with the EN 1015-10 standard [48].

Figure 3. (**a**) Workability measurement, (**b**) testing of the dry bulk density of hardened mortars.

2.3. Mechanical Characterization

For the mechanical characterization of the hardened cementitious composites, flexural and compressive strength tests were carried out according to the EN 1015-11 standard [49]. A COINSA Controls Industrial servo-controlled testing machine with two different heads, one for flexure testing and the other for compression testing, as shown in Figure 4, was used to test the reinforced cementitious mortars. To assess the bending strength, a three-point flexion test was carried out with a load rate of 50 ± 10 N/s. The compressive strength tests were carried out following the flexion tests using one of the two resulting fragments with a loading rate of 2400 ± 200 N/s. The two tests were carried out until the fracture of the prismatic specimen and the breaking load were recorded.

Figure 4. COINSA Controls Industrial servo-controlled testing machine (**a**) flexion test (**b**) compression test.

2.4. Thermal Characterization

2.4.1. Analytical Prediction of Composite Thermal Conductivity

As a bulk property, the thermal conductivity of composite material can be predicted by several theoretical models. Starting with Maxwell [50], numerous analytical investigations [51–53] have been performed to solve the problem of thermal conduction in heterogeneous materials. The suggested models are mathematical expressions that determine the effective thermal conductivity of a composite material based on the thermal conductivities of its components as well as the shape and proportion of the reinforcing material. The suitability of a particular model is established by the underlying assumptions made during its development. Thus far, various analytical solutions are available to predict thermal conductivity for the most common composite structures, including dispersion compounds. In this work, several theoretical models were presented, which assume that the reinforcing material consists of randomly dispersed spherical units within the matrix, for comparison with the experimental results. In all of the models, we considered a composite material with thermal conductivity Kc, consisting of spherical units with thermal conductivity Krf embedded in a continuous matrix with thermal conductivity Kmx at a volume fraction φ.

- Series and Parallel models: Figure 5 shows the two plain theoretical methodologies that have been followed to predict the effect of adding a reinforcing material on the thermal conductivity of a matrix. The first approach focuses on individually considering the contribution of each component to model the thermal conductivity of the composite through the application of the percolation theory [54]. Based on the electrical analogy, this model is called a series model. In this case, the effective thermal conductivity of the composite material is given by [55–57]:

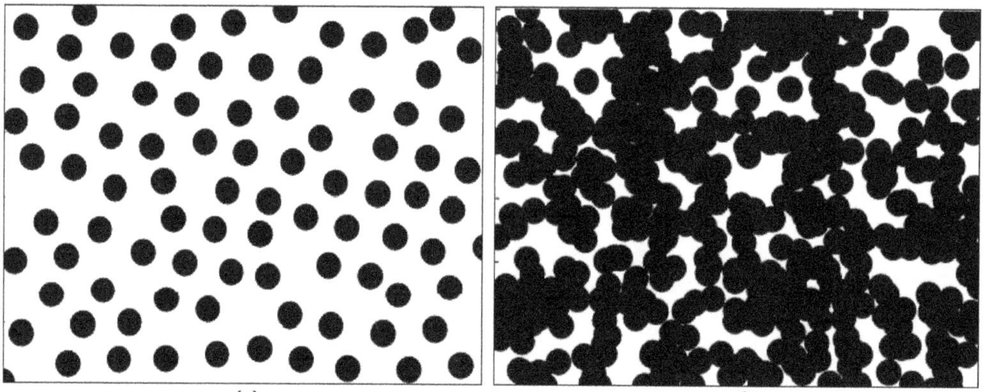

Figure 5. Basic models used to predict the effective thermal conductivity of composite materials: (**a**) non-penetrating dispersed particles, (**b**) interpenetrating dispersed particles.

$$\mathrm{Kc} = (1-\varphi)\,\mathrm{Kmx} + \varphi\,\mathrm{Krf} \quad (1)$$

The second approach assumes that the composite reacts as a homogeneous material. This model follows the arrangement of parallel electric circuits, which is represented by the following expression:

$$\mathrm{Kc} = 1/(((1-\varphi))/\mathrm{Kmx} + \varphi/\mathrm{Krf}) \quad (2)$$

The aforementioned models present upper (interactive model) and lower bounds (non-interactive model) of the effective thermal conductivity that the data encompass [55–57].

- Maxwell model: This model was developed to define the electrical conductivity of a heterogeneous medium composed of dispersed spheres. The development of this model provides an accurate solution for the effective thermal conductivity of arbitrarily distributed homogeneous spherical particles without interaction in a homogeneous matrix [56]:

$$Kc = Kmx\ (2\ Kmx + Krf - 2\ (Kmx - Krf)\ \varphi)/(2\ Kmx + Krf + (Kmx - Krf)\ \varphi) \quad (3)$$

The main limitation of this model is its limited applicability to only low concentrations of reinforcing particles. However, it has served as a base for creating new, advanced models that account for additional parameters related to the various components involved [57].

- Rayleigh model: This model was adapted to predict the effect of cylindrical reinforcement materials on the thermal conductivity of a matrix. The equation below allows for the effective thermal conductivity to be calculated [58]:

$$Kc = Kmx\ (Kmx + Krf - (Kmx - Krf)\ \varphi)/(Kmx + Krf + (Kmx - Krf)\ \varphi) \quad (4)$$

- Hashin and Shtrikman model: Following the approach of Maxwell and using the perturbation hypothesis, Hashin and Shtrikman developed a model to predict the thermal conductivity of randomly scattered units in a continuous matrix. This model provides upper and lower limits of the effective thermal conductivity rather than deriving an equation for it. Moreover, it is shown that, in the case of composite material, these limits are the most restrictive that one can obtain in terms of volume fractions of charge and conductivity [59]. Equations (5) and (6) indicate the lower and upper limits of the conductivity [55]:

$$Kc\_min = Kmx\ (1 + (p - 1)\ D\ \varphi)/(1 - D\ \varphi) \quad (5)$$

$$Kc\_max = Kmx\ (Krf\ (1 + (p - 1)\ (1 - \varphi)\ d)/(Kmx\ (1 - (1 - \varphi)\ d) \quad (6)$$

where

$$D = (Krf - Kmx)/(Krf + (p - 1)\ Kmx) \quad (7)$$

$$d = (Kmx - Krf)/(Kmx + (p - 1)\ Krf) \quad (8)$$

and p is a parameter involving the morphology of the particles, in the case of a spherical dispersion $p = 3$ and for a cylindrical dispersion $p = 2$.

- Hatta and Taya model: Hatta and Taya developed a model to predict the thermal conductivity of a composite consisting of short fibers with different orientations [60,61] based on the analogy of Eshelby [62]. This approach is based on predicting the steady-state equivalent thermal conductivity of the composite by considering the shape and interactions between the additions with different orientations. The equation they arrived at is [55]:

$$Kc = Kmx + (\varphi\ Kmx)/(S\ (1 - \varphi) + Kmx/((Krf - Kmx))) \quad (9)$$

where $S = 1/3$ for spherical particles.

- Nielsen and Lewis model: Nielsen and Lewis derived a semi-theoretical model for predicting thermal conductivity [63,64], based on the Halpin-Tsai equation [65]. They adjusted the model to handle non-spherical additives using a coefficient, which depends on the shape and orientation of the particles. Moreover, Nielsen and Lewis's model considers the effect of the maximum fraction of the additive, φmax. The semi-empirical model developed is as follows [66]:

$$Kc = Kmx\ (1 + C\ E\ \varphi)/(1 - \beta\ E\ \varphi) \quad (10)$$

where:

$$E = (Krf/Kmx - 1)/(Krf/Kmx + C) \quad (11)$$

$$\beta = 1 + ((1 - \varphi max)\ \varphi)/(\varphi max)^2 \quad (12)$$

For spherical additive randomly dispersed in the matrix, $C = 1.5$ and $\varphi max = 0.637$ [57].

2.4.2. Experimental Characterization

To study the thermal properties of reinforced mortars, The TEMPOS thermal properties analyzer was used on the remaining fragment from the flexion test. As shown in Figure 6, the samples must be drilled first with a rotary hammer and then cleaned with compressed air. Before inserting the sensor needle, the hole must be filled with thermal grease. Good thermal contact between the sensor and the sample is critical for accurate measurements; therefore, it is important to ensure that the sensor fits tightly into the hole. Finally, the thermal conductivity and thermal resistance were measured in the climatic chamber at a temperature of around 20 °C.

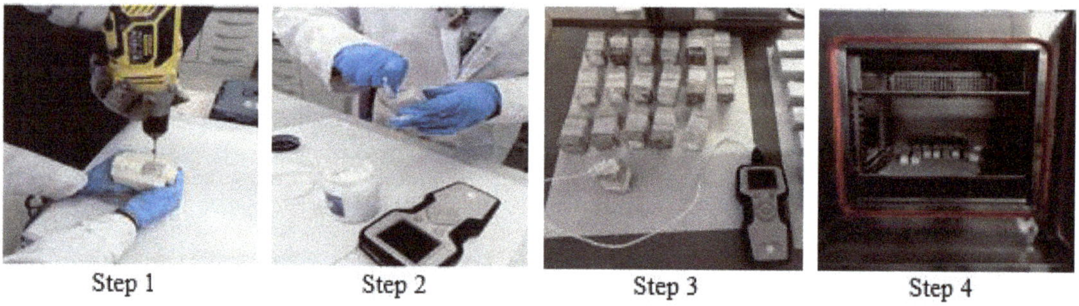

Figure 6. Thermal conductivity test process using the TEMPOS thermal properties analyzer.

2.4.3. Numerical Characterization

A numerical study was performed to investigate the thermal properties of the textile-reinforced mortars using COMSOL Multiphysics software. The first crucial step for a successful simulation is to generate an appropriate geometry for the problem. In this case, a three-dimensional model featuring a rectangular block defining the mortar matrix and spherical units representing the reinforcement material was used to conduct the thermal analysis on the composite materials with varying additive concentrations. In this steady-state heat conduction problem, the heat flux was only considered along the x-direction. The faces that were perpendicular to the direction of the heat flow were considered isothermal. One surface was set at 15 °C and the other at 40 °C. The other faces that were parallel to the x-direction were all considered adiabatic. The thermal conductivities that were predicted by this model were compared and validated with the experimental results. After validation of the model, it was be used to predict the thermal diffusivity and the volumetric heat capacity of the textile-reinforced mortars.

3. Evaluation of Characterization Results

In order to assess the influence of textile fibers on cement-based mortars, several tests were carried out on the fresh and hardened states of the mixtures. Each measurement was repeated three times to guarantee repeatability, and the mean value was reported.

3.1. Workability Testing

The amount of water added to the dry mix was adjusted differently for each percentage of textile waste added to achieve acceptable workability. The details of the modified cementitious mortar mix proportions are summarized in Figure 1. The Water to Cement (W/C) ratio values with different fiber percentages are illustrated in Figure 7. The measured flow diameter of the control mixture without fibers was 145 mm. In this test program, the flow rate for all of the mortar mixtures was kept at a specified value of 140 ± 5 mm, in order to obtain a fluid and workable consistency. As can be seen, fiber-reinforced mortars showed a higher water requirement in order to achieve similar workability to plain mortar. This was due to the high water absorption of the fibers.

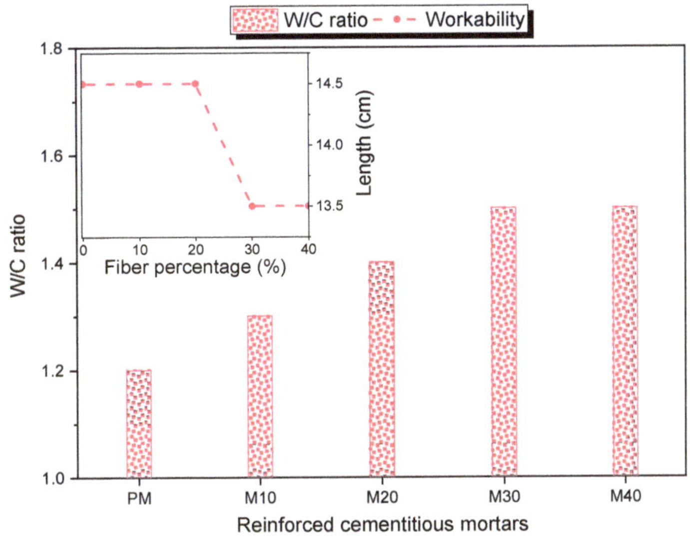

Figure 7. W/C ratio values with different textile fiber percentages.

3.2. Bulk Density Testing

Figure 8 shows the variation in apparent density of the samples in both fresh and dry states versus the reinforcing material percentage. The fiber-free cement mortar had a fresh apparent density of 1900 kg/m^3. After adding varying amounts of textile fibers, the fresh state apparent density dropped to a value of 1585 kg/m^3, exhibiting a difference of around 315 kg/m^3. For the M40 sample, in the hardened state, the density diminished to 1538 kg/m^3. with a difference of around 302 kg/m^3 compared to the plain mortar PM. Thus, it can be deduced that the incorporation of the fibers into the cementitious mortar is accompanied by a loss of density. This lightness is due to the low fiber density.

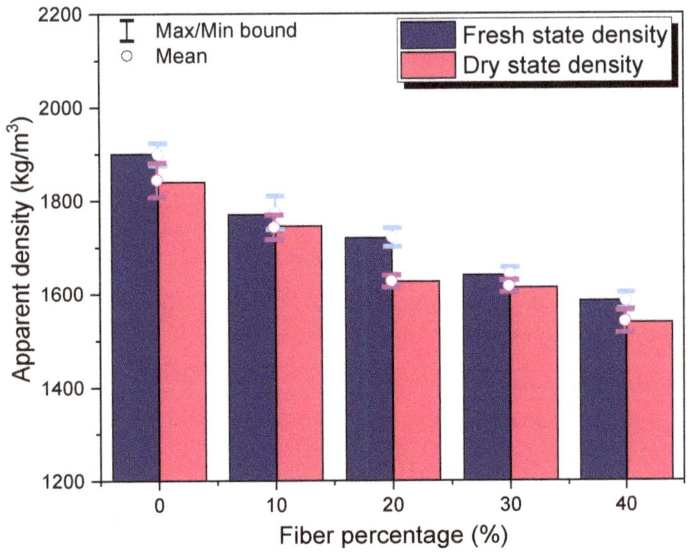

Figure 8. Dry and fresh bulk densities of the textile fiber reinforced mortars.

3.3. Mechanical Characterization

Coating mortars must have a sufficient mechanical strength to withstand the multiple impacts to which they may be subjected during their lifetime. Accordingly, the mechanical performance of the textile waste-reinforced mortars was investigated at 7 and 28 days of curing. Figure 9 illustrates the failure modes of the textil-ized mortars during the mechanical tests after 7 and 28 days, respectively. The visual inspection of the defective samples (Figure 9a) confirmed that the mortars containing more textile fibers showed more severe damage under the applied load. This predicts that increasing fiber content will cause a dramatic drop in sample stiffness and a subsequent reduction in compressive strength. Figures 10 and 11 show the average compressive and flexural strength results, respectively. Based on the graphs, increasing the curing period significantly enhances mechanical performance. For the compressive strength testing, the increasing percentage between the 7th day and the 28th day was approximately 35% for the M40 sample. Compared to the 7-day flexural strength, the increase in the 28-day strength was 34% for the PM sample and 22% for the M40 sample. It is worth noting that, as the amount of textile increased, the flexural strength gained over the curing period was affected. As can be seen, the plain mortar PM had the highest compressive and flexural strength regardless of the curing period. In adding the textile fibers, a decrease in the compressive strength of mortars occurs. Figure 10 shows that the seven-day compressive strength for the M40 sample decreased by approximately 33% compared to ordinary mortar PM. However, the compression resistance of the samples after 28 days of curing slightly decreased until reaching a difference of 15 % for the M40 mix compared to the PM sample. This diminution in mechanical strength for the textile-reinforced mortars could be attributed to the low strength and high compressibility of the textile fibers. Despite the reduction in compressive strength, all of the reinforced mortars met the requirement for use as a coating mortar in building construction according to EN 998-1 [67]. Concerning Figure 11, regardless of the curing period, the flexural strength of the mortars increased as the percentage of reinforcing material increased. The increase in the seven-day flexural strength was about 11%, 25%, 33%, and 34% for the M10, M20, M30, and M40 samples, respectively, compared to the control sample. The 28-day flexural strength increased to 3.7 MPa, which corresponds to an increase of approximately 22% compared to the textile-free mortar.

3.4. Thermal Characterization

Figure 12 illustrates the variation in thermal conductivity and thermal resistance of the reinforced mortars under climatic chamber conditions. As can be seen, the thermal conductivity shows a clear downward trend with an increasing proportion of reinforcement material. With 10% fibers incorporated, the thermal conductivity decreased by 15% compared to the plain mortar. By increasing the amount of textile, the thermal conductivity was further decreased until a value of about 0.87 W/m·K was reached, which is about 42% lower than the PM sample. Accordingly, the addition of textile fibers increased the thermal resistance of the cement mortar to approximately 1.27 m·K/W. This improvement in the thermal properties of mortars is attributed to the low thermal conductivity of the textile fiber.

Figure 13 shows a comparison between the experimentally measured thermal conductivities of the reinforced mortars and those predicted using the models mentioned above. Both the experimental data and theoretical data show a similar trend. An increase in the proportion of the reinforcing material corresponds to a decrease in the thermal conductivity of cementitious mortar. The results showed that the experimental outcomes fall within the bounds of the series and parallel models and Hashin–Strikman model. The other analytical models had similar results with minor variations. Since the reinforcing materials are assumed to be spherical particles dispersed in the cementitious matrix, the Hata and Taya model coincides with the Maxwell model. Although the experimental thermal conductivity was close to the results of these two models, a slight discrepancy can be observed. This difference is due to the fact that the model only considers the thermal conductivity of the constituent materials of the composite. The Rayleigh model exhibited the best agreement with the experimental data, especially for the samples with high percentages of reinforcing materials. These results demonstrate that the textile fibers were well-dispersed in the matrix, resulting in homogeneous composites.

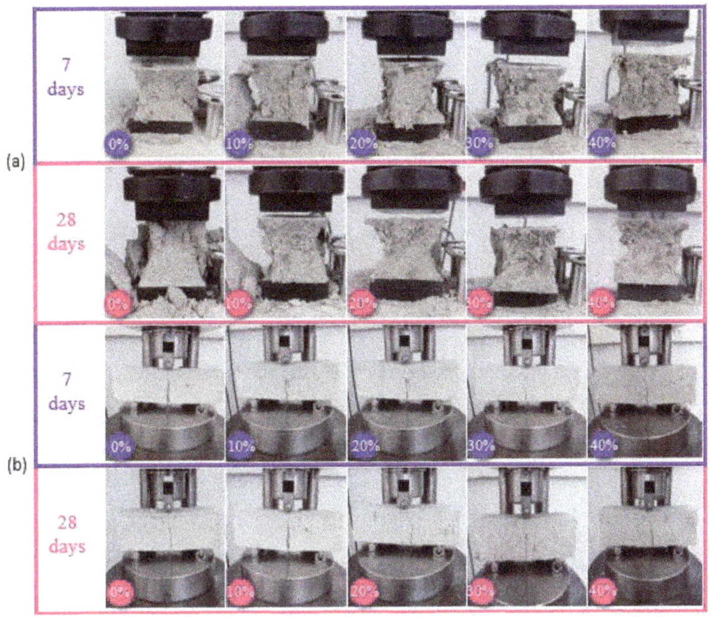

Figure 9. Failure behavior of the textile-reinforced mortars during (**a**) the compression and (**b**) bending tests.

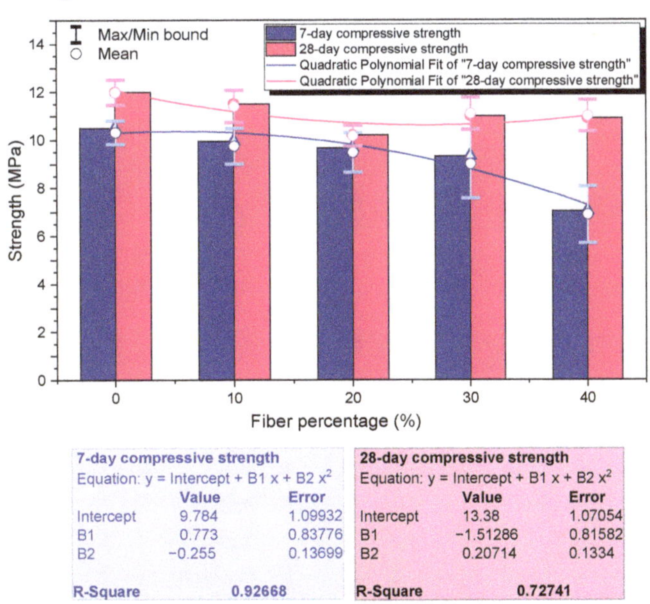

Figure 10. Compressive strength of textile waste-reinforced mortars after 7 and 28 days of water curing.

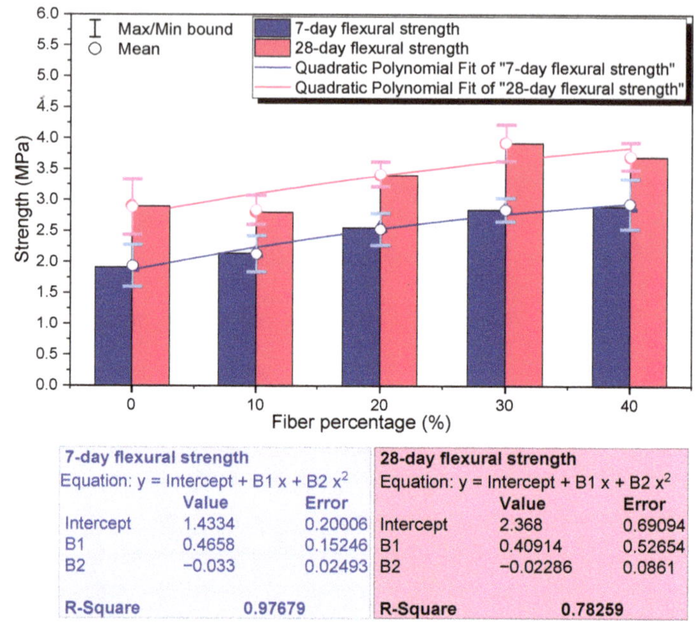

Figure 11. Flexural strength of textile waste-reinforced mortars after 7 and 28 days of water curing.

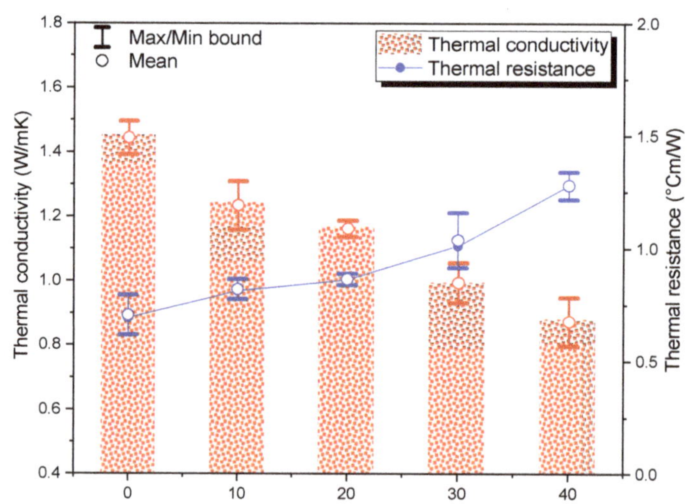

Figure 12. Thermal conductivity and thermal resistance of the textile waste-reinforced mortars.

Figure 14 shows a comparison of the numerical and experimental results of the thermal conductivity of textile-reinforced mortars. Both results show the same tendency, with an error interval marked by the orange area. For the samples with a higher textile content, the margin of error between the predicted and experimental results increased slightly. The experimental values of thermal conductivity are lower than those predicted. This can be explained by the uncontrolled phenomena, which can occur during the preparation and which are not taken into account by the model, such as the porosity and the contact resistance between components. In general, the results indicate the effectiveness of the computational model.

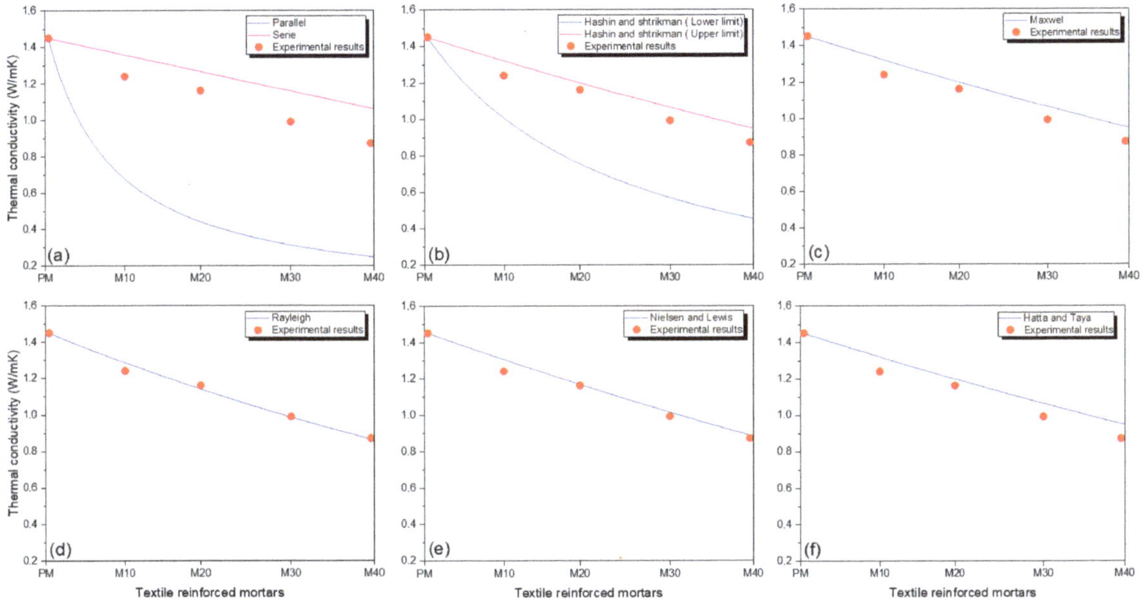

Figure 13. Thermal conductivity of textile-reinforced mortars predicted by the model of (**a**) Parallel and series, (**b**) Hashin and stickman, (**c**) Maxwel, (**d**) Rayleigh, (**e**) Nielson and Lewis, and (**f**) Hatta and Taya models compared to experimental results.

Thermal diffusivity quantifies the speed at which heat is transferred through materials from a hot to a cold end. It is calculated by dividing the thermal conductivity of the material by its density and specific heat capacity at a constant pressure. The mathematical expression of thermal diffusivity is as follows:

$$\alpha = K/\rho Cp \tag{13}$$

where K is the thermal conductivity; Cp is the specific heat capacity; ρ is the density. Moreover, ρCp is the volumetric heat capacity.

Figure 15 shows the numerically predicted thermal diffusivity and volumetric heat capacity of the mortars reinforced with textile waste. The addition of fibers to the cementitious mortar reduced both thermal diffusivity and volumetric heat capacity. With a low content of textile fibers, the thermal diffusivity and the volumetric heat capacity of textilized mortars marginally diminished until 1.13 mm^2/s and 1.12 $MJ/m^3{\cdot}K$, respectively. As well, by increasing the proportion of textile fibers to 40 %, the thermal diffusivity and volumetric heat capacity were decreased to 0.97 mm^2/s and 0.91 $MJ/m^3{\cdot}K$, respectively. The addition of the textile fiber waste into the cementitious mortar reduced the thermal diffusivity and the volumetric heat capacity in correlation with the drop in thermal conductivity. Compared to the PM sample, a 40% decrease in thermal conductivity for the M40 sample caused drops of 21% and 23% in thermal diffusivity and volumetric heat capacity, respectively.

4. Case Study: Numerical Investigation of a Hollow Brick Wall Coated with Textile-Reinforced Mortar

As the textile-reinforced mortars showed better thermal potential than an ordinary cementitious mortar with acceptable mechanical behavior, the application of these composites as a coating mortar appears to be a relevant solution to reducing energy consumption and carbon dioxide (CO_2) emissions. Thus, a transient heat transfer analysis was conducted using the COMSOL Multiphysics software to predict the effect of adding textile fibers on the thermal performance of buildings.

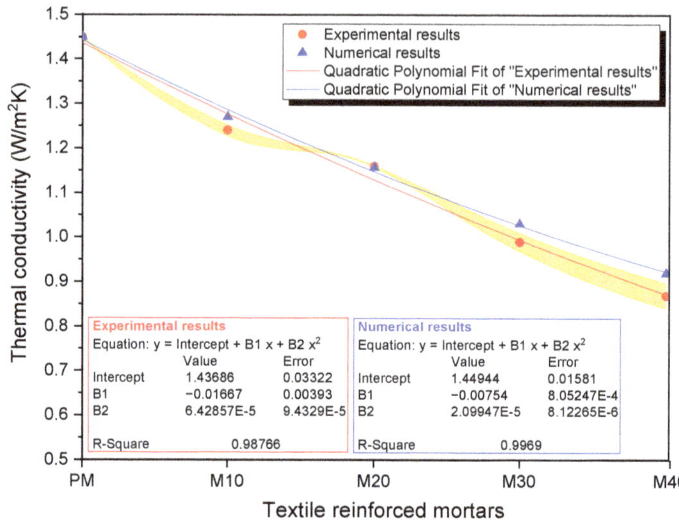

Figure 14. Numerical and experimental validation of the thermal conductivity of textile-reinforced mortars.

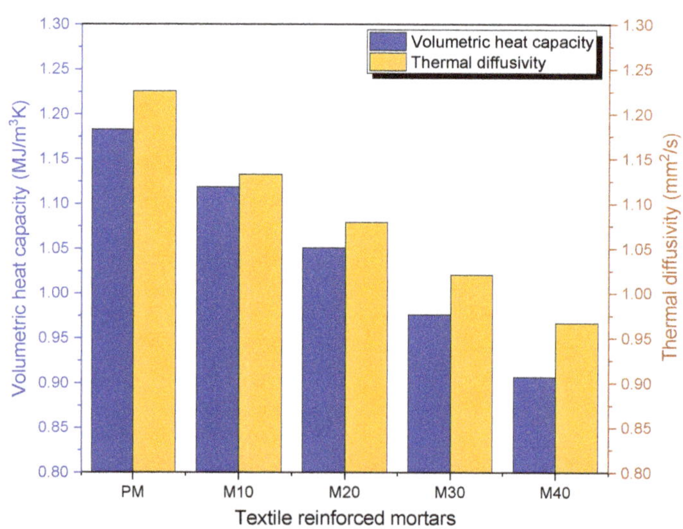

Figure 15. Numerically predicted thermal diffusivity and volumetric heat capacity of textile-reinforced mortars.

4.1. Numerical Model and Validation

The developed numerical model is intended to investigate the thermal response of a building wall after the application of thermal excitation on the outer surface. The investigated wall consisted of hollow bricks measuring 15 × 20 × 30 cm and contained 12 cavities. The bricks were joined horizontally and vertically with ordinary mortar (PM) with a thickness of 10 mm. Moreover, the wall was coated on both sides with a 20 mm-thick layer of cement mortar in three different scenarios, as shown in Figure 16. The impact of changing the type of coating mortar was analyzed numerically based on the temperature change on the inner surface of each wall configuration after imposing a temperature of 40 °C on the outer surface. To solve this heat transfer problem, the heat conduction in solids and heat convection in air cavities were considered, while the radiative heat transfer through air cavities was excluded. Before starting the simulation, the model was meshed using tetrahedral elements, resulting in a mesh with 492,357 elements. The validation of the aforementioned model involved comparing its predictions with the analytical solution of a transient heat transfer problem. The analytical solution was obtained by transforming the partial differential equation into an ordinary differential equation (Equation (14)) using the Gaussian error function [68].

$$T\,(x,t) = Tc - (Tc - Tin)\,(erf((x)/\sqrt{((4kt))})) \quad (14)$$

where T is the temperature, Tc is the imposed heat temperature, Tin is the initial temperature, x is the location of calculation, k is the thermal conductivity, and t is the time.

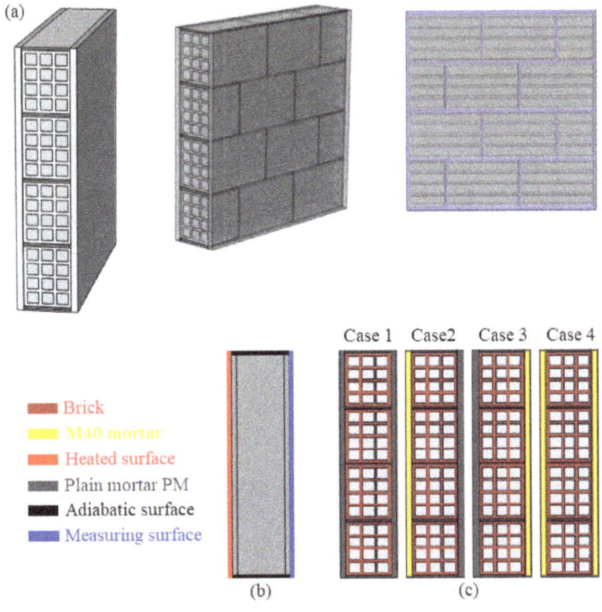

Figure 16. Coated wall designed with COMSOL Multiphysics software (**a**) 3D model (**b**) boundary conditions (**c**) wall configurations.

To evaluate the performance of the proposed model, the Root Mean Squared Error (RMSE) was calculated as follows:

$$\text{RMSE} = \sqrt{\sum_{j=1}^{n} (mj - pj)^2 / n} \quad (15)$$

where mj and pj are, respectively, the average of the measured and predicted parameters, and n is the number of variables.

In order to validate the numerical model mentioned above, the result of case study 1, where the wall was coated with ordinary mortar on both sides, was compared with the analytical results. Figure 17 shows a comparison of the numerically predicted and the analytically calculated change in the internal surface temperature as a function of the wall thickness. As can be seen, the numerical and analytical thermal behaviors illustrate a similar tendency, with a root mean square error (RMSE) of 0.59. This consistency of results demonstrates the effectiveness of the numerical model in studying heat transfer through the brick wall.

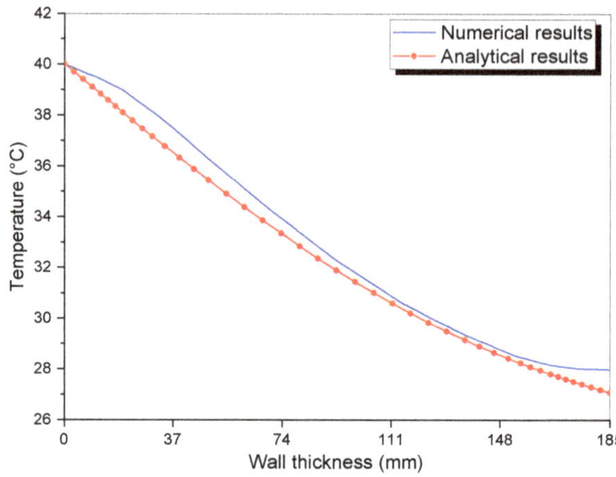

Figure 17. Numerical and analytical validation of a building wall temperature with textile-reinforced mortars.

4.2. Computational Assessment of a Textile Reinforced Wall

Figure 18 summarizes the numerical results of the spatial variation in temperature after 50 min of thermal excitation for the different wall configurations. Figure 18d shows the temperature change along the wall thickness for both the first and second cases. As the graph shows, the temperature propagation in the wall of case 2, coated with textile-reinforced mortar on the outside and plain mortar on the inside, was slower than in the wall coated with ordinary mortar on both sides. Comparing case 1 and case 3, the temperature propagation along the first parts of the walls shows the same tendency since the outer coating layers consist of the same material. However, the integration of the textile into the inner layer of the wall of case 3 slowed down the spread of temperature, as shown in Figure 18e. The application of a textile-reinforced mortar as an inner coating material leads to a greater decrease in temperature than insulation applied to the external mortar layer. Figure 18f illustrates the effect of coating a wall with fiber-reinforced mortar on both sides. The use of textile mortar as a coating mortar on both sides showed the best results for reducing the temperature spread.

Figures 19 and 20 present the variation in heat fluxes and inner surface temperatures versus time after the thermal excitation on the outer surface. As expected, the temperature profile of a wall depends on the thermal conductivity and, consequently, the thermal resistance of its components. As shown in Figure 19, the integration of the textile into the coating mortar reduced the temperature spread in the wall. In fact, the insulation of the outer surface caused a slight decrease in temperature, while the integration of the textile in the inner surface of the wall improved the thermal performance by almost 4% compared to case 1. However, the optimal values were obtained by the fourth case, where the wall was coated on both sides with textile-reinforced mortar. The temperature of the inner surface may decrease by approximately 1.5 °C using the textilized composite. The variation in heat flux, as shown in Figure 20b, also corroborates the results presented for the variation in inner surface temperature. Lower transmitted heat flux values were achieved when the reinforced mortar was incorporated into the investigated wall.

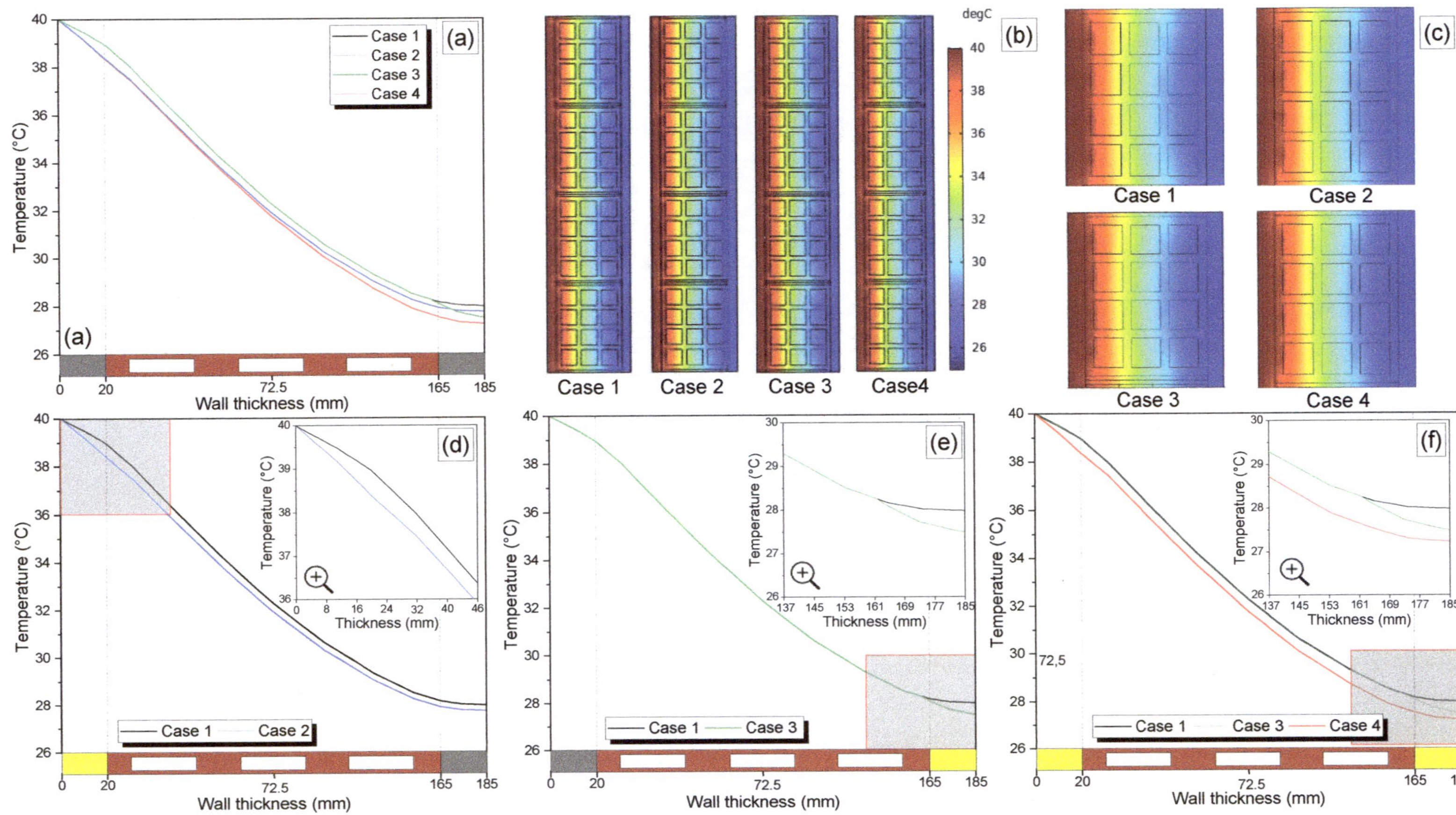

Figure 18. Predicted temperature (**a**) curves and (**b**,**c**) contours in the investigated wall in (**d**) case 2, (**e**) case 3 and (**f**) case 4 compared to case1.

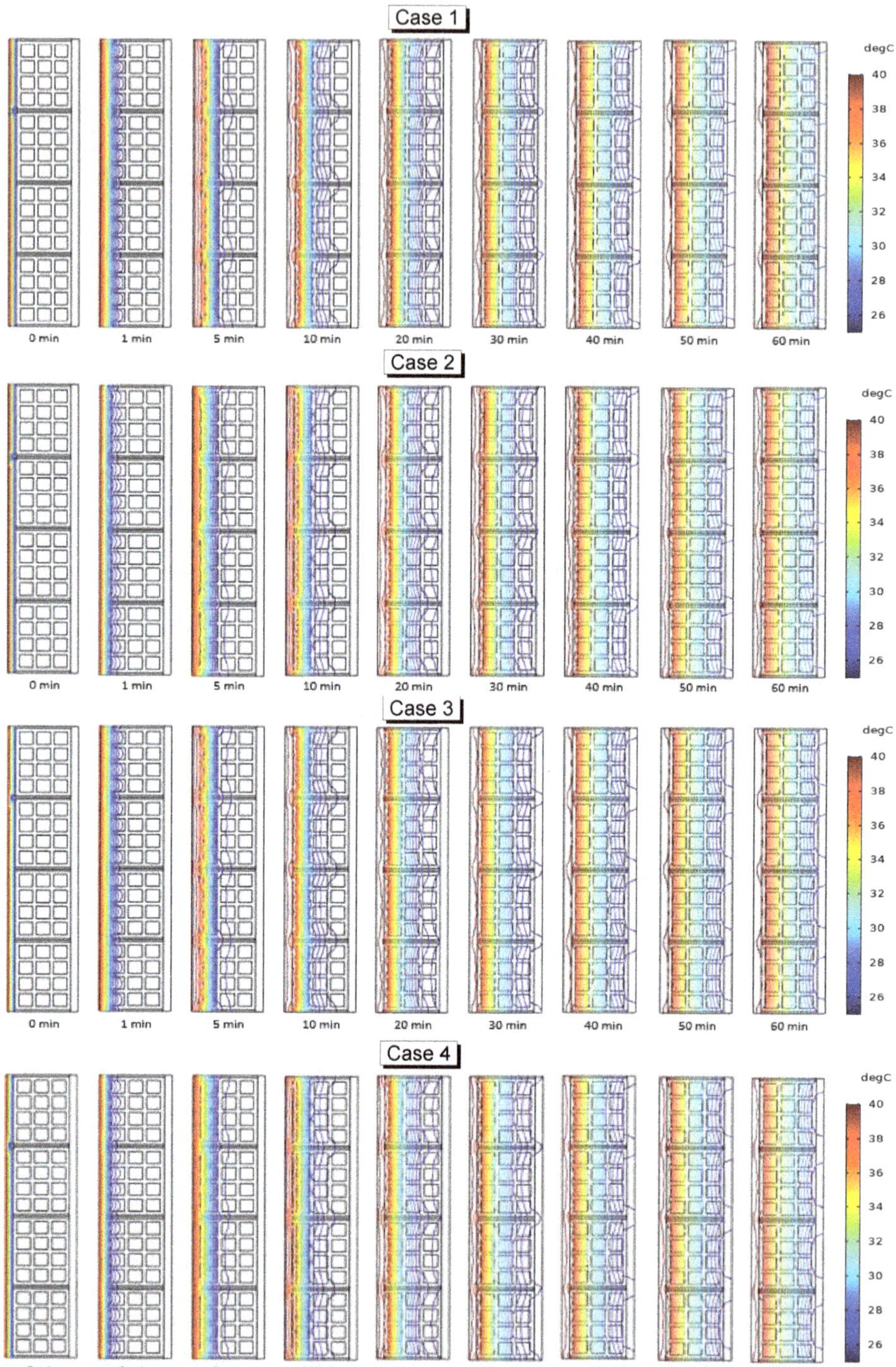

Figure 19. Comparison of the wall temperature fields with different locations of the textile-reinforced mortar.

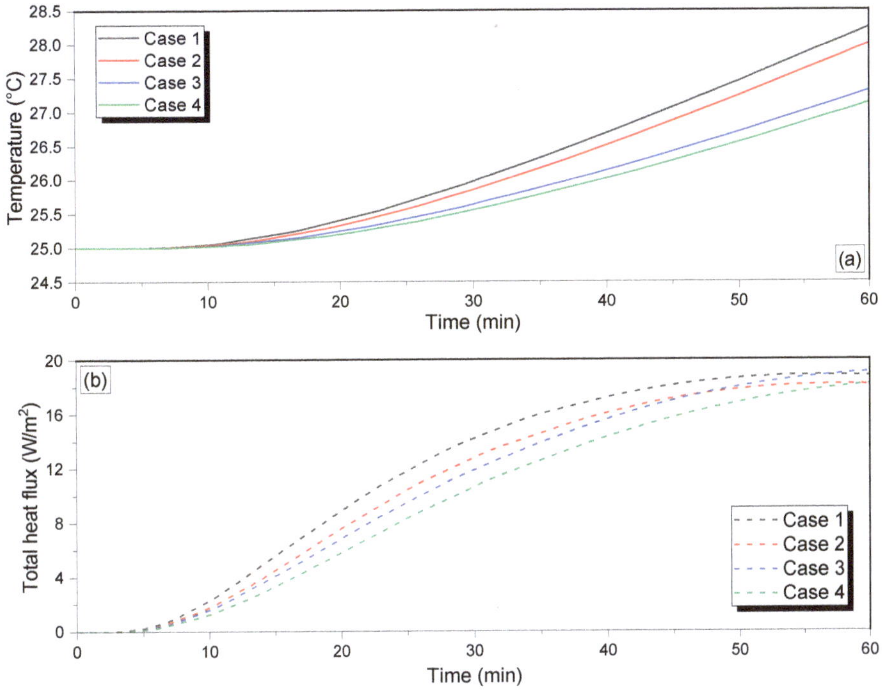

Figure 20. Variation in (**a**) the inner surface temperatures and (**b**) the heat fluxes versus time.

5. Conclusions and Perspective

The incorporation of recycled materials into building components is not only promoted for ecological reasons, but can also improve certain properties of building materials. Therefore, the objective of this study was to develop thermally enhanced cement-based materials by incorporating textile fiber waste as a replacement for sand in varying ratios. The results showed that textile fiber-reinforced mortars can be promising coating mortars due to their improved thermal properties and acceptable mechanical strength. In fact, replacing 40% of the sand with textile fibers caused drops in thermal conductivity, thermal diffusivity, and volumetric heat capacity by about 40%, 21%, and 23%, respectively, compared to ordinary cement mortar. Furthermore, the integration of textile fibers improved the flexural strength of cementitious mortar. Although the reinforcement of the mortar with textile fibers decreased the compressive strength, it still met the requirement for use as a coating mortar. Since the textile-reinforced mortar has a high potential to be used as a coating mortar, a numerical study was performed to study its impact on the thermal behavior of a hollow brick wall. The examined wall was coated with the reinforced mortar in different arrangements. The results showed that coating a wall with a 20 mm layer of textile-reinforced mortar on both sides improved the overall thermal performance.

This work sheds light on the influence of incorporating textile fiber waste as a thermal reinforcement material into cementitious mortar on both its thermal and mechanical properties and, thus, on its impact on a building element. The results highlighted the significant potential of using textile fibers in the development of more energy-efficient and environmentally-friendly building materials and practices as a means to minimize energy consumption and alleviate the environmental effects of widespread waste materials. Further research can be carried out to investigate the impact of textile fibers on the durability and long-term characteristics of cement mortars. Moreover, it would be beneficial to assess the influence of incorporating these materials on the thermal efficiency of entire buildings, beyond simply analyzing their effect on a single wall.

Author Contributions: Conceptualization, R.A. and S.B.; methodology, R.A., S.B. and L.B.; software, R.A.; validation, R.A., S.B. and S.S.; formal analysis, R.A., S.B. and S.S.; investigation, R.A., S.B. and L.B.; resources, L.F.C. and M.L.; data curation, R.A. and S.B.; writing—original draft preparation, R.A., S.B. and S.S.; writing—review and editing, R.A., S.B. and L.F.C.; visualization, R.A. and S.B.; supervision, L.F.C. and M.L.; project administration, L.F.C.; funding acquisition, L.F.C. All authors have read and agreed to the published version of the manuscript.

Funding: This study was supported by the Thermal Processes Laboratory of the Research and Technology Center of Energy (CRTEn) and the University of Tunis El Manar in Tunisia. This work is partially supported by ICREA under the ICREA Academia programme.

Data Availability Statement: Not Applicable.

Acknowledgments: Cabeza would like to thank the Catalan Government for the quality accreditation given to her research group (2017 SGR 1537). GREiA is certified agent TECNIO in the category of technology developers from the Government of Catalonia.

Conflicts of Interest: The authors declare no conflict of interest.

References

1. de Oliveira, E.M.; Machado de Oliveira, E.; de Oliveira, C.M.; Dal-Bó, A.G.; Peterson, M. Study of the Incorporation of Fabric Shavings from the Clothing Industry in Coating Mortars. *J. Clean. Prod.* **2021**, *279*, 123730. [CrossRef]
2. United Nations Environment Programme. *Global Status Report for Buildings and Construction*; United Nations Environment Programme: Washington, DC, USA, 2021.
3. Pour le développement industriel Cartographie Des Déchets Textiles En Tunisie et Maroc; Competitivite Pacc. 2021, Volume 9. Available online: https://switchmed.eu/wp-content/uploads/2021/07/MAROC_TUNISIE_DA_9_Cartographie_D%C3%A9chetsTextiles.pdf (accessed on 17 October 2022).
4. Agence de Promotion de l'Industrie et de l'Innovation. *INDUSTRIES DU TEXTILE ET DE L'HABILLEMENT EN TUNISIE*; Centre de Documentation et de l'Information Industrielle: Tunis, Tunisie, 2019; Volume 7.
5. Wang, Y.; Jiang, Z.; Li, L.; Qi, Y.; Sun, J.; Jiang, Z. A Bibliometric and Content Review of Carbon Emission Analysis for Building Construction. *Buildings* **2023**, *13*, 205. [CrossRef]
6. International Energy Agency. *Net Zero by 2050: A Roadmap for the Global Energy Sector*; International Energy Agency: Paris, France, 2021; Volume 224.
7. Paraschiv, S.; Paraschiv, L.S.; Serban, A. Increasing the Energy Efficiency of a Building by Thermal Insulation to Reduce the Thermal Load of the Micro-Combined Cooling, Heating and Power System. *Energy Rep.* **2021**, *7*, 286–298. [CrossRef]
8. Aste, N.; Leonforte, F.; Manfren, M.; Mazzon, M. Thermal Inertia and Energy Efficiency—Parametric Simulation Assessment on a Calibrated Case Study. *Appl. Energy* **2015**, *145*, 111–123. [CrossRef]
9. Zach, J.; Novak, V.; Peterkova, J.; Bubenik, J.; Kosir, M.; Bozicek, D.; Krejza, Z. The Use of Advanced Environmentally Friendly Systems in the Insulation and Reconstruction of Buildings. *Buildings* **2023**, *13*, 404. [CrossRef]
10. Balo, F.; Ulutas, A. Energy-Performance Evaluation with Revit Analysis of Mathematical-Model-Based Optimal Insulation Thickness. *Buildings* **2023**, *13*, 408. [CrossRef]
11. Majumder, A.; Canale, L.; Mastino, C.C.; Pacitto, A.; Frattolillo, A.; Dell'isola, M. Thermal Characterization of Recycled Materials for Building Insulation. *Energies* **2021**, *14*, 3564. [CrossRef]
12. Rosca, B.; Corobceanu, V. Structural Grade Concrete Containing Expanded Polystyrene Beads with Different Particle Distributions of Normal Weight Aggregate. *Mater. Today Proc.* **2020**, *42*, 548–554. [CrossRef]
13. Gencel, O.; Yavuz Bayraktar, O.; Kaplan, G.; Arslan, O.; Nodehi, M.; Benli, A.; Gholampour, A.; Ozbakkaloglu, T. Lightweight Foam Concrete Containing Expanded Perlite and Glass Sand: Physico-Mechanical, Durability, and Insulation Properties. *Constr. Build. Mater.* **2022**, *320*, 126187. [CrossRef]
14. Mo, K.H.; Lee, H.J.; Liu, M.Y.J.; Ling, T.C. Incorporation of Expanded Vermiculite Lightweight Aggregate in Cement Mortar. *Constr. Build. Mater.* **2018**, *179*, 302–306. [CrossRef]
15. Borinaga- Treviño, R.; Cuadrado, J.; Canales, J.; Rojí, E. Lime Mud Waste from the Paper Industry as a Partial Replacement of Cement in Mortars Used on Radiant Floor Heating Systems. *J. Build. Eng.* **2021**, *41*, 102408. [CrossRef]
16. Ayed, R.; Baddadi, S.; Bouadila, S.; Skouri, S.; Lazaar, M. Architectural Development of the Buildings' Envelope to Improve Energy Efficiency. In *Advances in Science, Technology and Innovation*; Springer International Publishing: Berlin/Heidelberg, Germany, 2022; pp. 459–461. ISBN 9783031008085.
17. Li, J.; Meng, X.; Gao, Y.; Mao, W.; Luo, T.; Zhang, L. Effect of the Insulation Materials Filling on the Thermal Performance of Sintered Hollow Bricks. *Case Stud. Therm. Eng.* **2018**, *11*, 62–70. [CrossRef]

18. Zukowski, M.; Haese, G. Experimental and Numerical Investigation of a Hollow Brick Filled with Perlite Insulation. *Energy Build.* **2010**, *42*, 1402–1408. [CrossRef]
19. Ayed, R.; Baddadi, S.; Dellagi, A.; Bouadila, S.; Lazaar, M. Thermal Behavior Improvement of Building Materials Using Expanded Polystyrene. In Proceedings of the 2022 13th International Renewable Energy Congress, IREC 2022, Hammamet, Tunisia, 13–15 December 2022; IEEE: New York, NY, USA, 2022; pp. 13–16.
20. Cardinale, T.; Arleo, G.; Bernardo, F.; Feo, A.; De Fazio, P. Thermal and Mechanical Characterization of Panels Made by Cement Mortar and Sheep's Wool Fibres. *Energy Procedia* **2017**, *140*, 159–169. [CrossRef]
21. de Azevedo, A.R.G.; Marvila, M.T.; Tayeh, B.A.; Cecchin, D.; Pereira, A.C.; Monteiro, S.N. Technological Performance of Açaí Natural Fibre Reinforced Cement-Based Mortars. *J. Build. Eng.* **2021**, *33*, 101675. [CrossRef]
22. Rama Rao, P.; Ramakrishna, G. Experimental Investigation on Mechanical Properties of Oil Palm Empty Fruit Bunch Fiber Reinforced Cement Mortar. *Mater. Today Proc.* **2021**, *46*, 471–477. [CrossRef]
23. Benmansour, N.; Agoudjil, B.; Gherabli, A.; Kareche, A.; Boudenne, A. Thermal and Mechanical Performance of Natural Mortar Reinforced with Date Palm Fibers for Use as Insulating Materials in Building. *Energy Build.* **2014**, *81*, 98–104. [CrossRef]
24. Jiang, D.; Lv, S.; Cui, S.; Sun, S.; Song, X.; He, S.; Zhang, J.; An, P. Effect of Thermal Insulation Components on Physical and Mechanical Properties of Plant Fibre Composite Thermal Insulation Mortar. *J. Mater. Res. Technol.* **2020**, *9*, 12996–13013. [CrossRef]
25. Quiñones-Bolaños, E.; Gómez-Oviedo, M.; Mouthon-Bello, J.; Sierra-Vitola, L.; Berardi, U.; Bustillo-Lecompte, C. Potential Use of Coconut Fibre Modified Mortars to Enhance Thermal Comfort in Low-Income Housing. *J. Environ. Manag.* **2021**, *277*, 111503. [CrossRef] [PubMed]
26. Awoyera, P.O.; Akinrinade, A.D.; de Sousa Galdino, A.G.; Althoey, F.; Kirgiz, M.S.; Tayeh, B.A. Thermal Insulation and Mechanical Characteristics of Cement Mortar Reinforced with Mineral Wool and Rice Straw Fibers. *J. Build. Eng.* **2022**, *53*, 104568. [CrossRef]
27. Borinaga-treviño, R.; Orbe, A.; Canales, J.; Norambuena-contreras, J. Thermal and Mechanical Properties of Mortars Reinforced with Recycled Brass Fibres. *Constr. Build. Mater.* **2021**, *284*, 122832. [CrossRef]
28. Zhang, X.; He, W.; Zhang, Y.; Chen, C.; Wu, X. Tensile Behavior of Basalt-Fiber-Grid-Reinforced Mortar before and after Exposure to Elevated Temperature. *Buildings* **2022**, *12*, 2269. [CrossRef]
29. Koutas, L.N.; Papakonstantinou, C.G. Flexural Strengthening of RC Beams with Textile-Reinforced Mortar Composites Focusing on the Influence of the Mortar Type. *Eng. Struct.* **2021**, *246*, 113060. [CrossRef]
30. Zhang, H.Y.; Liu, H.Y.; Kodur, V.; Li, M.Y.; Zhou, Y. Flexural Behavior of Concrete Slabs Strengthened with Textile Reinforced Geopolymer Mortar. *Compos. Struct.* **2022**, *284*, 115220. [CrossRef]
31. Adheem, A.H.; Kadhim, M.M.A.; Jawdhari, A.; Fam, A. Confinement Model for Concrete Wrapped with Fiber Reinforced Cementitious Mortar. *Constr. Build. Mater.* **2021**, *312*, 125401. [CrossRef]
32. Dinh, N.H.; Van Tran, H.; Choi, K.K. Direct Shear Behavior of Cementitious Mortar Reinforced by Carbon Fiber Textile. *Constr. Build. Mater.* **2020**, *249*, 118760. [CrossRef]
33. Brazão Farinha, C.; de Brito, J.; Veiga, R. Incorporation of High Contents of Textile, Acrylic and Glass Waste Fibres in Cement-Based Mortars. Influence on Mortars' Fresh, Mechanical and Deformability Behaviour. *Constr. Build. Mater.* **2021**, *303*, 124424. [CrossRef]
34. Shamseldein, A.; ELgabbas, F.; Elshafie, H. Tensile Behavior of Basalt Textile-Reinforced Mortar (BTRM). *Ain Shams Eng. J.* **2022**, *13*, 101488. [CrossRef]
35. Mercimek, Ö.; Ghoroubi, R.; Özdemir, A.; Anil, Ö.; Erbaş, Y. Investigation of Strengthened Low Slenderness RC Column by Using Textile Reinforced Mortar Strip under Axial Load. *Eng. Struct.* **2022**, *259*, 114191. [CrossRef]
36. Guo, L.; Deng, M.; Chen, H.; Li, R.; Ma, X.; Zhang, Y. Experimental Study on Pre-Damaged RC Beams Shear-Strengthened with Textile-Reinforced Mortar (TRM). *Eng. Struct.* **2022**, *256*, 113956. [CrossRef]
37. Dong, Z.; Deng, M.; Dai, J.; Ma, P. Diagonal Compressive Behavior of Unreinforced Masonry Walls Strengthened with Textile Reinforced Mortar Added with Short PVA Fibers. *Eng. Struct.* **2021**, *246*, 113034. [CrossRef]
38. Pinto, J.; Peixoto, A.; Vieira, J.; Fernandes, L.; Morais, J.; Cunha, V.M.C.F.; Varum, H. Render Reinforced with Textile Threads. *Constr. Build. Mater.* **2013**, *40*, 26–32. [CrossRef]
39. Sadrolodabaee, P.; Claramunt, J.; Ardanuy, M.; de la Fuente, A. A Textile Waste Fiber-Reinforced Cement Composite: Comparison between Short Random Fiber and Textile Reinforcement. *Materials* **2021**, *14*, 3742. [CrossRef] [PubMed]
40. Gulinelli, P.; Aprile, A.; Rizzoni, R.; Grunevald, Y.H.; Lebon, F. Multiscale Numerical Analysis of TRM-Reinforced Masonry under Diagonal Compression Tests. *Buildings* **2020**, *10*, 196. [CrossRef]
41. Abbas, S.; Amjad, R.; Nehdi, M.L.; Ali, S. Material Characterization of Locally Available Textile Fabrics for Structural Applications. *Buildings* **2022**, *12*, 1589. [CrossRef]
42. Briga-Sá, A.; Nascimento, D.; Teixeira, N.; Pinto, J.; Caldeira, F.; Varum, H.; Paiva, A. Textile Waste as an Alternative Thermal Insulation Building Material Solution. *Constr. Build. Mater.* **2013**, *38*, 155–160. [CrossRef]
43. Briga-Sá, A.; Gaibor, N.; Magalhães, L.; Pinto, T.; Leitão, D. Thermal Performance Characterization of Cement-Based Lightweight Blocks Incorporating Textile Waste. *Constr. Build. Mater.* **2022**, *321*, 126330. [CrossRef]
44. *EN 197-1*; Cement—Part 1: Composition, Specifications and Conformity Criteria for Common Cements. European Commitee for Standarization: Brussels, Belgium, 1992.
45. *EN 1015-2*; Methods of Testing Mortars for Masonry—Part 2: Bulk Sampling of Mortars and Preparation of Mortars for Testing. European Commitee for Standarization: Brussels, Belgium, 1999.

46. *EN 1015-3*; Methods of Test for Mortar for Masonry—Part 3: Determination of Consistence of Fresh Mortar (by Flow Table). European Commitee for Standarization: Brussels, Belgium, 1999. Available online: https://standards.globalspec.com/std/733841/EN1015-3 (accessed on 17 October 2022).
47. *EN 1015-6*; Methods of Test for Mortar for Masonry—Part 6. Determination of Bulk Density of Fresh Mortar. European Commitee for Standarization: Brussels, Belgium, 1998.
48. *EN 1015-10*; Methods of Test for Mortar for Masonry—Part 10: Determination of Dry Bulk Density of Hardened Mortar. European Commitee for Standarization: Brussels, Belgium, 1999.
49. *EN 1015-11*; Methods of Test for Mortar for Masonry—Part 11: Determination of Flexural and Compressive Strength of Hardened Mortar 2020. European Commitee for Standarization: Brussels, Belgium, 2020.
50. Maxwell, J.C. *A Treatise on Electricity and Magnetism*; Dover Publications: New York, NY, USA, 1954.
51. Hamilton, R.L.; Crosser, O.K. Thermal Conductivity of Heterogeneous Two-Component Systems. *Ind. Eng. Chem. Fundam.* **1962**, *1*, 187–191. [CrossRef]
52. Agar, Y.; Uno, T. Estimation on Thermal Conductivities of Filled Polymers. *J. Appl. Polym. Sci.* **1986**, *32*, 5705–5712. [CrossRef]
53. Hasselman, D.P.; Johnson, L.F. Effective Thermal Conductivity of Composites with Interfacial Thermal Barrier Resistance. *J. Compos. Mater.* **1987**, *21*, 508–515. [CrossRef]
54. Sahimi, M.; Scriven, L.E.; Davis, H.T. On the Improvement of the Effective-Medium Approximation to the Percolation Conductivity Problem. *J. Phys. C Solid State Phys.* **1984**, *17*, 1941. [CrossRef]
55. Bigg, D.M. *Thermal Conductivity of Heterophase Polymer Compositions*; Springer: Berlin/Heidelberg, Germany, 1995; Volume 119.
56. Progelhof, R.C.; Throne, J.L.; Ruetsch, R.R. Methods for Predicting the Thermal Conductivity of Composite Systems: A Review. *Polym. Eng. Sci.* **1976**, *16*, 615–625. [CrossRef]
57. Kumlutas, D.; Tavman, I.H. A Numerical and Experimental Study on Thermal Conductivity of Particle Filled Polymer Composites. *J. Thermoplast. Compos. Mater.* **2006**, *19*, 441–455. [CrossRef]
58. Rayleigh, L. On the Influence of Obstacles Arranged in Rectangular Order upon the Properties of a Medium. *Lond. Edinb. Dublin Philos. Mag. J. Sci.* **2009**, 37–41. [CrossRef]
59. Hashin, Z.; Shtrikman, S. A Variational Approach to the Theory of the Effective Magnetic Permeability of Multiphase Materials. *J. Appl. Phys.* **1962**, *33*, 3125–3131. [CrossRef]
60. Hatta, H.; Taya, M. Effective Thermal Conductivity of a Misoriented Short Fiber Composite. *J. Appl. Phys.* **1985**, *58*, 2478–2486. [CrossRef]
61. Hatta, H.; Taya, M. Thermal Conductivity of Coated Filler Composites. *J. Appl. Phys.* **1986**, *59*, 1851–1860. [CrossRef]
62. Eshelby, J.D. The Determination of the Elastic Field of an Ellipsoidal Inclusion, and Related Problems. *Proc. R. Soc. London. Ser. A Math. Phys. Sci.* **1957**, *241*, 376–396. [CrossRef]
63. Lewis, T.B.; Nielsen, L.E. Dynamic Mechanical Properties of Particulate-Filled Composites. *J. Appl. Polym. Sci.* **1970**, *14*, 1449–1471. [CrossRef]
64. NIELSEN, L.E. Thermal Conductivity of Particulate-Filled Polymers. *J. Appl. Polym. Sci.* **1973**, *17*, 3819–3820. [CrossRef]
65. Halpin, J.C. Stiffness and Expansion Estimates for Oriented Short Fiber Composites. *J. Compos. Mater.* **1969**, 3–7. [CrossRef]
66. Bigg, D.M. Thermally Conductive Polymer Compositions. *Polym. Compos.* **1986**, *7*, 125–140. [CrossRef]
67. *EN 998-1:2017*; Specification for Mortar for Masonry—Part 1: Rendering and Plastering Mortar. European Commitee for Standarization: Brussels, Belgium, 2005.
68. Wendl, M.C. *Theoretical Foundations of Conduction and Convection Heat Transfer*; The Wendl Foundation: Saint Louis, MO, USA, 2012.

buildings

MDPI

Article

Thermal-Electric Modeling: A New Approach for Evaluating the Impact of Conservation Voltage Reduction on Cooling Equipment

Harshit Varshney, Himanshu Jain * and Ravi Tiwari

Indian Institute of Technology Roorkee, Roorkee 247667, Uttarakhand, India;
h_varshney@hre.iitr.ac.in (H.V.); r_tiwari@hre.iitr.ac.in (R.T.)
* Correspondence: himanshu.jain@hre.iitr.ac.in

Abstract: It has been suggested in the literature that by reducing the incoming voltage at a distribution feeder head or at the supply side of buildings, significant electricity savings can be achieved. This technique is called Conservation Voltage Reduction (CVR). Data-based analysis with and without CVR is primarily used to support such assertions, which does not explain the physics behind reduction in energy consumption with CVR. This paper presents a new approach for evaluating the impact of CVR on cooling equipment. In this approach, a thermal-electric model of the cooling process is developed in MATLAB's SIMSCAPE toolbox that can be used to explain the physics behind energy reduction with CVR. This model includes an accurate model of a compressor coupled to an induction motor whose supply voltage can be varied to simulate CVR. Simulations performed using this model show that the Coefficient of Performance (COP) of cooling equipment improves with a reduction in supply voltage. However, the energy lost in the motor windings may nullify the impact of the improvement in the COP and render the CVR programs ineffective if the range of speed change is small over the allowable voltage change. The simulation results show an increase in energy consumption of 4% at 90% rated voltage compared to the energy consumed at the rated voltage. However, if variable frequency drives-based cooling equipment is appropriately controlled, it is possible to reduce their net energy consumption using CVR. Simulations performed keeping the ratio of the supply voltage and the frequency constant showed a reduction in energy consumption of 2.5% at 90% rated voltage compared to the energy consumed at the rated voltage. Thermal-electric modeling of building cooling equipment is, therefore, vital to accurately evaluating the benefits of CVR as smart, power electronics-based end-use equipment is globally adopted.

Keywords: building energy efficiency; conservation voltage reduction; voltage optimization; energy conservation; demand response; climate change

check for updates

Citation: Varshney, H.; Jain, H.; Tiwari, R. Thermal-Electric Modeling: A New Approach for Evaluating the Impact of Conservation Voltage Reduction on Cooling Equipment. *Buildings* **2023**, *13*, 1287. https://doi.org/10.3390/buildings13051287

Academic Editors: Zhenjun Ma and Müslüm Arıcı

Received: 31 March 2023
Revised: 11 May 2023
Accepted: 13 May 2023
Published: 15 May 2023

1. Introduction

Energy conservation is one of the key strategies for combating climate change. There are several examples across the world where a significant reduction in energy consumption and greenhouse gas emissions has been achieved through energy conservation programs. In the United States (US), various energy efficiency programs have ensured around 23% less annual energy consumption [1]. In India, the number of LED bulbs being used between 2014 and 2018 increased from 5 million to 670 million resulting in energy savings of 30 terawatt hours (TWh) [2]. Energy conservation will continue to play a critical role in the future as the entire world strives to reduce greenhouse gas emissions and contain the average temperature rise from the beginning of the industrial age to less than 1.5 degrees by 2050 [3].

There are many tools and techniques to reduce power consumption. One such tool is Conversation Voltage Reduction (CVR), which has been used in the past and can be used in the future to reduce electricity consumption. As the name suggests, CVR involves reducing voltage at the feeder head of an electricity distribution network, or at the source-side of a commercial or industrial building to reduce electricity consumed by loads. Several researchers and utilities have used CVR to reduce power consumption [4–8].

In [9], an experiment was performed to evaluate the effects of CVR on electric home appliances. It was observed that in the case of constant resistive load, the power consumption decreased with a reduction in the supply voltage. However, in the case of constant power loads, a reduction in the supply voltage at the same power level resulted in an increased current demand leading to higher line losses. This highlights the importance of knowing the load model(s) to evaluate the effectiveness of CVR. In [10], the authors performed a comprehensive review of CVR implementation and discussed different methods to quantify its effects and arrived at the conclusion that while the simulation method can provide high precision, its accuracy depends on the load models used. However, no details were provided in this paper that would allow thermal-electric modeling of the cooling processes. In [11], an attempt was made to model cooling as a thermal-electric process for large-scale distribution simulations by expressing heat flow through a cooling unit as a function of voltage, ambient temperature, and Coefficient of Performance (COP). However, the explanation for the modeled thermal-electric coupling was not provided (e.g., how change in voltage affects mass flow rate and heat transfer). In [12], the authors carried out an experiment in which they operated equipment that worked at two voltage levels, 208 V and 480 V, and calculated energy savings and peak load savings for a 5% reduction in voltage. CVR and its implementation on the power systems of Washington EMC and the facilities of the Georgia Transmission Corporation were discussed in [13]. To measure, test, and verify the CVR system effects, and to obtain a reduction in peak demand and energy, an electric utility implemented a conservation voltage reduction system on its various distribution substations in [14].

The literature review on CVR reveals that while the benefits of using CVR on power consumption have been reported, modeling-based explanations for energy reduction by reducing voltage in cooling equipment are missing. Moreover, the results in the literature are for conventional loads that are increasingly being replaced with power electronics-based equipment such as inverter-air conditioners and refrigerators. The response of such electrical cooling equipment to CVR will be significantly different from those observed in conventional cooling equipment. To address these gaps, accurate thermal-electric models of cooling processes are required, where both conventional and power electronics-based cooling equipment can be simulated under CVR. This paper attempts to address these gaps by developing a MATLAB/SIMSCAPE-based thermal-electric model of the cooling process using a refrigerator as an example. Several experiments were performed using this model to identify the key reasons for reduction in energy consumption under CVR and the impact of CVR on conventional and power electronics-based cooling equipment performance. In the process, a foundation has been laid for a thorough CVR impact analysis in the future as appliance characteristics change.

The remainder of the paper is organized as follows. Section 2 presents the research methodology, where thermal-electric modeling of the refrigeration/air-conditioning process, compressor modeling, and three-phase induction motor modeling are discussed. Simulation results and major observations regarding the impact of CVR on cooling equipment performance are presented in Section 3. Experimental validation of simulation results is presented in Section 4 while Section 5 concludes this paper and identifies topics for future research.

2. Methodology

2.1. Thermal Model of Cooling Process

A thermal model of cooling process is shown in Figure 1, which is obtained from an inbuilt example in MATLAB-SIMSCAPE titled "ssc_refrigeration" [15]. This model explains the two-phase fluid refrigeration in which the refrigerant undergoes a phase change from liquid to vapor and back to liquid. The refrigerant fluid is compressed by the compressor, which increases its pressure and temperature. The high-pressure, high-temperature fluid then flows into the condenser, where it releases heat to the surrounding environment and condenses into a liquid. The liquid refrigerant then passes through the expansion valve, which reduces its pressure and causes it to undergo a phase change from liquid to gas (vapor). This cold, low-pressure vapor then flows into the evaporator, where it absorbs heat from the desired area and undergoes a phase change to become liquid again [16,17]. The mass and energy balance equations for the colling process are given below.

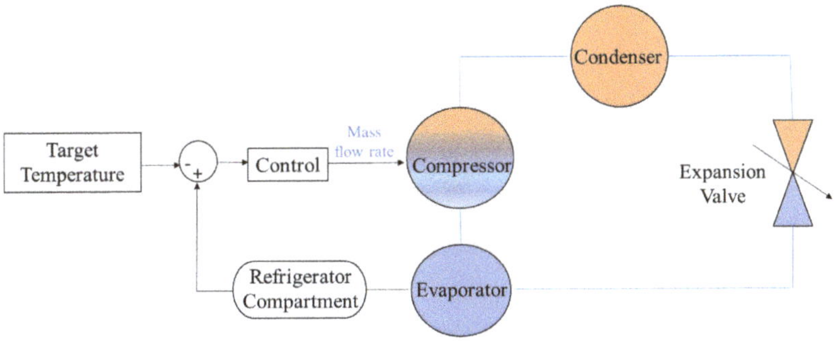

Figure 1. Block diagram of MATLAB simulation representation of two-phase fluid refrigeration system [15].

For the compressor

$$m_{comp\ in} = m_{comp\ out} \tag{1}$$

$$h_1 + W_{comp} = h_2 \tag{2}$$

For the condenser

$$m_{cond\ in} = m_{cond\ out} \tag{3}$$

$$h_2 = h_3 + Q_{cond} \tag{4}$$

For the expansion valve

$$m_{exp\ in} = m_{exp\ out} \tag{5}$$

$$h_3 = h_4 \tag{6}$$

For the evaporator

$$m_{evp\ in} = m_{evp\ out} \tag{7}$$

$$h_4 = h_1 + Q_{evp} \tag{8}$$

where

$m_{comp\ in}$, $m_{comp\ out}$ are the mass flow rates of refrigerant entering and leaving the compressor, respectively.

$m_{cond\ in}$, $m_{cond\ out}$ are the mass flow rates of refrigerant entering and leaving the condenser, respectively.

$m_{exp\ in}$, $m_{exp\ out}$ are the mass flow rates of refrigerant entering and leaving the expansion valve, respectively.

$m_{evp\ in}$, $m_{evp\ out}$ are the mass flow rates of refrigerant entering and leaving the evaporator, respectively.

h_1, h_2, h_3, h_4 are the enthalpies at the compressor inlet, compressor outlet, condenser outlet, and expansion valve outlet, respectively.

Q_{cond} is the heat rejected by the refrigerant during the condensation process.

Q_{evp} is the heat absorbed by the refrigerant during the evaporation process.

The refrigerant used in this model is R-134a. The entire refrigeration cycle is modeled along with a controller that turns the compressor ON or OFF as it tries to maintain the temperature within the desired range.

In this model, an ideal compressor is used. This means that the compressor maintains the same mass flow rate through it irrespective of the pressure difference between the compressor inlet and outlet. Moreover, the load torque that the compressor would encounter while compressing the vapor is not modeled. This compressor cannot be used to study the impact of CVR because changing the mass flow rate and speed changes the suction and discharge temperatures, pressures, and other thermodynamic parameters. To fix these gaps, a compressor model is developed as discussed next.

2.2. Compressor Modeling

A review of the literature showed that several authors have developed steady state thermodynamic models of reciprocating and centrifugal compressors. In this paper, we model the reciprocating compressor as it is widely used in refrigerators. In [18], authors suggested that in the modeling of the compressor, the effect of the compressor speed is large, and the effect of the refrigerant charge is much less on the COP of system. In [19], a compressor model was developed for a positive displacement compressor assuming that displacement, operational speed, volumetric efficiency, and isentropic efficiency are known. Modeling of the compressor motor was performed in [20] to investigate the air conditioner compressor stall. In [21], the temperature control model for coating plant air conditioner was explored, where the authors used a mass flow meter to measure the mass flow rate in the compressor, although in [22] they developed a method to instantaneously measure the mass flow rate inside the compressor using the p-v diagram.

Reference [19] was used as the reference to model the reciprocating compressor, where the refrigerant mass flow rate was a function of the piston displacement and the specific volume of the vapor entering the compressor (the evaporator temperature). The input work of the compressor was calculated using (9), while the refrigerant mass flow rate was calculated using (10).

$$\mathcal{W} = m_r \cdot \left(\frac{\mathcal{K}}{\mathcal{K}-1}\right) \cdot \mathcal{P}_3 \cdot \mathcal{v}_{g3} \cdot \left(\frac{\mathcal{P}_4}{\mathcal{P}_3}\right)^{(\mathcal{K}-1/\mathcal{K})} \quad (9)$$

$$m_r = \mathcal{N} \cdot \frac{\pi}{4} \cdot d^2 \cdot \mathcal{L} \cdot \gamma_v \cdot \frac{1}{v_{g3}} \quad (10)$$

where

m_r	refrigerant mass flow rate (kg/s)
$\mathcal{K}$	C_p/C_v = isentropic exponent
$\mathcal{P}_3$	compressor inlet pressure (kPa)
$\mathcal{P}_4$	compressors exit pressure (kPa)
$\mathcal{v}_{g3}$	refrigerant specific volume at compressor inlet (m^3/kg)
d	pipe diameter
$\mathcal{L}$	length of stroke
γ_v	volumetric efficiency
$\mathcal{N}$	speed (Hz)

As shown in (10), to obtain the mass flow rate specific volume at the compressor inlet is necessary, and it can be extracted from the compressor block of Figure 1. This specific volume value is sent to the product block through a state space block to avoid an algebraic loop.

The input of compressor is the mass flow rate, which depends on the pipe diameter, stroke length, volumetric efficiency, speed, and the specific volume of refrigerant at the compressor inlet. Accordingly, the two main variables are the specific volume of the refrigerant at the compressor inlet and the speed (represented by a constant block). These variables are passed through a product block, which is connected to a gain block to add the effects of the pipe diameter, stroke length, and volumetric efficiency. The modified thermal model of the cooling process with the reciprocating compressor is shown in Figure 2.

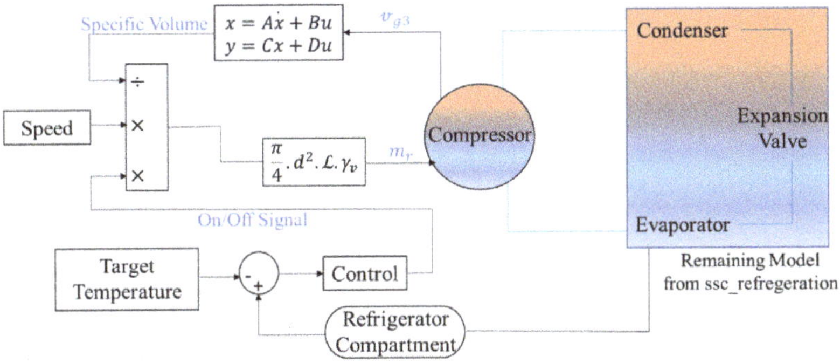

Figure 2. SIMSCAPE thermal model of cooling process with reciprocating compressor.

2.3. Induction Motor (IM) Modeling

To develop the thermal-electric model of the cooling process, a model of an induction motor (IM) is needed. The in-built IM models of SIMSCAPE were considered for modeling the electrical characteristics of the three-phase induction motor. However, the models were too detailed for the CVR study because only the steady state behavior of motor was required, which is reached very quickly (in few seconds or less). Therefore, it was decided to use Equation (11) for modeling the torque vs. speed characteristics of the three-phase IM [23]. This equation neglects the magnetizing reactance because it is very large and has a negligible impact on the on-load performance of the IM. Solving (11) gives (13), which is the expression of k. Equation (13) can be used with (12) to obtain the slip and hence the rotor speed. The IM model defined by (11)–(13) is shown in Figure 3.

$$\tau = \left[\frac{n \cdot v^2 \cdot k}{(R_1 + k)^2 + (X_1 + X_2)^2}\right]\left(\frac{1}{\omega_s}\right) \tag{11}$$

$$k = \frac{R_2}{s} \tag{12}$$

$$k = \frac{A_1 - B_1 + C_1^{\frac{1}{2}}}{2(T_{mech} \cdot w_s)} \tag{13}$$

where,

τ, τ_{mech}	electrical torque, mechanical torque
ω_s	synchronous angular speed of the rotor
ω_r	actual rotor speed
V	voltage
R_1 *and* R_2	stator and rotor resistance, respectively
X_1 *and* X_2	stator and rotor reactance, respectively
n	phases
s	slip
A_1	$n \cdot v^2$
B_1	$2 \cdot T_{mech} \cdot R_1 \cdot \omega_s$
C_1	$c_1 - c_2 - c_3 - c_4 - c_5$
c_1	$n^2 \cdot v^4$
c_2	$4 \cdot R_1 \cdot \tau n \cdot v^2 \cdot \omega_s +$
c_3	$4 \cdot \tau^2 \cdot \omega_s^2 \cdot X_1^2$
c_4	$8 \cdot \tau^2 \cdot \omega_s^2 \cdot X_1 \cdot X_2$
c_5	$4 \cdot \tau^2 \cdot \omega_s^2 \cdot X_2^2$

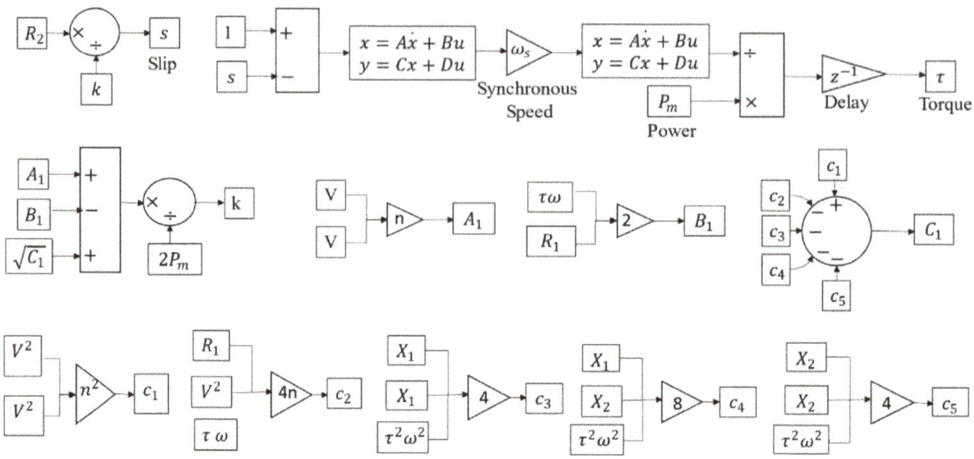

Figure 3. Developed model for induction motor.

2.4. Thermal-Electric Model of Cooling Process

To develop the thermal-electric model of the cooling process, the compressor model of "ssc_refrigeration" was replaced by the compressor model of Figure 2 and the IM model of Figure 3. Adding the IM model resulted in algebraic loops being formed because the speed was being used to convert the compressor power into torque that was being input to the motor, while the torque was being used to obtain the rotor speed that was changing the refrigerant mass flow rate. To break this algebraic loop, a unit delay (1 s) was added. The detailed thermal-electric model of the cooling process is shown in Figure 4.

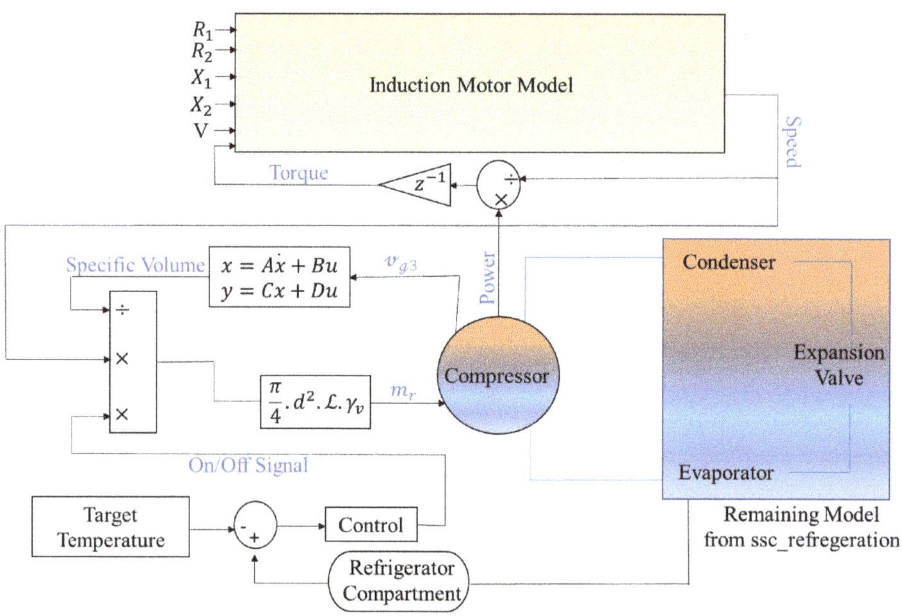

Figure 4. Detailed thermal-electric model of cooling process.

3. Results and Discussion

Several simulations were performed using the models developed in the previous section. The data used for performing these simulations were primarily obtained from the "ssc_refrigeration" model discussed earlier, and they are given in Table 1. Additional data required for modeling the reciprocating compressor and three-phase IM are given in Table 2.

Table 1. Parameters of ssc_refrigeration model.

Parameters	Value	Parameters	Value
Commanded mass flow	0.04 kg/s	Environment temperature	293 K
Condenser length	30 m	Target temperature	277 K
Evaporator length	30 m	Minimum throat temperature	270 K
Pipe diameter	0.01 m	Maximum throat temperature	250 K
Pipe thickness	0.0005 m	Minimum throat area	0.1 mm^2
Copper conductivity	400 W/m/K	Maximum throat area	1.5 mm^2
Foam conductivity	0.03 W/m/K	Fin area	1 m^2
Copper density	8940 kg/m^3	Refrigerant initial pressure	0.6 MPa
Copper specific heat	390 J/kg/K	Foam thickness	0.03 m
Compartment interior surface area	4.5 m^2	Compartment exterior surface area	6 m^2

Table 2. Parameters of developed models.

Parameters	Values
Pipe diameter, d *	0.025 m
Stroke length $\mathcal{L}$ *	0.0154 m
No. of poles	4
Rated speed	1500 rpm
Stator resistance, R_1	0.294 Ω
Rotor resistance, R_2	0.144 Ω
Stator reactance, X_1	0.503 Ω
Rotor reactance, X_2	13.25 Ω

* Parameters based on [24,25].

Two scenarios were simulated using the thermal-electric model of Figure 4. The first scenario simulated conventional cooling equipment, where the compressor motor was driven directly from the utility power supply. In the second scenario, power electronics-based cooling equipment was considered, where the motor was assumed to be driven by a variable frequency drive (VFD) that is controlled using the widely used V/f control technique. Because the VFD decouples the motor power supply from the utility power supply, it is assumed that the controller can change the motor supply frequency according to the utility side voltage to maintain a constant V/f ratio. This is a reasonable assumption because making such changes to the VFD controllers is easy and only requires changes in the VFD firmware. Two sensitivities were also performed to evaluate the changes in the impact of CVR under different ambient and refrigerator target compartment temperatures.

For both simulation scenarios, the supply voltage varied between 90% to 110% of the rated voltage (220 V). All simulations were run for a duration of one day (86,400 s), and the data were recorded at a sampling rate of 1 Hz.

3.1. Base Results for Scenario 1 (Conventional Cooling Equipment)

The effect of voltage on the COP of the compressor and energy savings is shown in Figure 5. The results show that the COP increases on reducing the voltage. This is because the compressor speed reduces on reducing the supply voltage, due to which the input energy of the compressor decreases resulting in a higher COP. This effect, however, is negligible (maximum variation of 0.09%) because the change in speed over the entire voltage range is very small (24.8 Hz to 24.75 Hz or 0.9%).

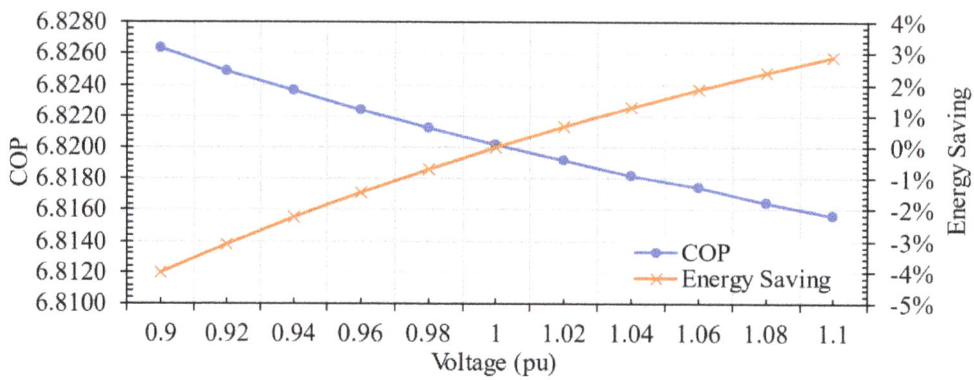

Figure 5. Effect of voltage on COP and energy saving (pu stands for per unit and it is calculated in this paper as actual voltage divided by 220 V).

Contrary to the increase in the COP, the total energy consumption (total energy drawn from the supply, which is the sum of the compressor work and motor ohmic losses) of the cooling process increases as voltage is reduced compared to the rated voltage. This is shown in Figure 5 as the negative energy savings (total energy at the rated voltage minus total energy at a specified voltage) is expressed as a percentage of the total energy consumption at the rated voltage. This occurred because of the increase in motor ohmic losses (increased by 25% compared to losses at the rated voltage) when the voltage was reduced from rated to 90%. This observation aligns with the approximately constant power behavior of induction motors observed in practice, where reducing the voltage increases the current drawn at the same load, resulting in higher losses and reduced motor efficiency.

3.2. Base Results for Scenario 2 (Power Electronics-Based Cooling Equipment)

The effect of voltage on the COP of the compressor and energy savings is shown in Figure 6. In this simulation, the V/f ratio is kept constant, and the COP is observed by changing the voltage. The results show that the COP increases on reducing the voltage. This increase in the COP (about 4%) is around 40 times higher than the increase in the COP observed earlier for conventional cooling equipment (only 0.09%) when the voltage was reduced from 100% to 90%. This is because a decrease in the frequency to keep the V/f ratio constant results in a significantly higher reduction in motor speed (22.3 Hz) at reduced voltages compared to scenario 1, where the speed decreased to 24.75 Hz for the same change in utility voltage from the rated voltage to 90% of the rated voltage. In other words, the reduction in speed was 10% with respect to 25 Hz synchronous speed in scenario 2 (power electronics-based cooling equipment) while it was only 0.1% in scenario 1 (conventional cooling equipment). As a result, the compressor work reduces significantly more under the V/f control compared to conventional cooling equipment for the same voltage reduction.

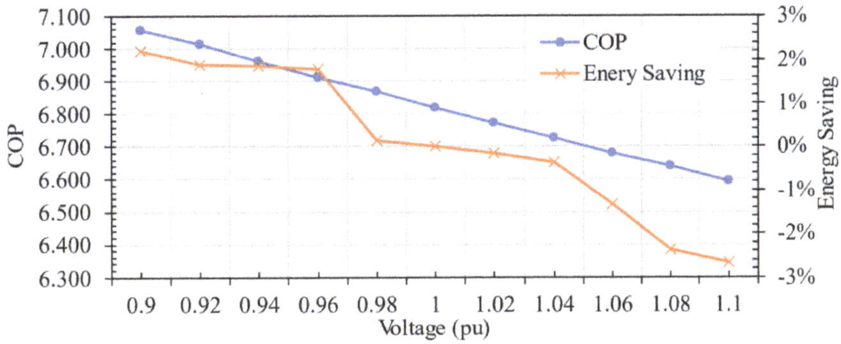

Figure 6. Effect of Voltage on COP and Energy Saving (with V/f control).

Unlike the conventional case, energy savings are positive at voltages lower than the rated voltage. This is because the compressor work decreases by as much as 3% compared to without the V/f control, where the maximum reduction was 0.045%. At the same time, the increase in motor ohmic losses was significantly lower (maximum increase of 4% compared to losses at the rated voltage) than those observed in scenario 1 with conventional cooling equipment (maximum increase of 25% compared to losses at the rated voltage).

3.3. *Sensitivity Results*

Because of the very small impact of CVR on conventional cooling equipment, sensitivity results are provided only for scenario 2 with power electronics-based cooling equipment and V/f control.

3.3.1. Effect of Ambient Temperature on COP and Total Energy

The COP of the developed model is calculated by changing the ambient temperature from 25 °C to 40 °C in steps of 5 °C and varying the voltage between 90% and 110%. The COP of the power electronics-based cooling equipment at different ambient temperatures with variations of voltage is given in Table 3. The table shows that, irrespective of the ambient temperature, the COP increases as the utility voltage is reduced. However, the COP of the power electronics-based refrigerator was very sensitive to changes in ambient temperature because the COP at 90% rated voltage decreased from 5.94 to 3.86 as the ambient temperature increased from 25 °C to 40 °C. The reduction in the COP at higher ambient temperatures can be explained by the higher heat ingress in the refrigerator compartment, thereby requiring the compressor to perform more work.

Table 3. COP at different ambient temperature with variation of voltage.

Voltage (pu)	COP			
	Ambient Temperature = 25 °C	Ambient Temperature = 30 °C	Ambient Temperature = 35 °C	Ambient Temperature = 40 °C
0.9	5.94	5.08	4.41	3.86
0.92	5.90	5.05	4.38	3.84
0.94	5.87	5.02	4.36	3.82
0.96	5.83	4.99	4.34	3.80
0.98	5.80	4.97	4.32	3.79
1	5.76	4.94	4.29	3.77
1.02	5.73	4.92	4.27	3.75
1.04	5.69	4.89	4.25	3.73
1.06	5.66	4.86	4.23	3.71
1.08	5.63	4.83	4.21	3.70

Total energy consumption at different ambient temperatures is given in Table 4. The table shows that, irrespective of the ambient temperature, the total energy consumption decreases as the utility voltage is reduced. However, like the COP, the total energy consumption of the power electronics-based refrigerator was very sensitive to changes in ambient temperature because the total energy consumption at 90% rated voltage increased from 432.64 Wh to 1196.67 Wh as the ambient temperature increased from 25 °C to 40 °C.

Table 4. Total energy consumption at different ambient temperature with variation of voltage.

Voltage (pu)	Total Energy Consumption (Wh)			
	Ambient Temperature = 25 °C	Ambient Temperature = 30 °C	Ambient Temperature = 35 °C	Ambient Temperature = 40 °C
0.9	432.64	638.82	891.43	1196.67
0.92	434.67	640.37	888.83	1198.49
0.94	436.72	642.29	894.12	1199.71
0.96	437.23	642.54	892.46	1200.33
0.98	439.99	642.44	893.56	1200.35
1	440.69	643.16	897.73	1199.96
1.02	442.46	646.78	900.43	1205.78
1.04	444.99	650.77	900.97	1206.95
1.06	444.49	650.19	902.17	1206.16
1.08	449.87	652.21	903.99	1212.43

3.3.2. Effect of Compartment Temperature on COP and Total Energy

In this sensitivity, the compartment temperature varied from 4 °C to −8 °C in steps of 4 °C. The utility voltage again varied from 110% to 90% of the rated voltage. The COP values obtained from this sensitivity are shown in Table 5. While the trend of increasing COP at reduced voltages was observed at all compartment temperatures, the difference between the COPs at the simulated voltages reduced as the compartment temperature was reduced. In other words, the effect of CVR in improving the COP reduced as the compartment temperature was reduced.

Table 5. COP at different compartment temperature with variation of voltage.

Voltage (pu)	COP			
	Compartment Temperature = 4 °C	Compartment Temperature = 0 °C	Compartment Temperature = −4 °C	Compartment Temperature = −8 °C
0.9	7.06	6.95	6.41	4.88
0.92	7.01	6.91	6.38	4.87
0.94	6.96	6.86	6.34	4.85
0.96	6.91	6.82	6.31	4.84
0.98	6.87	6.77	6.28	4.82
1	6.82	6.73	6.24	4.81
1.02	6.77	6.68	6.21	4.79
1.04	6.73	6.64	6.19	4.77
1.06	6.68	6.60	6.15	4.76
1.08	6.64	6.56	6.11	4.74

Similarly, the total energy consumption increased substantially as the compartment temperature was reduced but the reduction in total energy consumption at reduced voltages became much less as the compartment temperature decreased from 4 °C to −8 °C. For example, as shown in Table 6, the reduction in total energy consumption was 2.2% (at 90% rated voltage compared to the rated voltage) at the compartment temperature of 4 °C, but this decreased to −0.6% (or the total energy increased slightly at 90% rated voltage compared to the rated voltage) when the compartment temperature was reduced to −8 °C.

Table 6. Total energy consumption at different compartment temperature with variation of voltage.

Voltage (pu)	Total Energy Consumption (Wh)			
	Compartment Temperature = 4 °C	Compartment Temperature = 0 °C	Compartment Temperature = −4 °C	Compartment Temperature = −8 °C
0.9	274.26	343.15	444.79	693.32
0.92	275.15	344.85	446.01	693.82
0.94	275.25	345.56	446.87	692.16
0.96	275.47	348.35	447.26	690.89
0.98	280.08	347.94	447.15	691.09
1	280.47	352.32	448.98	688.93
1.02	280.91	352.13	451.24	688.91
1.04	281.45	353.01	453.78	688.16
1.06	284.20	356.75	453.42	689.84
1.08	287.12	356.90	454.43	690.37

4. Experimental Validation

Hardware validation of the simulation results was performed at the Renewable Grid Integration Laboratory of the Department of Hydro and Renewable Energy at IIT Roorkee. The experimental setup is shown in Figure 7, and it is comprised of the following equipment:

- **Programmable AC power supply**: This was a 6 kVA, Kikusui make power supply (PCR6000WEA2), which acts like a grid except that it does not allow reverse power flow from equipment to the supply. Individual phase voltage magnitudes and phase angles and three-phase supply frequency can be precisely controlled. This feature was used to simulate the CVR and maintain the constant voltage to the frequency ratio. Moreover, the supply also displays power and current measurements, which were used to observe the power consumed by the air conditioner during experiments.
- **Single-phase, conventional air conditioner**: This was a Voltas window type air conditioner (Voltas 123 LZF) of 1000 W power rating, which acts as cooling equipment. The refrigerant used in this air conditioner is R32.
- **Temperature sensor**: Testo 417 vane anemometer was used to measure the fan speed and ambient temperature.

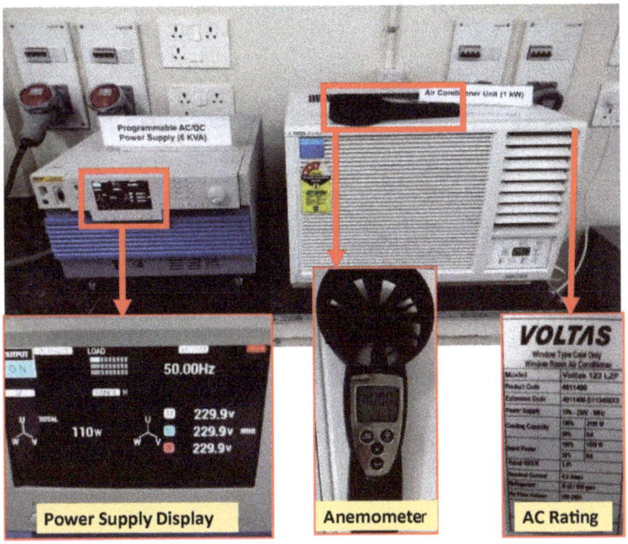

Figure 7. Experimental setup.

Experiments were performed for two scenarios, one for conventional cooling equipment and the other for power electronics-based equipment. The experimental setup is shown in Figure 7.

Two sets of the CVR experiments were performed. In one, only the supply voltage was reduced to 95% and 90% of the rated voltage of the air conditioner (230 V). In the second, the voltage to the frequency ratio was held constant as the voltage was reduced to 95% and 90% of the rated voltage. The wind speed of the fan and the room temperature were measured with the testo 417 vane anemometer. Each voltage level was maintained for 5 min and the steady value of power as displayed by the power supply was recorded.

In the simulation, only the compressor was modelled, whereas in the air conditioner a fan was also present. Therefore, to align the experiments with simulations, the power required by the air conditioner's fan was subtracted from the total power displayed on the power supply to obtain the estimate for the compressor's power consumption.

The average powers consumed by the compressor for the two sets of experiments are shown in Tables 7 and 8, respectively. For the first set of experiments, where only the supply voltage was reduced, the power required by the compressor increased by 1.1% on reducing the voltage by 10%, whereas for the second set of experiments, where both voltage and frequency were reduced to keep their ratio constant, the power required by the compressor decreased by 5.07% on reducing the voltage by 10%.

Table 7. Power saving with variation of voltage when only voltage was reduced.

Voltage (pu)	Power (W)			
	AC	Only AC Fan	Compressor	Power Saving
0.9	920	91.5	828.5	−1.1%
0.95	924	99.6	828.4	−0.66%
1	927	108	819	0%

Table 8. Power saving with variation of voltage when both voltage and frequency were reduced to keep their ratio constant.

Voltage (pu)	Power (W)			
	AC	Only AC Fan	Compressor	Power Saving
0.9	860	82.5	777.5	5.07%
0.95	892	94.7	797.3	2.65%
1	927	108	819	0%

These results validate the trends observed in the simulations. The percentage increase or decrease in energy consumption was different because of the differences in the characteristics of the cooling equipment (single-phase motor, R34 refrigerant, and different and unknown thermal characteristics of the air conditioner) and the inability to control the ambient conditions because of the unavailability of a climate chamber with the research team.

5. Conclusions and Next Steps

Conservation Voltage Reduction (CVR) has been used by industries to save electricity by reducing incoming voltage. Many electricity distribution companies in the United States have also deployed CVR to reduce both peak power demand and the total energy use. Despite the long and successful experience with CVR programs, a physics-based modelling framework that could explain the key factors for energy savings by CVR and enable evaluation of CVR impacts in the future with changing end-use equipment, building technology, and deployment of distributed generation that can increase local voltages was missing.

This research attempted to address this gap by developing a thermal-electric model of cooling processes in MATLAB/SIMSCAPE and using it to demonstrate the efficacy of CVR programs under various conditions. Through these simulations, it was shown that it is difficult to generalize the effectiveness of CVR programs and attribute some fixed energy savings to CVR programs without performing detailed thermal-electric modeling of the facility and its equipment. It was found that for conventional, single speed compressors, contrary to the expectation, energy consumption (motor ohmic losses plus compressor work) increased by up to 4% as the voltage was reduced to 90% of the rated value. The trend of increased total energy consumption with reduced voltages (lower than the rated voltage) was also observed during compartment and ambient air temperature sensitivities.

The research performed has presented an intriguing opportunity of applying CVR on power electronics-based cooling equipment, which is rapidly increasing in our society. By appropriately modifying their control algorithms, the voltage–speed relationship can be changed to provide more impact of CVR (energy savings around 2.5%) than is possible in conventional cooling appliances (negative energy savings of 4%) that have limited speed range under normal operations.

These observations were also confirmed during hardware experiments, where only reducing the voltage while keeping the supply frequency fixed at 50 Hz resulted in an increase in energy (increase of 1.1% at 90% rated voltage and increase of 0.66% at 95% rated voltage when compared to the energy consumed at the rated voltage). However, when the ratio of the voltage and frequency was fixed, the energy consumption decreased by 5% and 2.6% at 90% and 95% rated voltages, respectively, when compared to the energy consumed at the rated voltage.

In the future, the authors intend to expand the thermal-electric model by including more detailed thermal models of buildings and electrical models of other appliances and equipment to observe the effect of CVR and quantify its effectiveness as an energy conservation strategy. Moreover, the authors intend to perform hardware validation of the impact of CVR using a climate-controlled chamber (not presently available with the research team), where the ambient temperature and humidity can be precisely controlled.

Author Contributions: Conceptualization: H.J.; methodology, H.J., H.V. and R.T.; investigation, H.V., R.T. and H.J.; writing—original draft preparation, H.V., R.T. and H.J.; writing—review and editing, H.J. and H.V.; visualization, H.V. All authors have read and agreed to the published version of the manuscript.

Funding: This research received no external funding.

Institutional Review Board Statement: Not applicable.

Informed Consent Statement: Not applicable.

Data Availability Statement: The data required to run the simulations performed in this study is provided within the body of the paper. The data analyzed here is available on request.

Conflicts of Interest: The authors declare no conflict of interest.

References

1. Energy Efficiency Impact Report—Why Energy Efficiency Matters. Available online: https://energyefficiencyimpact.org/ (accessed on 8 January 2023).
2. Kamat, A.S.; Khosla, R.; Narayanamurti, V. Illuminating homes with LEDs in India: Rapid market creation towards low-carbon technology transition in a developing country. *Energy Res. Soc. Sci.* **2020**, *66*, 101488. [CrossRef]
3. The Paris Agreement | UNFCCC. Available online: https://unfccc.int/process-and-meetings/the-paris-agreement/the-paris-agreement (accessed on 8 January 2023).
4. Chen, M.S.; Shoults, R.; Fitzer, J.; Songster, H. The effects of reduced voltages on the efficiency of electric loads. *IEEE Trans. Power Appar. Syst.* **1982**, *PAS-101*, 2158–2166. [CrossRef]

5. Kennedy, B.W.; Fletcher, R.H. Conservation voltage reduction (CVR) at snohomish county PUD. *IEEE Trans. Power Syst.* **1991**, *6*, 986–998. [CrossRef]
6. Bokhari, A.; Alkan, A.; Dogan, R.; Diaz-Aguiló, M. Experimental determination of the ZIP coefficients for modern residential, commercial, and industrial loads. *IEEE Trans. Power Deliv.* **2014**, *29*, 1372–1381. [CrossRef]
7. Diaz-Aguiló, M.; Sandraz, J.; Macwan, R.; De Leon, F.; Czarkowski, D.; Comack, C.; Wang, D. Field-validated load model for the analysis of CVR in distribution secondary networks: Energy conservation. *IEEE Trans. Power Deliv.* **2013**, *28*, 2428–2436. [CrossRef]
8. Paul, S.; Jewell, W. Impact of load type on power consumption and line loss in voltage reduction program. In Proceedings of the 2013 North American Power Symposium (NAPS), Manhattan, KS, USA, 22–24 September 2013. [CrossRef]
9. Singaravelan, A.; Kowsalya, M. A Practical Investigation on Conservation Voltage Reduction for its Efficiency with Electric Home Appliances. *Energy Procedia* **2017**, *117*, 724–730. [CrossRef]
10. Wang, Z.; Wang, J. Review on implementation and assessment of conservation voltage reduction. *IEEE Trans. Power Syst.* **2014**, *29*, 1306–1315. [CrossRef]
11. Schneider, K.P.; Fuller, J.C.; Chassin, D.P. Multi-state load models for distribution system analysis. *IEEE Trans. Power Syst.* **2011**, *26*, 2425–2433. [CrossRef]
12. Kampezidou, S.; Wiegman, H. Energy and power savings assessment in buildings via conservation voltage reduction. In Proceedings of the 2017 IEEE Power and Energy Society Innovative Smart Grid Technologies Conference, ISGT 2017, Washington, DC, USA, 23–26 April 2017. [CrossRef]
13. El-Shahat, A.; Haddad, R.J.; Alba-Flores, R.; Rios, F.; Helton, Z. Conservation voltage reduction case study. *IEEE Access* **2020**, *8*, 55383–55397. [CrossRef]
14. Sunderman, W.G. Conservation voltage reduction system modeling, measurement, and verification. In Proceedings of the PES T&D 2012, Orlando, FL, USA, 7–10 May 2012. [CrossRef]
15. Two-Phase Fluid Refrigeration—MATLAB & Simulink—MathWorks India. Available online: https://in.mathworks.com/help/simscape/ug/two-phase-fluid-refrigeration.html;jsessionid=e45a89a857a5384060dc04f30bac (accessed on 8 January 2023).
16. Incropera, F.P.; DeWitt, D.P.; Bergman, T.L.; Lavine, A.S. *Fundamentals of Heat and Mass Transfer*, 6th ed.; Wiley: Hoboken, NJ, USA, 2007.
17. Çengel, Y.A.; Boles, M.A. *Thermodynamics: An Engineering Approach*, 7th ed.; McGraw-Hill: New York, NY, USA, 2011.
18. Macagnan, M.H.; Copetti, J.B.; De Souza, R.B.; Reichert, R.K.; Amaro, M. Analysis of the Influence of Refrigerant Charge and Compressor Duty Cycle in an Automotive Air Conditioning System. In Proceedings of the 22nd International Congress of Mechanical Engineering (COBEM 2013), Ribeirão Preto, SP, Brazil, 3–7 November 2013.
19. Arthur, J.H.; Beard, J.T.; Bolton, C. Simplified analytical modeling of an air conditioner with a positive displacement compressor. In Proceedings of the Intersociety Energy Conversion Engineering Conference, Washington, DC, USA, 11–16 August 1996; Volume 3, pp. 2009–2014. [CrossRef]
20. Liu, Y.; Vittal, V.; Undrill, J.; Eto, J.H. Transient model of air-conditioner compressor single phase induction motor. *IEEE Trans. Power Syst.* **2013**, *28*, 4528–4536. [CrossRef]
21. Jie, D. Modeling and Simulation of Temperature Control System of Coating Plant Air Conditioner. *Procedia Comput. Sci.* **2017**, *107*, 196–201. [CrossRef]
22. Negrao, C.O.; Erthal, R.; Andrade, D.E.; Silva, L. An Algebraic Model for Transient Simulation of Reciprocating Compressors. International Compressor Engineering Conference. January 2010. Available online: https://docs.lib.purdue.edu/icec/2032 (accessed on 30 March 2023).
23. Umans, S.D.; Fitzgerald, A.E. *Fitzgerald & Kingsley's Electric Machinery*; McGraw-Hill Companies: New York, NY, USA, 2014; p. 706.
24. Compressor Model GS26TB_T 220-240V 50Hz ~ 1 Refrigerant. Technical Data Sheet. Available online: https://lightcommercialrefrigeration.danfoss.com/pdf/danfoss_GS26TB_T_R134a_220_50.pdf (accessed on 30 March 2023).
25. Lin, S.; He, Z.; Guo, J. Study of aerodynamic noise in hermetic refrigerator compressor. In Proceedings of the 8th International Conference on Compressors and Their Systems, London, UK, 9–10 September 2013; Institution of Mechanical Engineers: London, UK, 2013; pp. 397–403. [CrossRef]

buildings

Review

An Overview of Emerging and Sustainable Technologies for Increased Energy Efficiency and Carbon Emission Mitigation in Buildings

Zhenjun Ma *, Muhammad Bilal Awan, Menglong Lu, Shengteng Li, Muhammad Shahbaz Aziz, Xinlei Zhou, Han Du, Xinyi Sha and Yixuan Li

Sustainable Buildings Research Centre, University of Wollongong, Wollongong 2522, Australia
* Correspondence: zhenjun@uow.edu.au

Abstract: The building sector accounts for a significant proportion of global energy usage and carbon dioxide emissions. It is important to explore technological advances to curtail building energy usage to support the transition to a sustainable energy future. This study provides an overview of emerging and sustainable technologies and strategies that can assist in achieving building decarbonization. The main technologies reviewed include uncertainty-based design, renewable integration in buildings, thermal energy storage, heat pump technologies, thermal energy sharing, building retrofits, demand flexibility, data-driven modeling, improved control, and grid-buildings integrated control. The review results indicated that these emerging and sustainable technologies showed great potential in reducing building operating costs and carbon footprint. The synergy among these technologies is an important area that should be explored. An appropriate combination of these technologies can help achieve grid-responsive net-zero energy buildings, which is anticipated to be one of the best options to simultaneously reduce building emissions, energy consumption, and operating costs, as well as support dynamic supply conditions of the renewable energy-powered grids. However, to unlock the full potential of these technologies, collaborative efforts between different stakeholders are needed to facilitate their integration and deployment on a larger and wider scale.

Keywords: energy savings; building decarbonization; data-driven; energy sharing; optimal control; grid-buildings integrated control

check for updates

Citation: Ma, Z.; Awan, M.B.; Lu, M.; Li, S.; Aziz, M.S.; Zhou, X.; Du, H.; Sha, X.; Li, Y. An Overview of Emerging and Sustainable Technologies for Increased Energy Efficiency and Carbon Emission Mitigation in Buildings. *Buildings* **2023**, *13*, 2658. https://doi.org/10.3390/buildings13102658

Academic Editor: Elena Lucchi

Received: 13 September 2023
Revised: 11 October 2023
Accepted: 19 October 2023
Published: 22 October 2023

1. Introduction

The undeniable evidence of climate change, coupled with the worldwide dedication to environmental decarbonization, has given rise to an imperative need for sustainable transformation of the power sector. Globally, different policies have been developed to tackle the challenge of climate change through the development of greenhouse gas emission reduction targets, deployment of renewable energy systems, the use of energy efficiency measures, and implementation of international agreements such as the Paris Agreement and carbon pricing. Energy efficiency is a central element of such policies because it can significantly reduce greenhouse gas emissions, increase energy security, and promote economic growth. Buildings are among the major players in the energy transition, as they consume around one-third of global energy usage and are responsible for 39% of global carbon emissions [1]. The emissions from buildings are projected to increase in the coming years with the increase in the world population and urbanization, as well as improved energy equality. A number of global commitments and programs are in place to enhance building energy efficiency and sustainability throughout their life cycle [2–5]. One of the leading concepts to achieve energy and carbon reductions in buildings is low-energy or net/nearly-zero energy buildings (NZEBs).

The concept of NZEBs has shown high potential to increase the sustainability of the building sector [6,7]. NZEBs are structures that strive to achieve a balance between their power import and export over the course of a defined period (usually one year) by utilizing innovative design, energy-efficient technologies, and renewable energy resources to significantly reduce their carbon footprint while maintaining a comfortable indoor environment [8]. A key benefit of such buildings is to support the integration of distributed renewable energy resources and minimize power import during peak demand hours.

Characterized by the advantage of reducing the dependence on traditional energy sources, more attention has been paid to NZEBs [6,7]. Wei and Skye [6] categorized different methods and technologies used to achieve NZEBs. It was reported that there are three features, as shown in Figure 1, that can define an NZEB, including the connection of a building with energy infrastructure, a lower-demand building with implemented energy conservation measures, and a building with onsite renewable energy generation sources.

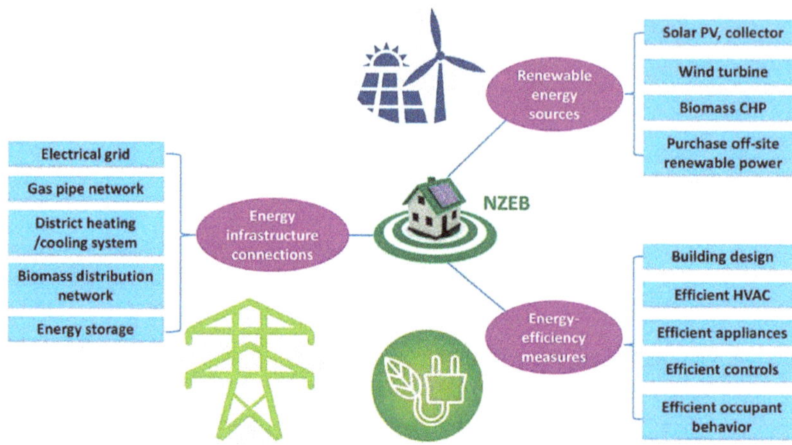

Figure 1. Methods to achieve NZEBs [6].

To achieve NZEBs, a range of energy-efficient solutions and technologies are often needed, which should offer increased energy efficiency and improved demand flexibility. To further support achieving the global decarbonization goals within the built environment, this study aims to explore advances in several emerging and sustainable technologies (Figure 2) for building energy savings and carbon emissions mitigation to enhance building sustainability. The key elements of each strategy and technology are also highlighted, along with the benefits reported in the available literature. It is worthwhile to note that the selection of these emerging and sustainable technologies was based on several key factors and considerations, such as technology maturity, market availability, interoperability, scalability, applicability, long-term viability, and most importantly, carbon reduction and energy reduction potential. Moreover, a systematic approach was adopted to review these technologies by first summarizing the importance of near-optimal sizing of building energy systems using uncertainty-based design. Emerging and sustainable technologies and strategies, along with control schemes, were then discussed, and lastly, grid-buildings integrated control was summarized for decarbonization of the building sector while supporting dynamic supply conditions of renewable-integrated grids.

Figure 2. Illustration of the emerging and sustainable technologies reviewed in this study.

2. Uncertainty-Based Design of Building Energy Systems

Appropriate design of building energy systems is critical to achieving an energy-efficient built environment. The fundamental objective for building decarbonization is to achieve a reduction in energy consumption through energy efficiency measures while meeting energy demand by using renewable energy resources and storage technologies. Therefore, appropriate design of energy systems plays an important role, but current literature shows that most energy systems in buildings were designed based on deterministic conditions [9,10], which may lead to an oversized capacity, and thus, unnecessarily extra initial investment [11].

In traditional design methods, two simplified design approaches, known as the worst scenario and safety factor, are often used [12]. In the worst scenario method, the maximum demand under the worst scenario is used to determine the equipment capacity, which often leads to over-evaluated demand and consequently, the oversized capacity. In the safety factor method, a safety factor is added to the calculated demand at the design condition to determine the equipment capacity, and the reliability of the selected capacity highly depends on the experience of the designers. Despite long practice in engineering applications, these traditional design approaches always lead to oversized systems, which increase initial cost, and, most importantly, energy waste due to low operating efficiency [13].

With the wide deployment of distributed renewable energy generation, the energy systems in buildings are characterized by uncertainties. Discrepancies are always found between the expected performance and field observation due to uncertainties from different factors such as weather conditions and occupancy behaviors [14–16]. These uncertain parameters could be categorized into three different groups [17,18], including outdoor conditions, building construction characterization, and indoor conditions, as shown in Figure 3. These uncertainties directly affect the estimation of both energy demand and generation [19]. Due to the lack of systematic consideration of the impact of uncertainties, inappropriate system configuration could occur in engineering applications [11]. To mitigate this issue, uncertainties should be carefully considered in designing and sizing building energy systems, in particular for NZEBs. Generally, the uncertainty-based design of building energy systems consists of three steps: (1) uncertainty characterization; (2) performance assessment under uncertainties, and (3) multi-objective decision-making. The uncertainty characterization includes the identification of uncertain parameters and the qualification of identified uncertainties. Subsequently, simulations are needed to evaluate the performance of the developed system models under uncertain conditions, enabling the generation of a Pareto solution set based on multiple objectives, such as life cycle costs, energy efficiency, environmental influence, and thermal comfort. The final step involves a comprehensive evaluation and comparison of different design alternatives. The optimal design solution should be determined according to the predefined objectives and alignment of the goals and requirements of stakeholders.

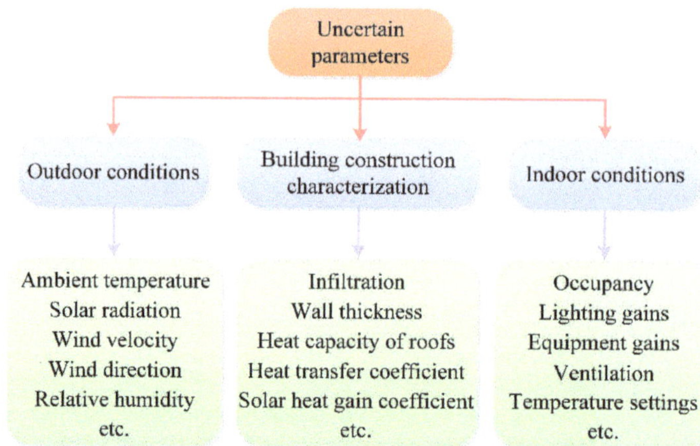

Figure 3. Uncertainties in evaluating building energy demand.

Numerous studies have been reported to optimize the energy systems in buildings by considering uncertainties. Huang et al. [18], for instance, analyzed the life cycle performance of a NZEB under uncertainties. It was concluded that the traditional method cannot achieve the target of nearly zero energy, and 12.61% of life cycle costs can be reduced by using their proposed method. A discrete Markov chain model was proposed in [20] to characterize the uncertainties of electric vehicles (EVs) when determining the capacities and positions of the PV system in a residential building cluster. The results demonstrated that energy sharing greatly improved renewable energy self-consumption, achieving 77% self-consumption and over 20% self-sufficiency. Fan et al. [21] introduced a two-layer collaborative approach using the Monte Carlo simulation for the design and operation optimization of a renewable energy system with hybrid energy storage. The energy system designed for a nearly zero energy community exhibited improved economic and environmental performance, resulting in a reduction of 39.5% and 25.6% in the annual carbon emissions and total annual costs, including investment, maintenance, operation, carbon tax, and renewable energy revenue, respectively. Combining Monte Carlo simulation with the K-means clustering method, Guo et al. [22] proposed a load forecasting method to facilitate the design of a distributed energy system for a nearly zero energy community. It was shown that this method can accurately predict the actual energy demand of end-users. In addition, the designed energy system showed good grid independence with a 36.7% energy import rate only. Zhang et al. [23] developed an uncertainty-based method to size renewable energy systems by taking energy balance, grid stress, and initial investment into consideration. This method demonstrated a 44% improvement in the overall performance of a NZEB when compared to the use of the traditional worst scenario method.

EVs have been identified as a profitable solution to increase building demand flexibility [24]. Despite the increasing share of EVs, their energy demand was only considered in a few studies at the design stage of building energy systems [22,25–27]. EVs are inherently characterized by strong uncertainties [28]. To improve energy flexibility and further investigate potential benefits, more efforts should be made to analyze the effect of EVs on the optimal configuration and sizing of energy systems in buildings. Furthermore, EVs can also be used to support demand management, which can further improve the bidirectional exchange of energy between buildings and the grid [29].

Uncertainty-based design methods can enhance system robustness, reduce unnecessary initial investment, and improve the overall performance of the designed energy systems. Therefore, uncertainties should be considered during the design phase of building energy systems.

3. Sustainable Energy Technologies

3.1. Renewable Energy Integration in Buildings

Renewable energy is a key solution for climate change and building decarbonization. In the roadmap developed by the International Renewable Energy Agency (IRENA), it was suggested that renewables can make up to 60% or more of many countries' total final energy consumption in 2050 [30]. As shown in Figure 4, different renewable energy technologies can be integrated with buildings to meet electricity and heating and cooling demands. Among different technologies, solar energy holds paramount importance due to its abundance and versatility. Its widespread application has been facilitated by favorable policies, significantly reducing the implementation costs of photovoltaic (PV) panels from 4.6 cents/kWh in 2021 to 2 cents/kWh in 2030 [31]. The diversification of solar energy applications has evolved from active to passive technologies. PV systems, prevalent in mitigating power supply pollution, are anticipated to constitute over half of the total renewable energy electricity generation by 2050 [32]. The share of distributed PV generation is continuously increasing. For instance, in 2021, China's distributed PV capacity was 108,580 MW, whereas it was 17,037, 48,559, 48,292, and 42,677 MW for Australia, Germany, Japan, and the United States, respectively. All these countries also have a high capacity for centralized PV generation [33]. PV and photovoltaic/thermal (PVT) systems can be integrated into diverse building elements, including roofs, atriums, skylights, ventilated facades, double-skin facades, and window curtains. Hybrid solar systems that combine active and passive technologies, such as Trombe walls combined with PVs, solar chimneys paired with PVs or phase change materials (PCMs), and PV-integrated ventilation blinds, have also been studied [34]. Although wind energy typically requires expansive open spaces and distance from urban areas, the proliferation of tall buildings in densely populated urban settings has opened new opportunities. Properly spaced tall buildings can harness and transfer wind energy over short distances. Wind energy has the potential to fulfill around 10%–20% of urban tall building energy demand, and urban wind energy systems predominantly comprise wind turbines, typically ranging from 1 to 20 kW, placed on buildings and adjacent grounds [35]. However, careful consideration of local environmental conditions and urban boundary layers is essential for appropriate design and efficient deployment of wind energy systems [36]. Due to the limited initial exploration, many wind turbine projects failed to meet output expectations after a few years of operation [37]. Biomass energy, accounting for approximately 10% of global primary energy demand, represents a vital renewable energy source [38]. Biomass-fired heaters, in conjunction with heat pumps, can provide an economically viable solution for building heating needs. These heaters can cover up to 65% of space heating demand, effectively decreasing reliance on the grid and heat pumps [39]. Furthermore, biomass-derived porous carbon materials resulting from pyrolysis can serve as favorable supporting materials for shape-stabilized PCMs, exhibiting beneficial attributes such as micropore structure, chemical stability, and the potential for integration into construction materials like cement bricks/blocks [40]. There are other types of renewable energy, such as micro-hydro and hydrogen fuel cells, that have been tried in buildings. Micro-hydroelectric systems can harness the energy of flowing water to drive turbines, producing electricity, and are often employed by homeowners in remote areas. While the utilization of hydrogen energy in buildings has not yet been widespread, it is being investigated as a potential means of decarbonizing heating and cooling in buildings. The prevalent approach involves fuel cells, which can generate both electricity and heat for buildings. It is still in its infancy stage, with ongoing projects to explore its potential [41]. However, the adoption of hydrogen energy in buildings faces challenges such as high production costs, limited infrastructure, and safety concerns associated with handling hydrogen, which restricts its widespread integration in buildings [42].

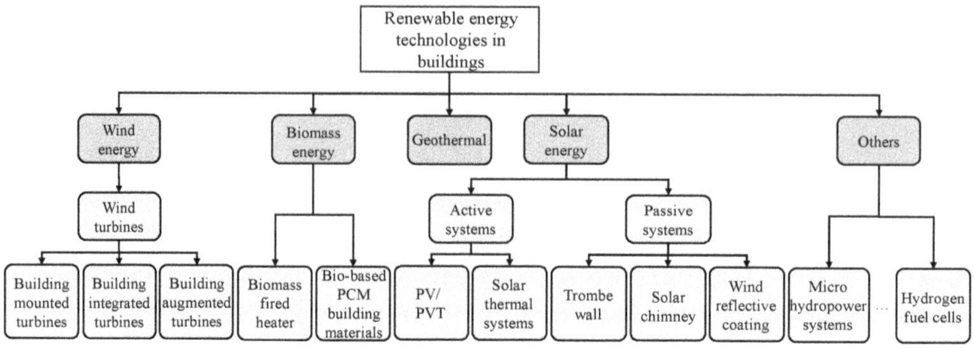

Figure 4. Integration of renewable energy in buildings.

3.2. Thermal Energy Storage

Ensuring a reliable energy supply is imperative due to the inconsistent availability and seasonal variations of renewable energy sources. To address the intermittence of renewable energy supply and temporal disparities, thermal energy storage (TES) has emerged as a promising solution. The ability of TES systems to provide flexible assets to wind and solar energy resources increases the overall value of TES in the modern energy supply chain. The 2020 innovation outlook on thermal energy storage published by the International Renewable Energy Agency (IRENA) [43] demonstrated that the TES global market will grow to triple the current size by 2030, which indicates an increase from 234 GWh in 2019 to 800 GWh within a decade. Moreover, it is expected that the investments in TES technologies for cooling and power applications will surge from 13 billion USD to 28 billion USD in the same period [43]. In buildings, TES can be mainly used for demand shifting and load modulation through integration with heat pumps and hot water storage systems as well as with building envelopes [44].

Sensible and latent energy storage systems are the two major TES systems used in buildings. Sensible energy storage systems require large volumes compared with that of latent energy storage systems. However, currently, sensible energy storage systems in the form of hot water tanks are among the most used TES systems in buildings because of their low cost and low complexities [45]. Research has shown that latent energy storage systems are more effective than sensible energy storage systems, and they can be used in different ways to enhance the overall thermal performance of buildings and building energy systems. Phase change materials (PCMs) have demonstrated high performance and practicality to be used as the leading latent energy storage medium in building applications. Figure 5 illustrates different applications of PCMs in buildings, where PCMs can be integrated with building envelope and building energy systems for enhanced performance.

Various benefits and limitations associated with the use of PCMs for different applications in buildings have been reported. For instance, in [46], it was illustrated that the integration of PCMs with gypsum and wallboards can result in enhanced reliability and durability over extended durations, though with drawbacks, such as potential deposition of volatile pollutants on surfaces and limited fire resistance. Another study demonstrated that the excessive use of hydrated salt PCMs in concrete can jeopardize material strength, but incorporating paraffin can enhance thermal storage while maintaining compressive strength [47]. Recommendations for improving thermal efficiency and mitigating supercooling involve the introduction of an internal ventilation system within PCM roof and floor systems to enhance convective heat transfer across the PCM surface [48]. For optimal energy efficiency, lighter mass PCMs are suggested when combined with air conditioning systems to minimize undesired heat transfer between upper and lower PCM layers. Furthermore, a well-designed operational strategy for the combined air conditioning-PCM system can significantly improve the overall energy performance [49]. The effective integration of PCMs into bricks requires the selection of PCMs with melting temperatures closely aligned with room temperature to maximize performance enhancement [50]. Thin PCMs with expansive heat transfer areas are preferred over thicker alternatives within bricks, and a staggered arrangement of PCMs can enhance the heat exchange between the PCM and brick materials [51].

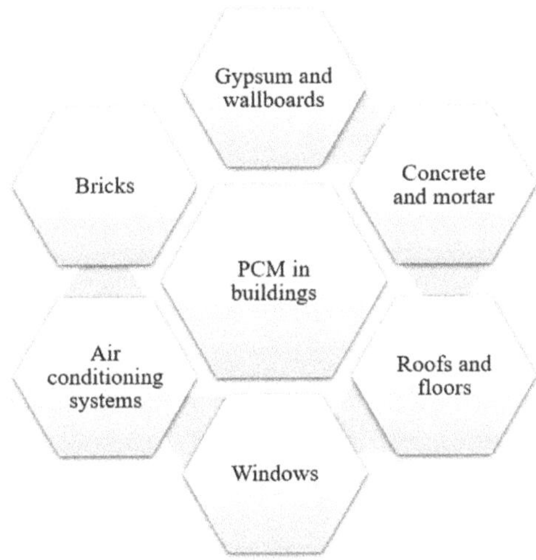

Figure 5. Integration of PCMs in buildings.

A range of studies have also demonstrated the high value of TES systems in increasing the efficiency and demand flexibility of buildings and building energy systems, contributing to the reduction in overall building emissions. For instance, in [52], a TES system filled with a PCM was used to shift the peak demand of an HVAC system to off-peak demand hours. The results showed that the TES system integration with the HVAC system helped reduce peak demand by 73% and provided 37.1% cost savings for the analyzed period. Tuncbilek et al. [44] reviewed a number of studies involving the use of TES for enhanced flexibility of buildings and building energy systems. The review demonstrated that the use of TES can reduce overall energy consumption, operational costs, and emissions of buildings. Overall, the TES systems will be a central part of decarbonizing the building and power sectors, along with providing opportunities to reduce overall operating costs and increase energy efficiency. However, lack of technology readiness, awareness, policy, and market mechanisms are among the main bottlenecks that are important to be addressed to explore the full potential of TES technologies in building applications [43].

3.3. Heat Pump Technologies

Heat pumps are increasingly recognized as a key and cost-effective solution for the electrification of heating systems and tackling carbon emissions, supporting the goal of achieving a net zero energy future [53]. The International Energy Agency (IEA) estimated that the installed number of heat pumps will rise from 180 million globally in 2020 to around 600 million in 2030 [54], and heat pumps could reduce global CO_2 emissions by at least 500 million tons in 2030 [55]. A recent report published by the Australian Energy Efficiency Council and the Australian Alliance for Energy Productivity concluded that ambitious deployment of heat pumps in different buildings and industrial processes could reduce energy usage by up to 14,391 PJ and save 746 Mt of greenhouse gas (GHG) emissions by 2050 [56].

Heat pumps are generally energy-efficient and can be powered by renewable energy resources. They can be easily integrated with thermal energy storage systems for load management, such as load shifting and peak demand reduction. Different types of heat pumps are now available in the market. According to the types of heat sources used, heat pumps can be classified into air-source heat pumps, water-source heat pumps, and ground-source heat pumps. It is noted that ground-source heat pumps are simply a variation of water-source heat pumps that are capable of working with a broader range of source water temperatures. Other heat pumps include absorption heat pumps and desiccant heat pumps. An absorption heat pump is a heat pump driven by thermal energy instead of electricity. It is most cost-effective when low-cost thermal energy is available and in applications where access to electricity is expensive. Desiccant heat pumps are developed based on the vapor compression cycle but with desiccant materials coated on heat exchangers for dehumidification. Such heat pumps are suitable for applications with strict humidity control and in tropical and subtropical climate conditions.

Among different heat pump technologies, ground-source heat pump (GSHP) systems are receiving increasing attention [57,58] as they can take advantage of the relatively stable earth temperature to achieve heat extraction from or heat rejection to the ground through ground heat exchangers to meet different air-conditioning requirements. The long-term operation of stand-alone GSHP systems may lead to ground thermal imbalance, resulting in system performance degradation [59]. Hybrid GSHP systems assisted with supplementary devices have been considered to be a solution to alleviate soil and system performance deterioration [60]. In addition to environmental and energy performance improvements brought by the application of auxiliary devices, obvious economic benefits can also be achieved, especially in terms of cost savings in the upfront purchase and installation of ground heat exchangers [61,62]. As shown in Figure 6, different auxiliary devices can be coupled with GSHP systems [63] to increase the overall efficiency and mitigate ground thermal imbalance.

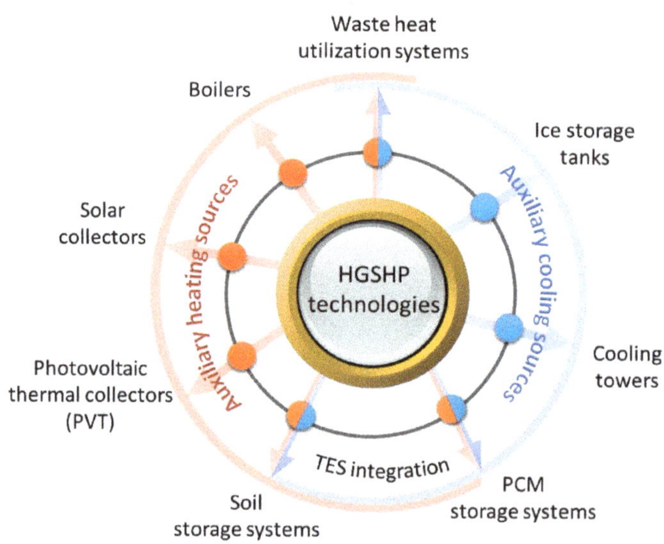

Figure 6. Hybrid GSHP systems integrated with different auxiliary devices.

Heat pumps will remain the main solution to provide heating and cooling to buildings in the coming decades, and cost-effective solutions/strategies to further improve their efficiency are needed. Solar heat pumps are also among the interesting solutions.

3.4. Thermal Energy Recovery and Sharing

Thermal energy recovery and sharing can capture waste energy from one system and redistribute the captured waste energy to nearby systems or end-users [64]. Figure 7 illustrates an example of a thermal energy recovery and sharing system. Such a system can provide simultaneous cooling and heating to interconnected networks of different buildings or utilities, leading to enhanced energy efficiency and reduced carbon emissions. Thermal energy recovery and sharing systems can effectively harness and re-utilize otherwise wasted energy and further improve resource utilization and sustainability [65].

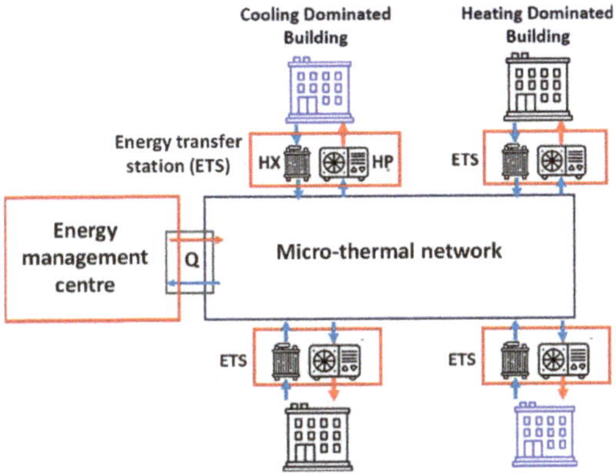

Figure 7. Illustration of a thermal energy sharing system, where HX represents the heat exchanger, and HP indicates the heat pump [64].

The cooling-dominated buildings or utilities in thermal energy sharing systems can include cooling systems of data centers, ice sports centers, and supermarkets, as well as industrial processes. Data centers can produce a substantial amount of waste heat during the cooling process of their equipment [66]. The temperature of this waste heat varies depending on the different types of cooling systems employed [67]. The cooling systems in supermarkets and ice rinks can also generate a significant amount of waste heat by providing the necessary cooling energy, and the waste heat from supermarkets can reach temperatures of more than 50 °C [68–70]. Many industrial processes also generate significant amounts of waste heat, often at high temperatures, making them a potential heat source [71]. These potential heat sources often exhibit stable and large heat generation characteristics, providing ample opportunities for integration into thermal energy recovery and sharing systems.

One notable utilization of waste heat is to supply thermal energy for heating purposes, such as integrating with district heating (DH) networks [72,73] and preheating swimming pools [74,75]. Recent studies have demonstrated the potential benefits of thermal energy sharing systems. Pan et al. [76] examined the performance of an adsorption chiller, which was powered by thermal energy recovered from a data center. Araya et al. [77] presented a lab-scale Organic Rankine Cycle (ORC) system, which was designed to operate using low-grade waste heat obtained from a server rack within a data center. Abdalla et al. [64] introduced a reduced model of a thermal energy sharing system, which effectively allowed thermal energy to be shared among different facilities. It was found that such a system could fulfill 48% of the overall heating demand, and by incorporating a water tank, an additional 12% of thermal energy could be covered, resulting in a remarkable 74% reduction in total carbon emissions. In another study by Wirtz et al. [78], an optimal design approach based on Linear Programming was developed for the design of the 5th-generation district heating system, as shown in Figure 8. It was found that the optimized system achieved a remarkable 42% cost reduction and an impressive 56% decrease in carbon emissions through component selection and optimal sizing. Similarly, a study described in [79] focused on thermal energy sharing between the cooling system of a data center and a multi-unit residential building. The findings indicated that energy sharing reduced the heating requirements of the building by 55% and the cooling requirements of the data center by 50%. Zhang et al. [80] developed an economic evaluation model for waste heat utilization of data centers in district heating networks. It was reported that reusing waste heat from data centers could achieve both economic and environmental benefits. Li et al. [81] examined two thermal energy storage methods in thermal energy sharing systems, including short-term storage using a water tank and long-term storage using a borehole thermal energy storage system. The water tank was found to effectively reduce peak loads by 31% and lower annual energy costs by 5%, and the borehole system can increase waste heat utilization from 77% to 96% and achieve an annual reduction of 8% in CO_2 emissions. By dynamically optimizing the parameters of the water tank, an optimal storage size was determined by considering the trade-off between the payback period and heating cost savings. It was found that prosumers could save up to 9% of their annual heating costs [82]. Wang et al. [83] proposed an efficient system using CO_2 direction-expansion GSHPs to harvest the waste heat from a data center for heating surrounding buildings. It was found that this system could reduce energy consumption by nearly 50% when compared to air-source heat pumps.

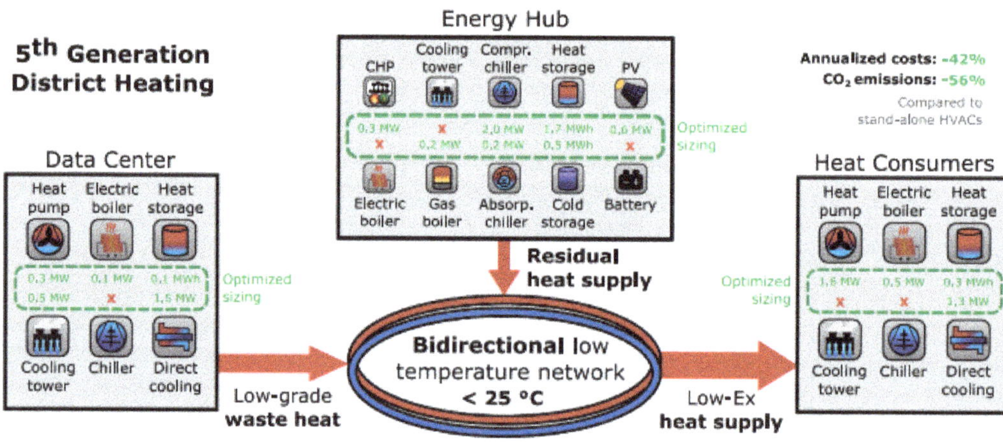

Figure 8. The 5th generation district heating system [78].

In addition to thermal energy sharing, electricity generated by distributed renewables can also be shared among different buildings. In summary, while thermal energy recovery and sharing systems showed significant advantages in terms of GHG reduction, cost savings, energy efficiency, and waste heat utilization, further research is needed to fully explore opportunities to increase their overall performance.

3.5. Retrofits for Increased Energy Efficiency

Most buildings are existing buildings, and retrofitting of existing buildings has been widely accepted as one of the effective measures to achieve building decarbonization. According to a report recently published by Market Research Future [84], the global energy retrofit systems market was valued at USD 150.9 billion in 2022, and it was projected to increase to USD 272.8 billion by 2032 with a compound annual growth rate of 6.8% from 2023 to 2032. In general, there are two types of retrofits, including shallow retrofits and deep retrofits. Shallow retrofits focus on small alternations by using low-cost or no-cost energy conservation measures, while deep retrofits focus on major renovations, and a holistic approach is often required to systematically explore co-benefits from energy retrofits and determine the best retrofit options. A recent study showed that stock-wide implementation of deep retrofits in existing buildings of eight cities (i.e., Braga, Cairo, Dublin, Florianopolis, Kiel, Middlebury, Montreal, and Singapore) can reduce energy use and carbon emissions by up to 66% and 84%, respectively, without additional grid decarbonization efforts [85]. Based on the review of energy retrofit policies for the UK's existing buildings, it was concluded that deep retrofits are needed for most UK's existing housing stock to bring more flexible financing schemes and reduce carbon emissions [86]. Figure 9 illustrates an example of deep retrofit of a typical Australian timber framed fibro house to a state-of-the-art net zero energy home through a combination of architectural alternation, envelop renovation, facilitating daylighting and natural ventilation, and integration of renewable energy systems and highly efficient air conditioning solutions.

Figure 9. Illustration of deep retrofit of a typical Australian timber framed fibro house.

In the past, the selection and quantification of co-benefits of retrofit measures were often achieved with the assistance of energy auditing, building performance simulations, and domain expertise. With the continued development of low-cost sensors, massive data can be readily available from existing buildings, and this presents significant opportunities to use machine learning and big data analytics to optimize the retrofit process and explore energy-saving potential. Figure 10 summarizes different retrofit methodologies for large-scale building retrofits [87]. Based on the data collected from existing buildings, the bottom-up approach focuses on retrofitting individual buildings by analyzing different retrofit strategies and then extrapolates the results to larger scales, while the top-down approach focuses on using the available datasets to explore retrofit potential, establishing benchmarks, and providing guidelines for the development of energy retrofit policies [88].

To maximize the benefits of existing building retrofits, significant efforts are still needed. More rigorous retrofit methods should be developed to optimize the selection of the most cost-effective retrofit options. Although many retrofit techniques can reduce building energy consumption and carbon footprint, their longevity has rarely been considered [89]. Circular economy is the key to sustainable development, and applying circular economy principles to building retrofits is an interesting direction to reduce the use of building materials and provide significant carbon benefits. The availability of large datasets from existing buildings will offer great opportunities to use state-of-the-art machine learning methods and big data analytics to explore useful information, providing new potentials to support the decision-making process and accelerating the selection and optimization of retrofit measures.

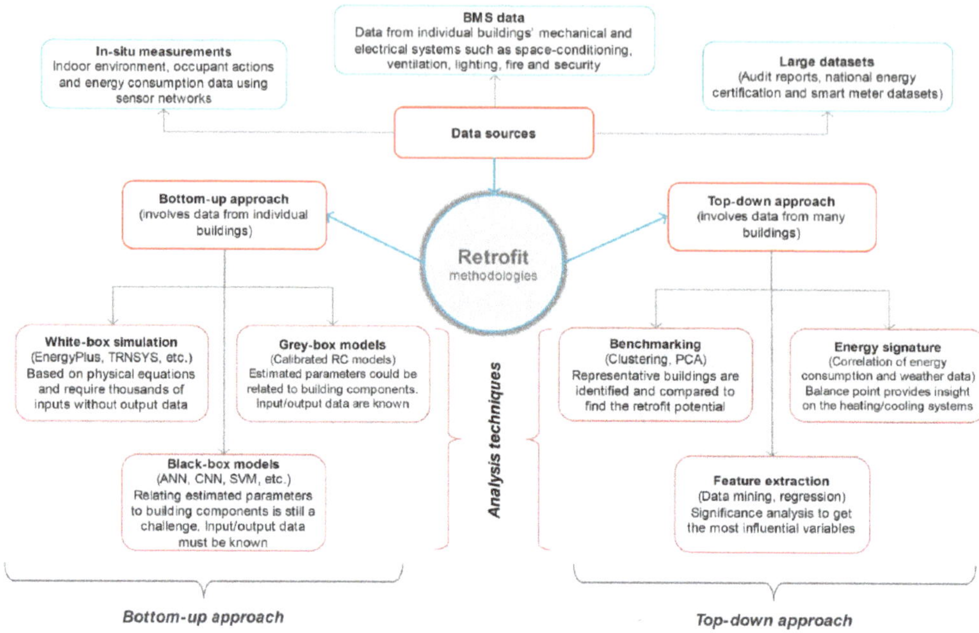

Figure 10. Illustration of different retrofit approaches [87].

4. Data-Driven Modeling, Demand Flexibility, and Integrated Control

4.1. Data-Driven Modeling

Data-driven modeling has evolved as one of the leading approaches for real-time control of buildings. Machine learning algorithms used for data-driven modeling can be generally categorized into two groups, including shallow structure machine learning and deep learning [90]. Generally, both categories of machine learning algorithms aim to identify correlations between model input and output variables, while deep learning employs more complex model structures and numerous parameters and can result in more powerful feature representation capabilities with lower interpretability and transparency [91]. A detailed comparison between the use of shallow structure machine learning and deep learning for building energy modeling can be found in [92]. Due to their superior ability to capture complex data features, two deep learning models, i.e., Recurrent Neural Networks (RNN) and Convolutional Neural Networks (CNN), have been frequently adopted to predict various building variables [93]. RNN models, such as LSTM networks, have shown outstanding performance in building energy consumption prediction due to their superior capacity to capture long-term temporal dependencies in sequential data [94–96]. CNN models have been highly successful in capturing spatial features, making them particularly well-suited for graphic data analysis. It was reported that a CNN model trained by sky image data can effectively improve the prediction accuracy of PV power generation as compared to conventional data-driven models [97].

In addition to the deployment of advanced deep learning algorithms, model fusion in data-driven modeling has emerged as a promising research area and has shown good performance in building energy management. In [98], a combination of an LSTM model and a CNN model was employed to simultaneously extract spatial and temporal features of model inputs, which led to increased accuracy in the predictive modeling of a GSHP system. In [99], building energy consumption was decomposed into a global part, a local part, and white noise. The global part was attributed to the typical user behavior of a building, while the local part was attributed to non-typical activities. A static deep learning model was used to predict the global part based on weather and time-related variables at a given time, while an auto-regression (AR) model was used to predict the local part using time series data of the local part as the input. The entire process is illustrated in Figure 11. This approach achieved superior performance when compared to the strategy that either used deep learning or AR models alone.

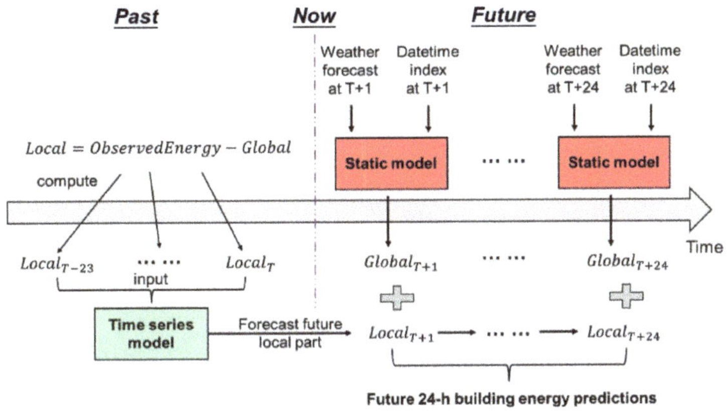

Figure 11. A hybrid machine learning model for building energy consumption prediction [99].

In [100], the integration of an LSTM model and a reinforcement learning (RL) model was explored to enhance the adaptability of deep learning prediction models. This integration allowed for dynamic updating of model parameters according to the most recent prediction errors during the prediction process, leading to improved accuracy in the prediction of building energy consumption. In [101], domain knowledge and decision trees were incorporated into an RL model to optimize building operations for improved energy flexibility. It was found that this method was effective in improving learning efficiency and model interpretability. Furthermore, the recent success of attention mechanisms has resulted in the increasing popularity of their integration with data-driven modeling. Attention mechanisms have been successfully used to improve modeling accuracy [102] and model interpretability [103] for data-driven modeling of building energy systems.

As the development of reliable data-driven models in practice is often limited by the scarcity of model training data, the solutions to address data scarcity for data-driven modeling have also been increasingly explored in recent years. A transfer learning scheme was developed for non-intrusive load monitoring in smart buildings [104]. The method reduced the data dependency of predictive modeling for energy consumption of indoor appliances. In [105], a meta-learning model, also known as the 'learning to learn' approach, effectively reduced the requirement of training data in the prediction of personalized thermal comfort conditions. In [106], Generative Adversarial Networks (GANs) were employed to generate synthetic data to train predictive models of building electricity demand and successfully alleviated data scarcity issues for data-driven model development.

The approaches explored in the above literature provided valuable references for the development of data-driven models for building energy management. Machine learning can learn patterns and relationships in complex datasets, enabling the development of accurate predictive models without explicit programming. However, they are prone to overfitting if they are not properly regularized or validated, and the results can be challenging to interpret and explain in comparison to the use of traditional statistical models. Appropriate selection, training, and optimization of machine learning models for a given application is critical.

4.2. Building Demand Flexibility

Demand flexibility of a building is its ability to manage its energy demand and generation based on user needs, grid requirements, and local climate conditions [4]. It can allow end-users to decrease, increase, modulate, or shift energy consumption to modify their load profiles using flexible load and/or onsite generation and storage sources, and is an attractive resource to provide grid support services and smooth out the so-called "duck curve" of power systems (Figure 12) while meeting interests and objectives of end-users such as energy availability, decreased energy costs, and reduced carbon emissions [107].

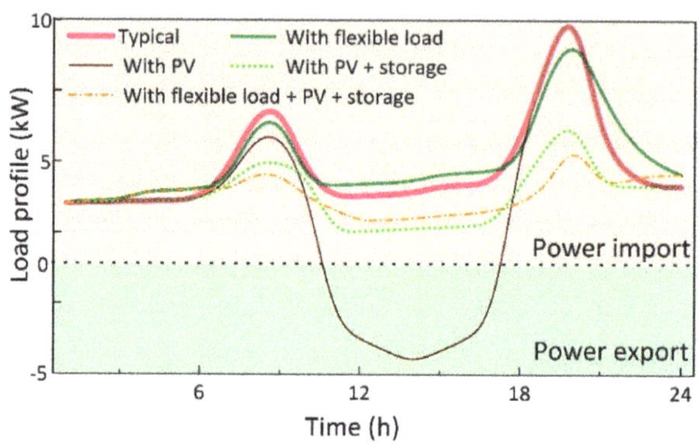

Figure 12. Indicative illustration of how load profiles can be modulated through demand flexibility.

Flexible demand is a viable component of "non-wires alternatives" and is a low-cost resource to effectively support power balance and accelerate renewables penetration in the energy mix. Demand flexibility in buildings can be greatly increased by using onsite generation, energy storage, and controllable loads, and can help reduce energy costs of end-users according to different flexibility types and demand management strategies (Figure 13) relating to flexibility capacity, duration, and response speed. By reducing and shifting the timing of electricity consumption in the grid-interactive efficient buildings, a 6% carbon emissions reduction in the US power sector can be achieved by 2030 [108]. The introduction of a Smart Readiness Indicator (SRI) to promote smart buildings with the capacity to provide demand flexibility and linking the SRI assessments to the Energy Performance Certificate in the building sector in Europe can result in final energy savings of up to 198 TWh by 2050 and 32 million tons of avoided carbon emissions per year [109]. A study for 498 residential homes showed that using battery and demand flexibility through home energy management systems along with energy efficiency upgrades can save households up to $590 in annual electricity costs [110].

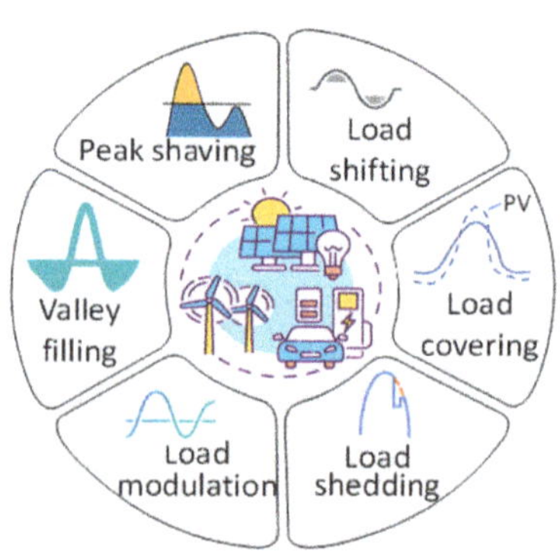

Figure 13. Various demand management strategies.

However, how to effectively use demand flexibility to manage energy consumption and how to actively participate in demand side management remain unclear for many residential customers. Many current studies have demonstrated the high value of using different flexible sources for demand management [111–113], but it remains critical to develop a generic characterization and optimization framework for enhanced demand flexibility and demand management. To ensure successful characterization and optimization of building demand flexibility, a holistic approach is required. This involves collaborations between architects, engineers, energy analysts, and stakeholders to design and implement energy-efficient building systems and practices. Continuous monitoring, data analysis, and performance evaluation are also vital for refining strategies and adapting to evolving energy landscapes.

Figure 14 illustrates a general process to characterize and optimize demand flexibility in buildings. Characterizing and optimizing building demand flexibility involves a comprehensive analysis and enhancement of a building's ability to dynamically adjust its energy consumption and generation patterns in response to external factors, grid conditions, and operational needs. It requires an in-depth understanding of a building's energy systems, envelope, occupant behavior, and the potential for integrating renewable energy sources and energy storage systems. Characterization involves the collection and interpretation of data related to occupancy patterns, thermal performance, equipment efficiency, weather data, energy demand profiles, and potential integration of smart technologies [113]. This information helps identify opportunities for demand response, such as load shifting and peak shaving. It also involves the use of different technologies and solutions for enhanced demand flexibility along with the quantification of the achieved flexibility. For instance, the integration with smart grid technologies enables a building to respond to signals that incentivize energy conservation during peak demand periods or take advantage of surplus energy during off-peak hours [114]. Furthermore, the implementation of energy-efficient HVAC systems, smart lighting, and automated shading systems can contribute to improved demand flexibility by reducing overall energy consumption and improving indoor thermal comfort [115–117]. Energy storage solutions, such as batteries or thermal storage systems, can play a crucial role in improving building demand flexibility through strategically managing stored energy, reducing their reliance on the grid during peak hours, and contributing to overall grid stability [111]. Characterization of demand management based on different sources and their potential benefits can help select optimal flexible systems for enhanced flexibility.

Optimization of building demand flexibility involves the development and deployment of advanced control strategies. This can be achieved through utilizing predictive algorithms, machine learning methods, and real-time monitoring to predict energy demand and supply variations [118–120]. Simulation models can also be opted for to optimize the flexibility potential of a building.

4.3. Improved Building Control

Appropriate building control is essential to ensuring service systems provide functional requirements with the least energy consumption. According to a study by Afroz et al. [121], appropriate control can reduce 25% of energy use of an HVAC system. Control strategies for buildings can be generally categorized into rule-based control, model predictive control, reinforcement learning control, and data-driven predictive control. The majority of the control strategies developed for buildings were focused on HVAC systems as HVAC systems are the major energy consumers in buildings.

Rule-based control is the most basic control strategy. It can provide safety limits and is easy to implement. Although rule-based control strategies have been widely used in practice, they largely rely on predefined rules and can only offer limited opportunities to enhance the performance and operating flexibility of building HVAC systems. Model predictive control (MPC), which uses models to predict future performance and make optimal control decisions, has attracted increasing attention in recent years due to its predictive feature. For instance, Fiorentini et al. [122] proposed a two-level MPC method to maintain indoor thermal comfort and reduce the energy use of an HVAC system. The high-level controller with a one-day prediction horizon was used to select the control modes of the system, and the low-level controller with a five-minute prediction horizon was used to make operational decisions. Merema et al. [123] developed a two-level MPC strategy to control the HVAC operation. The two-level MPC strategy included a black-box linear model with auto-regression and a grey-box resistor-capacitor nonlinear model to predict building thermal comfort. The method was tested in a real lecture room, and it was shown that this MPC strategy can reduce electricity consumption by 10–40%.

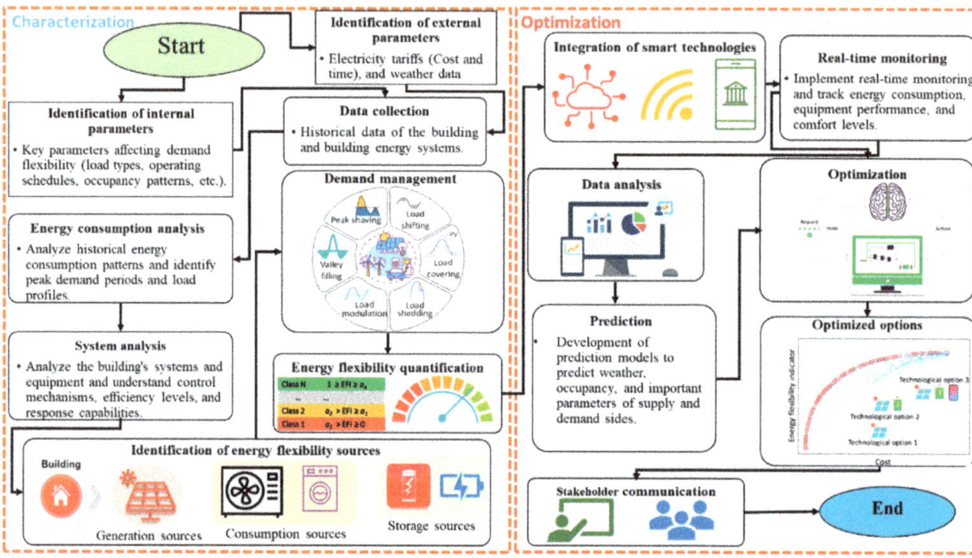

Figure 14. Characterization of demand flexibility and optimization of building demand management.

Reinforcement learning, which can directly establish a relationship between the inputs and outputs, has recently received increasing attention. Wei et al. [124] developed a control strategy by using deep reinforcement learning (DRL) to reduce energy costs and improve indoor thermal comfort. The DRL strategy was tested based on the EnergyPlus simulation of a physical house. Compared with rule-based and MPC strategies, this DRL strategy can reduce energy use by 27%. Azuatalam et al. [125] used a demand-response reinforcement learning method to control the operation of an HVAC system. In their study, proximal policy optimization was used to stabilize the hyper-parameters of RL agents in order to build a relationship between occupancy of the building and HVAC operation decisions. The method was tested in EnergyPlus, and it was shown that it can reduce 22% of energy use when compared with a handcrafted baseline controller. Yu et al. [126] proposed a DRL method using a multi-actor-attention-critic method to optimize the HVAC control of multiple-zone buildings. The method was tested using real-world data of energy prices, ambient air temperature, and zone-occupancy data. It was indicated that this control method was effective in controlling multiple zones' thermal comfort and reducing energy costs.

A data-enabled predictive control (DeePC) method was recently proposed to solve the high training cost of reinforcement learning [127]. The DeePC method used a non-parameter learning method to predict future trajectories. This method was not developed based on learning from a large dataset but on using trajectories for system identification. It can perform better in nonlinear stochastic systems with regularizations. Chinde et al. [128] used a DeePC method to optimize the operating parameters of a building HVAC system. The method was tested via an office prototype building in EnergyPlus simulation. Compared with a traditional MPC strategy, the operational cost of the HVAC system by using this method was reduced by 5% in a five-week test. The strengths and weaknesses of the commonly used strategies for building HVAC control are summarized in Table 1.

Table 1. Strengths and weaknesses of different control strategies for building HVAC systems.

Control Strategies	Strengths	Weaknesses
Rule-based control	Easy to implement and understand; and can enable safety rules to the control system.	Low flexibility to change setpoints to react to demand variation; and cannot provide global optimal solutions.
Model predictive control	It enables safety rules; reacts prior to variation to improve stability; and has high flexibility to meet control requirements.	High computational costs; and relies on accurate models for prediction.
Reinforcement learning	Model-free data-driven method; does not require much information for model development; and requires low reaction time for decision-making once trained.	Requires long training time and large training dataset; and doesn't enable safety rules.
Data-enabled predictive control	Does not require a large training dataset; and has high stability for nonlinear time-varying systems.	Has not been widely validated for HVAC control; and the efficiency of the model relies on the quality of the training data.

4.4. Grid-Buildings Integrated Control

To support grid operation, future buildings integrated with variable renewable generation sources should have the capability to regulate their demand according to the needs of the grid. To achieve this, a bidirectional interaction between the grid and buildings is required, which is different from the conventional unidirectional electricity flow from the grid to buildings. This bidirectional dynamic interaction can potentially result in a range of severe issues for the grid, including harmonic distortions, unbalances, voltage flicker, and other complications, and in extreme cases, it may even lead to grid collapse [129,130]. To address these issues, substantial efforts have been made to explore control strategies for grid integration of sustainable buildings. By adopting such strategies, buildings can effectively tackle the complexities associated with bidirectional energy flow. This approach enables seamless coordination between the building and the grid. It ensures efficient resource utilization while minimizing any adverse effects on the grid. Such control strategies can allow buildings to interact harmoniously with the grid, paving the way for optimized energy management and reduced environmental footprint [131]. In general, there is a relationship between energy efficiency and grid interactivity, which can be achieved by using smart control for energy management [132].

Grid-buildings integrated control systems are increasingly recognized as essential for optimized energy usage and improved reliability and efficiency of sustainable buildings. In contrast to building control, grid-buildings integrated control allows for real-time optimization of energy usage based on grid conditions, integration of renewable energy sources, effective utilization of energy storage systems, and provision of grid support services, leading to cost savings and increased resilience [133]. Grid-integrated buildings can be classified as microgrids. Microgrids are localized energy systems that operate either autonomously or in parallel with the main electrical grid [134]. They consist of interconnected loads, distributed energy resources, and advanced control strategies. In this context, grid-integrated buildings exhibit characteristics aligning with the microgrid concept. Control strategies developed for microgrids can be applied to grid-integrated buildings and vice versa.

A wide range of computational and experimental aspects related to the integration of buildings into the grid have been the subject of numerous studies. Huo et al. [135] proposed a two-level hybrid decentralized-centralized (HDC) strategy to control distributed energy resources in grid-interactive efficient buildings. This HDC algorithm can offer a scalable and bifurcated approach to managing a large number of grid-connected buildings and devices. The two-level design consisted of central aggregators that manage buildings and a system operator that coordinates the distribution network in a decentralized manner. Rastegarpour and Ferrarini [136] analyzed different modeling and control techniques to address building energy efficiency and management problems. The importance of adaptive control techniques and real-time predictive models in optimizing energy usage and integrating buildings into smart grids was highlighted. Figure 15 shows the control hierarchy and levels of integration of building components to the grid over a time period ranging from a week to one second. The yellow zone shows the area where the system behaved dynamically.

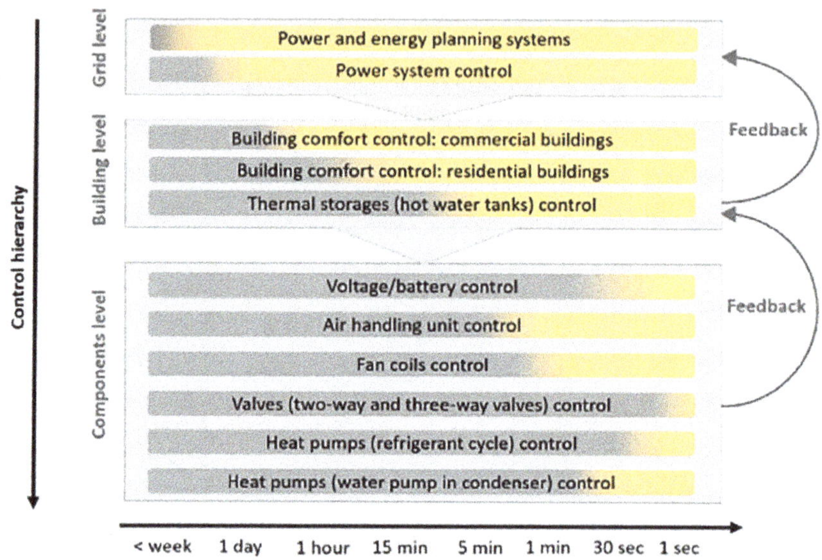

Figure 15. Illustration of control hierarchy and levels of integration [136].

An experimental architecture was proposed in [137], aimed to enable smart buildings by integrating HVAC systems with the grid. A bi-level optimization approach for integrating commercial buildings with a distribution grid was presented in [138]. It can reduce building operating costs while maintaining thermal comfort for occupants. For smart homes, a strategy including integration of renewable energy systems and scheduling and arranging the power flow during peak and off-peak periods was studied in [139]. More commonly, intelligent methods, including neural networks [140], linear programming [141], and differential evolution [142], have been used to optimize energy interactions for grid stress reduction.

In addition to the integration with the bulk grid, building energy management strategies have also been established for microgrids [143–145]. A management strategy for battery energy storage was proposed for commercial building microgrids, considering the operational costs and grid resilience. This strategy was able to increase the microgrid's resilience while maintaining the operational cost at a low level [143]. A framework for smart transactive energy in residential microgrids was proposed in [144]. An existing home-microgrid coalition formation scenario was used to verify the viability of the proposed method. Based on the work in [144], an enhanced transactive energy framework was further created to enable multiple residential microgrids to cooperate with each other by forming alliances in order to increase competitiveness [145]. In [146], the supply of regulation services by smart buildings was examined. Price signals were exchanged between the grid and building operators to adjust building energy usage. Additional research on buildings-to-grid (B2G) integration showed that HVAC control in grid-aware buildings can provide frequency management or ancillary services to the grid while maintaining occupants' thermal comfort to an adequate level. In [147], the control actions of both the power grid and buildings were coupled to optimize the performance. An MPC-based algorithm was developed to formulate the B2G integration, accounting for the time-scale discrepancy. To solve the duck-curve challenges by reducing the load-ramp rate, a predictive bidirectional control framework was presented in [138] for power flow control of B2G systems. Mirakhorli et al. [148] proposed a unique load aggregation method that used model predictive controlled loads and integrated power generation, grid constraints, and human behavior in a large cluster of buildings. The proposed method improved the control over multiple appliances and storage systems by introducing a behavior-driven price-based MPC for residential building energy management systems. Extensive testing on a large distribution network showed significant reductions in nodal voltage drop and peak loads with a 21% reduction in generation costs. Fan et al. [149] proposed a collaborative control optimization approach, as shown in Figure 16, for grid-connected NZEBs to improve their performance at the building group level. This approach used a game theoretic framework to coordinate the control of multiple NZEBs in order to maximize their overall performance. The results showed that the proposed approach can reduce peak demand, improve grid stability, and increase the utilization of renewable energy sources. Clastres et al. [150] proposed a mathematical model for optimizing the operation of a domestic PV system to provide ancillary services to the electricity grid. The model considered the intermittent nature of solar generation, the cost of electricity, and the value of ancillary services. It was found that PV systems can provide significant ancillary services to the grid, such as frequency regulation and voltage control. This could lead to increased reliability of the grid, reduced need for conventional generators, and lower electricity prices.

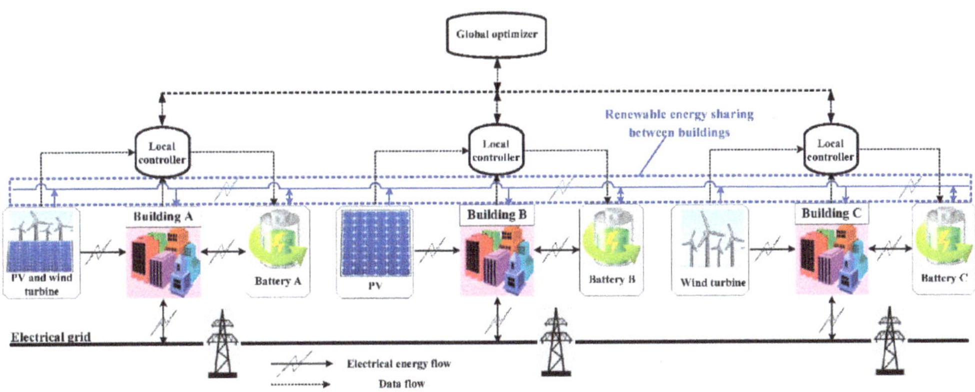

Figure 16. Illustration of a collaborative control approach [149].

Overall, these studies demonstrated the need for grid-integrated control for optimal demand response of buildings. Such control will be a critical aspect to support grid operation through the utilization of building integrated distributed energy sources. Further research is still needed to fully explore the benefits of grid-integrated control of buildings.

5. Conclusions, Future Direction, and Barriers

This paper provided a review of several emerging and sustainable technologies that can help achieve increased energy savings and reduced carbon emissions from buildings. These technologies ranged from onsite energy generation, thermal energy storage, heat pumps, and thermal energy sharing to advanced building design, retrofits, and data-driven modeling as well as improved building control and grid-buildings integrated control. Each technology showcased the potential to improve operating efficiency and increase demand flexibility to significantly reduce building energy consumption and carbon emissions while supporting grid operation.

However, how to rationally select technology combinations to achieve low energy operation is one of the key challenges during the design stage as this process is highly impacted by factors such as available budget, design objectives, and technological complexity. The uncertainty-based design has been considered in recent studies and has demonstrated its advances in improved robustness and increased resilience of the designed systems for buildings if circular economy principles are also considered during the design stage. To achieve building decarbonization, renewable energy and storage technologies are among the main solutions that are often considered, and they can potentially help take buildings off the grid. As a significant amount of electricity used in buildings is for heating and cooling, thermal energy recovery and energy sharing will play a key role in the development of smart communities and smart cities. Heat pumps will continue to be the main technology to provide heating and cooling to buildings in the coming decades. Research on the appropriate integration of heat pumps with renewable energy systems and energy storage technologies is needed.

For a particular application, the selection of appropriate energy technologies is quite complex, subject to the impact of many factors such as technology maturity, market availability, interoperability, scalability, applicability, long-term viability, and carbon reduction and energy reduction potential. A detailed cost-benefit analysis is often desirable to assess the short and long-term economic, social, and environmental impacts, enabling businesses and policymakers to make informed decisions. However, there often lacks sufficient historical data for sustainable energy technologies, making it challenging to estimate their future costs and benefits accurately. Moreover, the rapid development of technological advancements can make cost-benefit analyses quickly outdated, necessitating regular technology reviews and updates to ensure analyses remain reflective of the latest advancements and economic conditions. Hence, such analyses should also emphasize advanced location-specific assessments that consider regional variations in energy costs, incentives, and climate conditions. Expanding analyses to include life-cycle assessments can help assess the full environmental impact of technologies, including manufacturing, transportation, and disposal, enabling the identification of more sustainable options and leading to a circular economy.

On the technological side, exploring integrated technological solutions and models, in which multiple technologies can work synergistically to maximize their energy savings and carbon reduction, presents an exciting direction for research and implementation. However, the diversity of building types, sizes, and locations requires adaptable models to overcome these variations and provide generic assessments. High upfront costs associated with technologies like energy-efficient retrofits can impact their adoption and deployment, and thus, innovative financing models will be vital in lowering these initial investment barriers. Furthermore, proactive maintenance and monitoring strategies are required to optimize the system to ensure the long-term cost-effectiveness of these technologies.

To maximize the benefits of using these emerging and sustainable technologies in buildings, appropriate control, including building-level control and grid-buildings integrated control, is needed. Such control should have embedded intelligence and can enable the seamless optimization of building energy systems for increased demand flexibility and reduced operating costs while, at the same time, supporting flexible grid services with increased resilience. With massive data readily available from buildings, data-driven prediction and data-driven control will offer new opportunities to develop more advanced control solutions. Moreover, reinforcement learning and model predictive control can optimize building energy consumption through adaptive learning and near-optimal prediction, respectively. A combination of both controllers with embedded intelligence can enable buildings to continually learn, adapt, and operate in an energy-efficient manner. Grid-integrated control of buildings is predicted to be one of the leading solutions to support renewable energy-powered grids.

While numerous advancements have been made in recent years, ongoing research and development efforts remain crucial to improve the performance and effectiveness of these emerging and sustainable technologies and unlock their full potential in practical applications. The scalability and affordability of these technologies must be considered to ensure that they are accessible to a diverse range of building types and climatic conditions worldwide. In summary, by embracing and actively integrating these emerging and sustainable technologies into building design and operation, grid-responsive net zero energy buildings can be achieved.

Author Contributions: Conceptualization, Z.M. and M.B.A.; methodology, Z.M. and M.B.A.; resources, Z.M., M.B.A., M.L., S.L., M.S.A., X.Z., H.D., X.S. and Y.L.; writing—original draft preparation, Z.M., M.B.A., M.L., S.L., M.S.A., X.Z., H.D., X.S. and Y.L.; writing—review and editing, Z.M. and M.B.A.; supervision, Z.M.; All authors have read and agreed to the published version of the manuscript.

Funding: This research received no external funding.

Data Availability Statement: Not applicable to this review article.

Conflicts of Interest: The authors declare no conflict of interest.

References

1. Embodied Carbon—World Green Building Council. Available online: https://worldgbc.org/advancing-net-zero/embodied-carbon/ (accessed on 28 September 2023).
2. Laski, J.; Burrows, V. *From Thousands to Billions. Coordinated Action towards 100% Net Zero Carbon Buildings By 2050*; World Green Building Council: London, UK, 2017.
3. Czerwinska, D. Green Building: Improving the Lives of Billions by Helping to Achieve the UN Sustainable Development Goals —World Green Building Council. Available online: https://worldgbc.org/article/green-building-improving-the-lives-of-billions-by-helping-to-achieve-the-un-sustainable-development-goals/ (accessed on 3 July 2023).
4. Jensen, S.Ø.; Marszal-Pomianowska, A.; Lollini, R.; Pasut, W.; Knotzer, A.; Engelmann, P.; Stafford, A.; Reynders, G. IEA EBC Annex 67 Energy Flexible Buildings. *Energy Build.* **2017**, *155*, 25–34. [CrossRef]
5. Sustainable Buildings—The United Nations Environment Programme. Available online: https://www.unep.org/explore-topics/resource-efficiency/what-we-do/cities/sustainable-buildings (accessed on 3 July 2023).
6. Wei, W.; Skye, H.M. Residential Net-Zero Energy Buildings: Review and Perspective. *Renew. Sustain. Energy Rev.* **2021**, *142*, 110859.
7. Souley Agbodjan, Y.; Wang, J.; Cui, Y.; Liu, Z.; Luo, Z. Bibliometric Analysis of Zero Energy Building Research, Challenges and Solutions. *Sol. Energy* **2022**, *244*, 414–433. [CrossRef]
8. Sartori, I.; Napolitano, A.; Voss, K. Net Zero Energy Buildings: A Consistent Definition Framework. *Energy Build.* **2012**, *48*, 220–232. [CrossRef]
9. Sami, S.; Gholizadeh, M.; Dadpour, D.; Deymi-Dashtebayaz, M. Design and Optimization of a CCHDP System Integrated with NZEB from Energy, Exergy and Exergoeconomic Perspectives. *Energy Convers. Manag.* **2022**, *271*, 116347. [CrossRef]
10. Chegari, B.; Tabaa, M.; Simeu, E.; Moutaouakkil, F.; Medromi, H. An Optimal Surrogate-Model-Based Approach to Support Comfortable and Nearly Zero Energy Buildings Design. *Energy* **2022**, *248*, 123584. [CrossRef]
11. Djunaedy, E.; Van Den Wymelenberg, K.; Acker, B.; Thimmana, H. Oversizing of HVAC System: Signatures and Penalties. *Energy Build.* **2011**, *43*, 468–475. [CrossRef]
12. Wang, S.K. *Handbook of Air Conditioning and Refrigeration*, 2nd ed.; McGraw-Hill Education: New York, NY, USA, 2001.
13. Woradechjumroen, D.; Yu, Y.; Li, H.; Yu, D.; Yang, H. Analysis of HVAC System Oversizing in Commercial Buildings through Field Measurements. *Energy Build.* **2014**, *69*, 131–143. [CrossRef]
14. Kang, J.; Wang, S. Robust Optimal Design of Distributed Energy Systems Based on Life-Cycle Performance Analysis Using a Probabilistic Approach Considering Uncertainties of Design Inputs and Equipment Degradations. *Appl. Energy* **2018**, *231*, 615–627. [CrossRef]
15. Li, H.; Wang, S. Coordinated Robust Optimal Design of Building Envelope and Energy Systems for Zero/Low Energy Buildings Considering Uncertainties. *Appl. Energy* **2020**, *265*, 114779. [CrossRef]
16. Zhang, S.; Sun, Y.; Cheng, Y.; Huang, P.; Oladokun, M.O.; Lin, Z. Response-Surface-Model-Based System Sizing for Nearly/Net Zero Energy Buildings under Uncertainty. *Appl. Energy* **2018**, *228*, 1020–1031. [CrossRef]
17. Zou, B.; Peng, J.; Yin, R.; Li, H.; Li, S.; Yan, J.; Yang, H. Capacity Configuration of Distributed Photovoltaic and Battery System for Office Buildings Considering Uncertainties. *Appl. Energy* **2022**, *319*, 119243. [CrossRef]
18. Huang, P.; Huang, G.; Sun, Y. Uncertainty-Based Life-Cycle Analysis of near-Zero Energy Buildings for Performance Improvements. *Appl. Energy* **2018**, *213*, 486–498. [CrossRef]

19. Shen, L.; Sun, Y. Performance Comparisons of Two System Sizing Approaches for Net Zero Energy Building Clusters under Uncertainties. *Energy Build.* **2016**, *127*, 10–21. [CrossRef]
20. Huang, P.; Lovati, M.; Zhang, X.; Bales, C.; Hallbeck, S.; Becker, A.; Bergqvist, H.; Hedberg, J.; Maturi, L. Transforming a Residential Building Cluster into Electricity Prosumers in Sweden: Optimal Design of a Coupled PV-Heat Pump-Thermal Storage-Electric Vehicle System. *Appl. Energy* **2019**, *255*, 113864. [CrossRef]
21. Fan, G.; Liu, Z.; Liu, X.; Shi, Y.; Wu, D.; Guo, J.; Zhang, S.; Yang, X.; Zhang, Y. Two-Layer Collaborative Optimization for a Renewable Energy System Combining Electricity Storage, Hydrogen Storage, and Heat Storage. *Energy* **2022**, *259*, 125047. [CrossRef]
22. Guo, J.; Zhang, P.; Wu, D.; Liu, Z.; Liu, X.; Zhang, S.; Yang, X.; Ge, H. Multi-Objective Optimization Design and Multi-Attribute Decision-Making Method of a Distributed Energy System Based on Nearly Zero-Energy Community Load Forecasting. *Energy* **2022**, *239*, 122124. [CrossRef]
23. Zhang, S.; Huang, P.; Sun, Y. A Multi-Criterion Renewable Energy System Design Optimization for Net Zero Energy Buildings under Uncertainties. *Energy* **2016**, *94*, 654–665. [CrossRef]
24. Doroudchi, E.; Alanne, K.; Okur, Ö.; Kyyrä, J.; Lehtonen, M. Approaching Net Zero Energy Housing through Integrated EV. *Sustain. Cities Soc.* **2018**, *38*, 534–542. [CrossRef]
25. Liu, Z.; Li, Y.; Fan, G.; Wu, D.; Guo, J.; Jin, G.; Zhang, S.; Yang, X. Co-Optimization of a Novel Distributed Energy System Integrated with Hybrid Energy Storage in Different Nearly Zero Energy Community Scenarios. *Energy* **2022**, *247*. [CrossRef]
26. Mohammadi, F.; Faghihi, F.; Kazemi, A.; Salemi, A.H. The Effect of Multi -Uncertainties on Battery Energy Storage System Sizing in Smart Homes. *J. Energy Storage* **2022**, *52*, 104765. [CrossRef]
27. Park, M.; Wang, Z.; Li, L.; Wang, X. Multi-Objective Building Energy System Optimization Considering EV Infrastructure. *Appl. Energy* **2023**, *332*, 120504. [CrossRef]
28. Liu, Z.; Guo, J.; Li, Y.; Wu, D.; Zhang, S.; Yang, X.; Ge, H.; Cai, Z. Multi-Scenario Analysis and Collaborative Optimization of a Novel Distributed Energy System Coupled with Hybrid Energy Storage for a Nearly Zero-Energy Community. *J. Energy Storage* **2021**, *41*, 102992. [CrossRef]
29. Elkadeem, M.R.; Abido, M.A. Optimal Planning and Operation of Grid-Connected PV/CHP/Battery Energy System Considering Demand Response and Electric Vehicles for a Multi-Residential Complex Building. *J. Energy Storage* **2023**, *72*, 108198. [CrossRef]
30. International Renewable Energy Agency. *Global Energy Transformation: A Roadmap to 2050*; International Renewable Energy Agency: Masdar City, United Arab Emirates, 2018.
31. IEA. Funding for Thin Film Technologies for Solar PV—Policies. Available online: https://www.iea.org/policies/13275-funding-for-thin-film-technologies-for-solar-pv (accessed on 9 September 2023).
32. Kuşkaya, S.; Bilgili, F.; Muğaloğlu, E.; Khan, K.; Hoque, M.E.; Toguç, N. The Role of Solar Energy Usage in Environmental Sustainability: Fresh Evidence through Time-Frequency Analyses. *Renew. Energy* **2023**, *206*, 858–871. [CrossRef]
33. Statista. Grid-Connected Solar PV Capacity by Select Country. Available online: https://www.statista.com/statistics/665864/solar-capacity-in-selected-countries-by-grid-connection/ (accessed on 28 September 2023).
34. Bosu, I.; Mahmoud, H.; Ookawara, S.; Hassan, H. Applied Single and Hybrid Solar Energy Techniques for Building Energy Consumption and Thermal Comfort: A Comprehensive Review. *Sol. Energy* **2023**, *259*, 188–228. [CrossRef]
35. Kwok, K.C.S.; Hu, G. Wind Energy System for Buildings in an Urban Environment. *J. Wind Eng. Ind. Aerodyn.* **2023**, *234*, 105349. [CrossRef]
36. Arnfield, A.J. Two Decades of Urban Climate Research: A Review of Turbulence, Exchanges of Energy and Water, and the Urban Heat Island. *Int. J. Climatol.* **2003**, *23*, 1–26. [CrossRef]
37. NBC10 Philadelphia. What Happened to the Wind Turbines that Twirled above Philadelphia Eagles' Lincoln Financial Field? Available online: https://www.nbcphiladelphia.com/news/sports/nfl/philadelphia-eagles/what-happened-to-the-wind-turbines-that-twirled-above-philadelphia-eagles-lincoln-financial-field/169875/ (accessed on 30 August 2023).
38. Zhu, Y.; Li, W.; Li, J.; Li, H.; Wang, Y.; Li, S. Thermodynamic Analysis and Economic Assessment of Biomass-Fired Organic Rankine Cycle Combined Heat and Power System Integrated with CO_2 Capture. *Energy Convers. Manag.* **2020**, *204*, 112310. [CrossRef]
39. Behzadi, A.; Thorin, E.; Duwig, C.; Sadrizadeh, S. Supply-Demand Side Management of a Building Energy System Driven by Solar and Biomass in Stockholm: A Smart Integration with Minimal Cost and Emission. *Energy Convers. Manag.* **2023**, *292*, 117420. [CrossRef]
40. Jiang, T.; Zhang, Y.; Olayiwola, S.; Lau, C.K.; Fan, M.; Ng, K.; Tan, G. Biomass-Derived Porous Carbons Support in Phase Change Materials for Building Energy Efficiency: A Review. *Mater. Today Energy* **2022**, *23*, 100905. [CrossRef]
41. Lancaster University. Lancaster Hydrogen Hub. Available online: https://www.lancaster.ac.uk/energy-lancaster/research/hydrogen-hub/ (accessed on 30 August 2023).
42. Wu, Y.; Zhong, L. An Integrated Energy Analysis Framework for Evaluating the Application of Hydrogen-Based Energy Storage Systems in Achieving Net Zero Energy Buildings and Cities in Canada. *Energy Convers. Manag.* **2023**, *286*, 117066. [CrossRef]
43. International Renewable Energy Agency. *International Renewable Energy Agency Innovation Outlook: Thermal Energy Storage*; International Renewable Energy Agency: Abu Dhabi, United Arab Emirates, 2020; p. 144.
44. Tunçbilek, E.; Yıldız, Ç.; Arıcı, M.; Ma, Z.; Awan, M.B. Thermal Energy Storage for Enhanced Building Energy Flexibility. *Build. Energy Flex. Demand Manag.* **2023**, 89–119. [CrossRef]

45. Faisal Ahmed, S.; Khalid, M.; Vaka, M.; Walvekar, R.; Numan, A.; Khaliq Rasheed, A.; Mujawar Mubarak, N. Recent Progress in Solar Water Heaters and Solar Collectors: A Comprehensive Review. *Therm. Sci. Eng. Prog.* **2021**, *25*, 100981. [CrossRef]
46. Zhan, H.; Mahyuddin, N.; Sulaiman, R.; Khayatian, F. Phase Change Material (PCM) Integrations into Buildings in Hot Climates with Simulation Access for Energy Performance and Thermal Comfort: A Review. *Constr. Build. Mater.* **2023**, *397*, 132312. [CrossRef]
47. Asadi, I.; Baghban, M.H.; Hashemi, M.; Izadyar, N.; Sajadi, B. Phase Change Materials Incorporated into Geopolymer Concrete for Enhancing Energy Efficiency and Sustainability of Buildings: A Review. *Case Stud. Constr. Mater.* **2022**, *17*, e01162. [CrossRef]
48. Li, C.; Wen, X.; Cai, W.; Yu, H.; Liu, D. Phase Change Material for Passive Cooling in Building Envelopes: A Comprehensive Review. *J. Build. Eng.* **2023**, *65*, 105763. [CrossRef]
49. Qiao, X.; Kong, X.; Fan, M. Phase Change Material Applied in Solar Heating for Buildings: A Review. *J. Energy Storage* **2022**, *55*. [CrossRef]
50. Schmerse, E.; Ikutegbe, C.A.; Auckaili, A.; Farid, M.M. Using PCM in Two Proposed Residential Buildings in Christchurch, New Zealand. *Energies* **2020**, *13*, 6025. [CrossRef]
51. Gao, Y.; Meng, X. A Comprehensive Review of Integrating Phase Change Materials in Building Bricks: Methods, Performance and Applications. *J. Energy Storage* **2023**, *62*, 106913. [CrossRef]
52. Awan, M.B.; Ma, Z. Building Energy Flexibility: Definitions, Sources, Indicators, and Quantification Methods. *Build. Energy Flex. Demand Manag.* **2023**, 17–40. [CrossRef]
53. Rosenow, J.; Gibb, D.; Nowak, T.; Lowes, R. Heating up the Global Heat Pump Market. *Nat. Energy* **2022**, *7*, 901–904. [CrossRef]
54. IEA. Installation of about 600 Million Heat Pumps Covering 20% of Buildings Heating Needs Required by 2030—Analysis. Available online: https://www.iea.org/reports/installation-of-about-600-million-heat-pumps-covering-20-of-buildings-heating-needs-required-by-2030 (accessed on 13 August 2023).
55. IEA. *The Future of Heat Pumps*; IEA: Paris, France, 2022.
56. Energy Efficiency Council. *Harnessing Heat Pumps for Net Zero The Role of Heat Pumps in Saving Energy and Cutting Emissions*; Energy Efficiency Council: Melbourne, Australia, 2023.
57. Gaur, A.S.; Fitiwi, D.Z.; Curtis, J. Heat Pumps and Our Low-Carbon Future: A Comprehensive Review. *Energy Res. Soc. Sci.* **2021**, *71*, 101764. [CrossRef]
58. Ma, Z.; Xia, L.; Gong, X.; Kokogiannakis, G.; Wang, S.; Zhou, X. Recent Advances and Development in Optimal Design and Control of Ground Source Heat Pump Systems. *Renew. Sustain. Energy Rev.* **2020**, *131*, 110001. [CrossRef]
59. Cai, W.; Wang, F.; Chen, S.; Chen, C.; Zhang, Y.; Kolditz, O.; Shao, H. Importance of Long-Term Ground-Loop Temperature Variation in Performance Optimization of Ground Source Heat Pump System. *Appl. Therm. Eng.* **2022**, *204*, 117945. [CrossRef]
60. Lee, M.; Cha, D.; Yun, S.; Yoon, S.M.; Kim, Y. Comparative Heating Performance Evaluation of Hybrid Ground-Source Heat Pumps Using Serial and Parallel Configurations with the Application of Ground Heat Exchanger. *Energy Convers. Manag.* **2021**, *229*, 113743. [CrossRef]
61. Bae, S.; Nam, Y. Economic and Environmental Analysis of Ground Source Heat Pump System According to Operation Methods. *Geothermics* **2022**, *101*, 102373. [CrossRef]
62. Lee, M.; Kim, J.; Shin, H.H.; Cho, W.; Kim, Y. CO_2 Emissions and Energy Performance Analysis of Ground-Source and Solar-Assisted Ground-Source Heat Pumps Using Low-GWP Refrigerants. *Energy* **2022**, *261*, 125198. [CrossRef]
63. Xu, L.; Pu, L.; Zhang, S.; Li, Y. Hybrid Ground Source Heat Pump System for Overcoming Soil Thermal Imbalance: A Review. *Sustain. Energy Technol. Assess.* **2021**, *44*, 101098. [CrossRef]
64. Abdalla, A.; Mohamed, S.; Bucking, S.; Cotton, J.S. Modeling of Thermal Energy Sharing in Integrated Energy Communities with Micro-Thermal Networks. *Energy Build.* **2021**, *248*, 111170. [CrossRef]
65. Lindhe, J.; Javed, S.; Johansson, D.; Bagge, H. A Review of the Current Status and Development of 5GDHC and Characterization of a Novel Shared Energy System. *Sci. Technol. Built Environ.* **2022**, *28*, 595–609. [CrossRef]
66. Huang, P.; Copertaro, B.; Zhang, X.; Shen, J.; Löfgren, I.; Rönnelid, M.; Fahlen, J.; Andersson, D.; Svanfeldt, M. A Review of Data Centers as Prosumers in District Energy Systems: Renewable Energy Integration and Waste Heat Reuse for District Heating. *Appl. Energy* **2020**, *258*, 114109. [CrossRef]
67. Ebrahimi, K.; Jones, G.F.; Fleischer, A.S. A Review of Data Center Cooling Technology, Operating Conditions and the Corresponding Low-Grade Waste Heat Recovery Opportunities. *Renew. Sustain. Energy Rev.* **2014**, *31*, 622–638. [CrossRef]
68. Giunta, F.; Sawalha, S. Techno-Economic Analysis of Heat Recovery from Supermarket's CO_2 Refrigeration Systems to District Heating Networks. *Appl. Therm. Eng.* **2021**, *193*, 117000. [CrossRef]
69. Mateu-Royo, C.; Sawalha, S.; Mota-Babiloni, A.; Navarro-Esbrí, J. High Temperature Heat Pump Integration into District Heating Network. *Energy Convers. Manag.* **2020**, *210*, 112719. [CrossRef]
70. Reclaim Waste Heat from Rink—Arena Guide. Available online: https://arena-guide.com/go-green/heat-re-claim-heat-recovery/ (accessed on 26 June 2023).
71. Pourfarzad, H.; Saremia, M.; Ganjali, M.R. A Novel Tri-Generation Energy System Integrating Solar Energy and Industrial Waste Heat. *J. Therm. Eng.* **2021**, *7*, 1067–1078. [CrossRef]
72. Davies, G.F.; Maidment, G.G.; Tozer, R.M. Using Data Centres for Combined Heating and Cooling: An Investigation for London. *Appl. Therm. Eng.* **2016**, *94*, 296–304. [CrossRef]

73. Khosravi, A.; Laukkanen, T.; Vuorinen, V.; Syri, S. Waste Heat Recovery from a Data Centre and 5G Smart Poles for Low-Temperature District Heating Network. *Energy* **2021**, *218*, 119468. [CrossRef]
74. Kuyumcu, M.E.; Tutumlu, H.; Yumrutaş, R. Performance of a Swimming Pool Heating System by Utilizing Waste Energy Rejected from an Ice Rink with an Energy Storage Tank. *Energy Convers. Manag.* **2016**, *121*, 349–357. [CrossRef]
75. Oró, E.; Allepuz, R.; Martorell, I.; Salom, J. Design and Economic Analysis of Liquid Cooled Data Centres for Waste Heat Recovery: A Case Study for an Indoor Swimming Pool. *Sustain. Cities Soc.* **2018**, *36*, 185–203. [CrossRef]
76. Pan, Q.; Peng, J.; Wang, R. Experimental Study of an Adsorption Chiller for Extra Low Temperature Waste Heat Utilization. *Appl. Therm. Eng.* **2019**, *163*, 114341. [CrossRef]
77. Araya, S.; Wemhoff, A.P.; Jones, G.F.; Fleischer, A.S. Study of a Lab-Scale Organic Rankine Cycle for the Ultra-Low-Temperature Waste Heat Recovery Associated with Data Centers. *J. Electron. Packag. Trans. ASME* **2021**, *143*, 021001. [CrossRef]
78. Wirtz, M.; Kivilip, L.; Remmen, P.; Müller, D. 5th Generation District Heating: A Novel Design Approach Based on Mathematical Optimization. *Appl. Energy* **2020**, *260*, 114158. [CrossRef]
79. Murphy, A.R.; Fung, A.S. Techno-Economic Study of an Energy Sharing Network Comprised of a Data Centre and Multi-Unit Residential Buildings for Cold Climate. *Energy Build.* **2019**, *186*, 261–275. [CrossRef]
80. Zhang, C.; Luo, H.; Wang, Z. An Economic Analysis of Waste Heat Recovery and Utilization in Data Centers Considering Environmental Benefits. *Sustain. Prod. Consum.* **2022**, *31*, 127–138. [CrossRef]
81. Li, H.; Hou, J.; Hong, T.; Ding, Y.; Nord, N. Energy, Economic, and Environmental Analysis of Integration of Thermal Energy Storage into District Heating Systems Using Waste Heat from Data Centres. *Energy* **2021**, *219*. [CrossRef]
82. Li, H.; Hou, J.; Tian, Z.; Hong, T.; Nord, N.; Rohde, D. Optimize Heat Prosumers' Economic Performance under Current Heating Price Models by Using Water Tank Thermal Energy Storage. *Energy* **2022**, *239*, 122103. [CrossRef]
83. Wang, X.; Li, H.; Wang, Y.; Zhao, J.; Zhu, J.; Zhong, S.; Li, Y. Energy, Exergy, and Economic Analysis of a Data Center Energy System Driven by the CO_2 Ground Source Heat Pump: Prosumer Perspective. *Energy Convers. Manag.* **2021**, *232*, 113877. [CrossRef]
84. Energy Retrofit Systems Market Size, Share, Forecast. 2023. Available online: https://www.marketresearchfuture.com/reports/energy-retrofit-systems-market-11758 (accessed on 13 August 2023).
85. Ang, Y.Q.; Berzolla, Z.M.; Letellier-Duchesne, S.; Reinhart, C.F. Carbon Reduction Technology Pathways for Existing Buildings in Eight Cities. *Nat. Commun.* **2023**, *14*, 1689. [CrossRef]
86. Alabid, J.; Bennadji, A.; Seddiki, M. A Review on the Energy Retrofit Policies and Improvements of the UK Existing Buildings, Challenges and Benefits. *Renew. Sustain. Energy Rev.* **2022**, *159*, 112161. [CrossRef]
87. Deb, C.; Schlueter, A. Review of Data-Driven Energy Modelling Techniques for Building Retrofit. *Renew. Sustain. Energy Rev.* **2021**, *144*, 110990. [CrossRef]
88. Thrampoulidis, E.; Hug, G.; Orehounig, K. Approximating Optimal Building Retrofit Solutions for Large-Scale Retrofit Analysis. *Appl. Energy* **2023**, *333*, 120566. [CrossRef]
89. Densley Tingley, D. Embed Circular Economy Thinking into Building Retrofit. *Commun. Eng.* **2022**, *1*, 28. [CrossRef]
90. Xu, Y.; Zhou, Y.; Sekula, P.; Ding, L. Machine Learning in Construction: From Shallow to Deep Learning. *Dev. Built Environ.* **2021**, *6*, 100045. [CrossRef]
91. Janiesch, C.; Zschech, P.; Heinrich, K. Machine Learning and Deep Learning. *Electron. Mark.* **2021**, *31*, 685–695. [CrossRef]
92. Wang, Z.; Hong, T.; Piette, M.A. Building Thermal Load Prediction through Shallow Machine Learning and Deep Learning. *Appl. Energy* **2020**, *263*. [CrossRef]
93. Tien, P.W.; Wei, S.; Darkwa, J.; Wood, C.; Calautit, J.K. Machine Learning and Deep Learning Methods for Enhancing Building Energy Efficiency and Indoor Environmental Quality—A Review. *Energy AI* **2022**, *10*, 100198. [CrossRef]
94. Muzaffar, S.; Afshari, A. Short-Term Load Forecasts Using LSTM Networks. *Energy Procedia* **2019**, *158*, 2922–2927. [CrossRef]
95. Zhou, Y.; Wang, J.; Liu, Y.; Yan, R.; Ma, Y. Incorporating Deep Learning of Load Predictions to Enhance the Optimal Active Energy Management of Combined Cooling, Heating and Power System. *Energy* **2021**, *233*, 121134. [CrossRef]
96. Guo, W.; Che, L.; Shahidehpour, M.; Wan, X. Machine-Learning Based Methods in Short-Term Load Forecasting. *Electr. J.* **2021**, *34*, 106884. [CrossRef]
97. Feng, C.; Zhang, J.; Zhang, W.; Hodge, B.M. Convolutional Neural Networks for Intra-Hour Solar Forecasting Based on Sky Image Sequences. *Appl. Energy* **2022**, *310*, 118438. [CrossRef]
98. Zhang, W.; Zhou, H.; Bao, X.; Cui, H. Outlet Water Temperature Prediction of Energy Pile Based on Spatial-Temporal Feature Extraction through CNN–LSTM Hybrid Model. *Energy* **2023**, *264*, 126190. [CrossRef]
99. Liang, X.; Chen, S.; Zhu, X.; Jin, X.; Du, Z. Domain Knowledge Decomposition of Building Energy Consumption and a Hybrid Data-Driven Model for 24-h Ahead Predictions. *Appl. Energy* **2023**, *344*. [CrossRef]
100. Zhou, X.; Lin, W.; Kumar, R.; Cui, P.; Ma, Z. A Data-Driven Strategy Using Long Short Term Memory Models and Reinforcement Learning to Predict Building Electricity Consumption. *Appl. Energy* **2022**, *306*, 118078. [CrossRef]
101. Zhou, X.; Du, H.; Sun, Y.; Ren, H.; Cui, P.; Ma, Z. A New Framework Integrating Reinforcement Learning, a Rule-Based Expert System, and Decision Tree Analysis to Improve Building Energy Flexibility. *J. Build. Eng.* **2023**, *71*, 106536. [CrossRef]
102. Li, J.; Niu, H.; Meng, F.; Li, R. Prediction of Short-Term Photovoltaic Power Via Self-Attention-Based Deep Learning Approach. *J. Energy Resour. Technol. Trans. ASME* **2022**, *144*, 101301. [CrossRef]

103. Gangopadhyay, T.; Tan, S.Y.; Jiang, Z.; Sarkar, S. Interpretable Deep Attention Model for Multivariate Time Series Prediction in Building Energy Systems. In *Lecture Notes in Computer Science*; Including Subseries Lecture Notes in Artificial Intelligence and Lecture Notes in Bioinformatics; Springer Nature: Cham, Switzerland, 2020; Volume 12312 LNCS, pp. 93–101.
104. Li, D.; Li, J.; Zeng, X.; Stankovic, V.; Stankovic, L.; Xiao, C.; Shi, Q. Transfer Learning for Multi-Objective Non-Intrusive Load Monitoring in Smart Building. *Appl. Energy* **2023**, *329*, 120223. [CrossRef]
105. Chen, L.; Ermis, A.; Meng, F.; Zhang, Y. Meta-Learning of Personalized Thermal Comfort Model and Fast Identification of the Best Personalized Thermal Environmental Conditions. *Build. Environ.* **2023**, *235*, 110201. [CrossRef]
106. Moon, J.; Jung, S.; Park, S.; Hwang, E. Conditional Tabular GaN-Based Two-Stage Data Generation Scheme for Short-Term Load Forecasting. *IEEE Access* **2020**, *8*, 205327–205339. [CrossRef]
107. Ma, Z.; Arıcı, M.; Shahsavar, A. *Building Energy Flexibilty and Demand Management*; Elsevier: Amsterdam, The Netherlands, 2023; ISBN 9780323995894.
108. Satchwell, A.; Ann Piette, M.; Khandekar, A.; Granderson, J.; Mims Frick, N.; Hledik, R.; Faruqui, A.; Lam, L.; Ross, S.; Cohen, J.; et al. *A National Roadmap for Grid-Interactive Efficient Buildings*; Lawrence Berkeley National Lab: Berkeley, CA, USA, 2021.
109. Verbeke, S.; Aerts, D.; Reynders, G.; Ma, Y.; Waide, P. *Final Report on the Technical Support to the Development of A Smart Readiness Indicator for Buildings*; European Commission: Brussels, Belgium, 2020.
110. Munankarmi, P.; Maguire, J.; Balamurugan, S.P.; Blonsky, M.; Roberts, D.; Jin, X. Community-Scale Interaction of Energy Efficiency and Demand Flexibility in Residential Buildings. *Appl. Energy* **2021**, *298*, 117149. [CrossRef]
111. Ren, H.; Sun, Y.; Albdoor, A.K.; Tyagi, V.V.; Pandey, A.K.; Ma, Z. Improving Energy Flexibility of a Net-Zero Energy House Using a Solar-Assisted Air Conditioning System with Thermal Energy Storage and Demand-Side Management. *Appl. Energy* **2021**, *285*. [CrossRef]
112. Awan, M.B.; Sun, Y.; Lin, W.; Ma, Z. A Framework to Formulate and Aggregate Performance Indicators to Quantify Building Energy Flexibility. *Appl. Energy* **2023**, *349*, 121590. [CrossRef]
113. Jensen, S.Ø.; Marszal, A.J.; Johra, H.; Weiss, T.; Knotzer, A.; Kazmi, H.; Vigna, I.; Pernetti, R.; Le Dréau, J.; Zhang, K.; et al. *Characterization of Energy Flexibility in Buildings*; Energy in Buildings and Communities Programme Annex 67 Energy Flexible Buildings; International Energy Agency: Paris, France, 2019.
114. Langner, R.; Granderson, J.; Crowe, E. *Quantifying the Value of Grid-Interactive Efficient Buildings through Field Study*; National Renewable Energy Lab: Golden, CO, USA, 2022.
115. Ruggiero, S.; Francesca De Masi, R.; Assimakopoulos, M.N.; Peter Vanoli, G. Energy Saving through Building Automation Systems: Experimental and Numerical Study of a Smart Glass with Liquid Crystal and Its Control Logics in Summertime. *Energy Build.* **2022**, *273*, 112403. [CrossRef]
116. Langevin, J.; Harris, C.B.; Satre-Meloy, A.; Chandra-Putra, H.; Speake, A.; Present, E.; Adhikari, R.; Wilson, E.J.H.; Satchwell, A.J. US Building Energy Efficiency and Flexibility as an Electric Grid Resource. *Joule* **2021**, *5*, 2102–2128. [CrossRef]
117. Ochs, M. How Lighting Control Systems Contribute to Flexible, Future-Proof Buildings. Available online: https://www.csemag.com/articles/how-lighting-control-systems-contribute-to-flexible-future-proof-buildings/ (accessed on 12 September 2023).
118. Tang, H.; Wang, S. Game-Theoretic Optimization of Demand-Side Flexibility Engagement Considering the Perspectives of Different Stakeholders and Multiple Flexibility Services. *Appl. Energy* **2023**, *332*, 120550. [CrossRef]
119. Zhou, Y.; Wang, J.; Dong, F.; Qin, Y.; Ma, Z.; Ma, Y.; Li, J. Novel Flexibility Evaluation of Hybrid Combined Cooling, Heating and Power System with an Improved Operation Strategy. *Appl. Energy* **2021**, *300*, 117358. [CrossRef]
120. Finck, C.; Li, R.; Zeiler, W. Optimal Control of Demand Flexibility under Real-Time Pricing for Heating Systems in Buildings: A Real-Life Demonstration. *Appl. Energy* **2020**, *263*, 114671. [CrossRef]
121. Afroz, Z.; Shafiullah, G.M.; Urmee, T.; Higgins, G. Modeling Techniques Used in Building HVAC Control Systems: A Review. *Renew. Sustain. Energy Rev.* **2018**, *83*, 64–84. [CrossRef]
122. Fiorentini, M.; Wall, J.; Ma, Z.; Braslavsky, J.H.; Cooper, P. Hybrid model predictive control of a residential HVAC system with on-site thermal energy generation and storage. *Appl. Energy* **2017**, *187*, 465–479. [CrossRef]
123. Merema, B.; Saelens, D.; Breesch, H. Demonstration of an MPC Framework for All-Air Systems in Non-Residential Buildings. *Build. Environ.* **2022**, *217*, 109053. [CrossRef]
124. Wei, T.; Wang, Y.; Zhu, Q. Deep Reinforcement Learning for Building HVAC Control. In Proceedings of the Design Automation Conference, Austin, TX, USA, 18 June 2017; Institute of Electrical and Electronics Engineers Inc.: Piscataway, NJ, USA, 2017; Volume 12828.
125. Azuatalam, D.; Lee, W.L.; de Nijs, F.; Liebman, A. Reinforcement Learning for Whole-Building HVAC Control and Demand Response. *Energy AI* **2020**, *2*, 100020. [CrossRef]
126. Yu, L.; Sun, Y.; Xu, Z.; Shen, C.; Yue, D.; Jiang, T.; Guan, X. Multi-Agent Deep Reinforcement Learning for HVAC Control in Commercial Buildings. *IEEE Trans. Smart Grid* **2020**, *12*, 407–419. [CrossRef]
127. Coulson, J.; Lygeros, J.; Dorfler, F. Data-Enabled Predictive Control: In the Shallows of the DeePC. In Proceedings of the 2019 18th European Control Conference, ECC 2019, Naples, Italy, 25–28 June 2019; pp. 307–312. [CrossRef]
128. Chinde, V.; Lin, Y.; Ellis, M.J. Data-Enabled Predictive Control for Building HVAC Systems. *J. Dyn. Syst. Meas. Control Trans. ASME* **2022**, *144*, 081001. [CrossRef]
129. Bajaj, M.; Singh, A.K. Grid Integrated Renewable DG Systems: A Review of Power Quality Challenges and State-of-the-Art Mitigation Techniques. *Int. J. Energy Res.* **2020**, *44*, 26–69. [CrossRef]

130. Naderi, Y.; Hosseini, S.H.; Ghassem Zadeh, S.; Mohammadi-Ivatloo, B.; Vasquez, J.C.; Guerrero, J.M. An Overview of Power Quality Enhancement Techniques Applied to Distributed Generation in Electrical Distribution Networks. *Renew. Sustain. Energy Rev.* **2018**, *93*, 201–214. [CrossRef]
131. Norouzi, F.; Hoppe, T.; Elizondo, L.R.; Bauer, P. A Review of Socio-Technical Barriers to Smart Microgrid Development. *Renew. Sustain. Energy Rev.* **2022**, *167*, 112674. [CrossRef]
132. ACEEE. Grid-Interactive Efficient Buildings Are the Future, and Utilities Can Help Lead the Way. Available online: https://www.aceee.org/blog/2019/11/grid-interactive-efficient-buildings (accessed on 11 July 2023).
133. Tumminia, G.; Guarino, F.; Longo, S.; Aloisio, D.; Cellura, S.; Sergi, F.; Brunaccini, G.; Antonucci, V.; Ferraro, M. Grid Interaction and Environmental Impact of a Net Zero Energy Building. *Energy Convers. Manag.* **2020**, *203*. [CrossRef]
134. Lagrange, A.; de Simón-Martín, M.; González-Martínez, A.; Bracco, S.; Rosales-Asensio, E. Sustainable Microgrids with Energy Storage as a Means to Increase Power Resilience in Critical Facilities: An Application to a Hospital. *Int. J. Electr. Power Energy Syst.* **2020**, *119*, 105865. [CrossRef]
135. Huo, X.; Dong, J.; Cui, B.; Liu, B.; Lian, J.; Liu, M. Two-Level Decentralized-Centralized Control of Distributed Energy Resources in Grid-Interactive Efficient Buildings. *IEEE Control Syst. Lett.* **2022**, *7*, 997–1002. [CrossRef]
136. Rastegarpour, S.; Ferrarini, L. Energy Management in Buildings: Lessons Learnt for Modeling and Advanced Control Design. *Front. Energy Res.* **2022**, *10*, 899866. [CrossRef]
137. Stamatescu, G.; Stamatescu, I.; Arghira, N.; Calofir, V.; Făgărăşan, I. Building Cyber-Physical Energy Systems. *arXiv* **2016**, arXiv:1605.06903.
138. Razmara, M.; Bharati, G.R.; Hanover, D.; Shahbakhti, M.; Paudyal, S.; Robinett, R.D. Building-to-Grid Predictive Power Flow Control for Demand Response and Demand Flexibility Programs. *Appl. Energy* **2017**, *203*, 128–141. [CrossRef]
139. Al-Ali, A.R.; El-Hag, A.; Bahadiri, M.; Harbaji, M.; Ali El Haj, Y. Smart Home Renewable Energy Management System. *Energy Procedia* **2011**, *12*, 120–126. [CrossRef]
140. Palma-Behnke, R.; Benavides, C.; Aranda, E.; Llanos, J.; Sáez, D. Energy Management System for a Renewable Based Microgrid with a Demand Side Management Mechanism. In Proceedings of the IEEE SSCI 2011—Symposium Series on Computational Intelligence—CIASG 2011: 2011 IEEE Symposium on Computational Intelligence Applications in Smart Grid, Paris, France, 11–15 April 2011; pp. 131–138. [CrossRef]
141. Giorgos, G.S.; Christodoulides, P.; Kalogirou, S.A. Optimizing the energy storage schedule of a battery in a PV grid-connected nZEB using linear programming. *Energy* **2020**, *208*, 118177.
142. Abedi, S.; Alimardani, A.; Gharehpetian, G.B.; Riahy, G.H.; Hosseinian, S.H. A Comprehensive Method for Optimal Power Management and Design of Hybrid RES-Based Autonomous Energy Systems. *Renew. Sustain. Energy Rev.* **2012**, *16*, 1577–1587. [CrossRef]
143. Tavakoli, M.; Shokridehaki, F.; Funsho Akorede, M.; Marzband, M.; Vechiu, I.; Pouresmaeil, E. CVaR-Based Energy Management Scheme for Optimal Resilience and Operational Cost in Commercial Building Microgrids. *Int. J. Electr. Power Energy Syst.* **2018**, *100*, 1–9. [CrossRef]
144. Marzband, M.; Fouladfar, M.H.; Akorede, M.F.; Lightbody, G.; Pouresmaeil, E. Framework for Smart Transactive Energy in Home-Microgrids Considering Coalition Formation and Demand Side Management. *Sustain. Cities Soc.* **2018**, *40*, 136–154. [CrossRef]
145. Marzband, M.; Azarinejadian, F.; Savaghebi, M.; Pouresmaeil, E.; Guerrero, J.M.; Lightbody, G. Smart Transactive Energy Framework in Grid-Connected Multiple Home Microgrids under Independent and Coalition Operations. *Renew. Energy* **2018**, *126*, 95–106. [CrossRef]
146. Bilgin, E.; Caramanis, M.C.; Paschalidis, I.C.; Cassandras, C.G. Provision of Regulation Service by Smart Buildings. *IEEE Trans. Smart Grid* **2016**, *7*, 1683–1693. [CrossRef]
147. Taha, A.F.; Gatsis, N.; Dong, B.; Pipri, A.; Li, Z. Buildings-to-Grid Integration Framework. *IEEE Trans. Smart Grid* **2019**, *10*, 1237–1249. [CrossRef]
148. Mirakhorli, A.; Dong, B. Model Predictive Control for Building Loads Connected with a Residential Distribution Grid. *Appl. Energy* **2018**, *230*, 627–642. [CrossRef]
149. Fan, C.; Huang, G.; Sun, Y. A Collaborative Control Optimization of Grid-Connected Net Zero Energy Buildings for Performance Improvements at Building Group Level. *Energy* **2018**, *164*, 536–549. [CrossRef]
150. Clastres, C.; Ha Pham, T.T.; Wurtz, F.; Bacha, S. Ancillary Services and Optimal Household Energy Management with Photovoltaic Production. *Energy* **2010**, *35*, 55–64. [CrossRef]

buildings

MDPI

Review

Hybrid Photovoltaic Thermal Systems: Present and Future Feasibilities for Industrial and Building Applications

Mahendran Samykano

Faculty of Mechanical & Automotive Engineering Technology, Universiti Malaysia Pahang, Pekan 26600, Pahang, Malaysia; mahendran@ump.edu.my

Abstract: The growing demands of modern life, industrialization, and technological progress have significantly increased energy requirements. However, this heightened need for energy has raised concerns about its impact on the environment and the rising costs associated with it. Therefore, the engineering sector is actively seeking sustainable and cost-effective energy solutions. Among the promising innovations in solving the problem is the photovoltaic thermal system (PVT), which aims to capture electrical and thermal energy from solar radiation. Despite its potential, the application of PVT systems is currently limited due to the unpredictable nature of solar energy and the absence of efficient thermal energy storage capabilities. To address these challenges, researchers have explored the use of phase change materials and nano-improved phase change materials (NEPCMs) to optimize energy extraction from solar systems. By incorporating these materials, the PVT system can maximize energy utilization. This article provides a comprehensive overview of the potential applications of PVT techniques in both industrial and building settings. It also offers a detailed assessment of their commercial and environmental aspects. The research findings highlight several advantages of PVT systems, including reduced electricity consumption, efficient utilization of cooling and heating loads during off-peak periods, improved temperature stability, and enhanced thermal comfort. Furthermore, the integration of NEPCMs in PVT systems has demonstrated superior thermal performance, enabling 8.3% more heat energy storage during charging and 25.1% more heat energy release during discharging. Additionally, the implementation of solar-assisted combined heating and power systems showed the potential to prevent the emission of 911 tons of CO_2 per year compared to conventional PV systems. These systems offer a promising pathway towards mitigating environmental impacts while meeting energy demands. Overall, this review article serves as a valuable resource for fellow researchers by providing detailed insights into the viability of PVT systems for various applications in the industrial and building sectors.

Keywords: thermal energy storage; phase change materials; hybrid photovoltaic thermal systems; buildings

check for updates

Citation: Samykano, M. Hybrid Photovoltaic Thermal Systems: Present and Future Feasibilities for Industrial and Building Applications. *Buildings* **2023**, *13*, 1950. https://doi.org/10.3390/buildings13081950

Academic Editor: Danny Hin Wa Li

Received: 21 June 2023
Revised: 17 July 2023
Accepted: 27 July 2023
Published: 31 July 2023

1. Introduction

The efficient utilization of solar energy, also known as insolation energy, holds the potential to significantly alleviate environmental conservation problems and address global energy demand [1,2]. Photovoltaic (PV) technology is specifically designed to directly convert solar energy into valuable electrical energy [3]. Despite having a higher initial investment than other renewable energy systems, solar PV is widely accepted worldwide due to its lower operational and maintenance costs [4]. Additionally, solar PV offers advantages such as higher efficiency and the generation of pollution-free electricity [5]. Figure 1 illustrates the global installation of solar PV systems, with an estimated installed capacity of 1523 GW by 2023. The installed capacities in 2022, 2021, and 2020 were 1208 GW, 940 GW, and 770 GW, respectively [6,7].

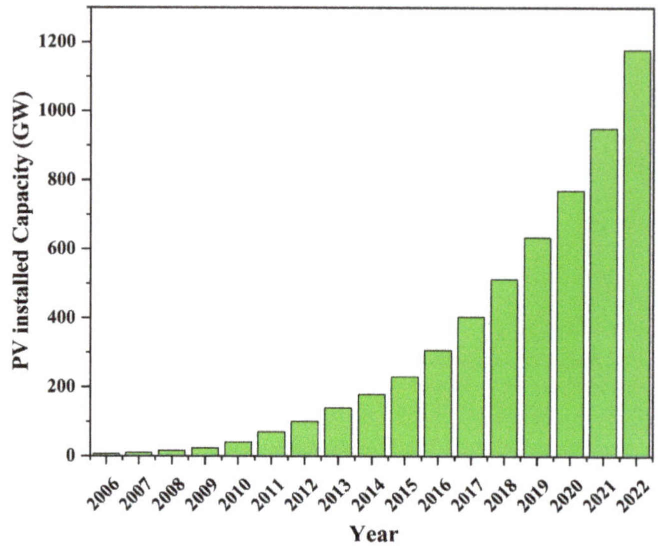

Figure 1. Global PV installed capacity [7].

The major limitation of photovoltaic panels is their capacity to convert a fraction of solar radiation into electrical energy, while the remaining energy increases the temperature of the PV cells, negatively impacting their performance. A mere 1 °C increase in the heat of PV cells can decrease electrical productivity by approximately 0.1% to 0.5%, significantly reducing the yield of PV cells, especially under continuous solar radiation [8,9]. Furthermore, increased circuit resistance caused by higher PV cell temperatures decreases electron velocity, affecting the open-circuit voltage and severely impacting the overall material performance of the cell. Therefore, the recommended operational temperature for PV cells is suggested to be within the range of 0 °C to 75 °C. The temperature co-efficient, which represents the rate of voltage decrease per unit temperature rise, defines the temperature dependence of the primary materials used in evaluating the performance of PV cells and highlights the strong relationship between output power and PV panel temperature [10].

For photovoltaic thermal (PVT) systems, where temperature control is crucial, it has been recommended that they incorporate cooling mechanisms into the solar collectors to improve their efficiency [11,12]. Thus, by introducing a heat transfer fluid (HTF) or cooling fluid into the solar collector, heat can be effectively removed from the PV cells, thereby increasing the photoelectric conversion efficiency and maximizing the utilization of waste heat [13,14]. PVT systems are designed to harvest electricity and heat energy simultaneously. The low-grade heat obtained can be used for various purposes, including space heating, industrial process heating, preheating household fluids, and crop drying [15–17]. Compared to conventional PV devices, PVT systems have lower manufacturing and installation costs per unit area and can be highly practical in efficient infrastructures such as buildings, hospitals, or factories [18,19]. This paper is focused on defining PVT systems based on different thermal transport strategies and exploring their integration into various potential frameworks [20]. The categorization of PVT systems for potential applications is based on different system parameters, including absorber plate configuration and fluid flow mechanisms (e.g., single-pass, double-pass, number of channels, normal and forced circulation) [21].

Given the significance of heat transfer and storage capacities in PVT systems, selecting suitable materials for storing heat energy during periods of solar power availability and releasing it when it is unavailable is crucial [22]. Phase change materials (PCMs), which are capable of efficiently storing and releasing heat energy through the process of melting and solidifying, are widely used in thermal energy storage (TES) applications [23,24]. However, the lower thermal conductivity of PCMs poses a challenge [25,26]. To overcome this limitation, researchers have suggested dispersing highly conductive nanomaterials, such as metals, metallic oxides, carbon nanotubes (CNTs), and graphene, within the PCMs. This new class of materials, known as nano-enriched phase change materials (NEPCMs), has become preferred for TES applications in PVT systems due to their higher heat transfer rates, higher storage density, and efficient energy release [27–31].

Multiple articles have been published that classify PVT systems and discuss experimental and numerical studies conducted on various PVT systems [32]. A comprehensive review has examined the performance, design, simulation, and experimental evaluation of different PVT systems, including air- and water-based collectors [33,34]. Studies have analyzed the seasonal performance of PVT systems under different climate conditions and found that simple cooling methods tend to outperform other systems [35]. Thorough analyses have highlighted the potential benefits of PVT systems in terms of energy efficiency, reliability, and long lifespan [36]. Extensive reviews have explored the application of PVT systems in solar water heaters and solar thermal systems with and without PCM for different temperature ranges and residential use [37–39]. Various cooling methods for PVT systems have also been reviewed, focusing on their accomplishments and applications [40,41]. The state-of-the-art solar-based heat pump PVT systems for meeting thermal energy demand in buildings have also been discussed, highlighting integration approaches, feasible arrangements, multiple foundations, and mechanism design [42]. Numerous ongoing scientific studies to enhance the reliability and application of this technology in decarbonizing buildings are also being reported [43]. However, based on the literature and Figure 2, there is a lack of review articles focusing on the industrial applications of PVT systems in terms of their environmental aspects. This article aimed to fill this gap by providing a general classification of PVT systems, discussing cooling methods, exploring industrial and building applications of PVT systems, and analyzing the environmental impacts based on previous studies conducted on various PVT devices. Moreover, this study aimed to assist engineers and researchers who are considering PVT systems for industrial and building facades, by providing them with a better understanding of the effects on electrical and thermal energy for heating and cooling applications.

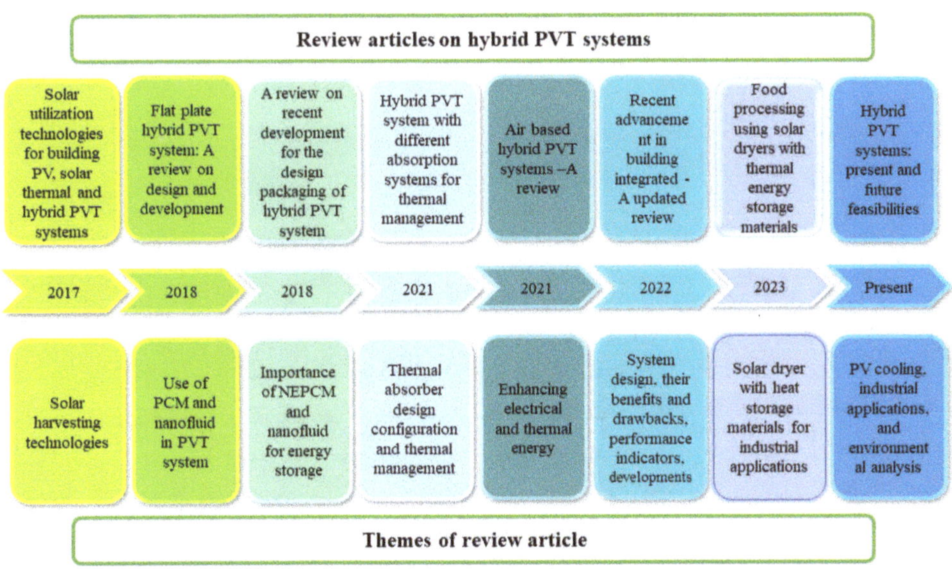

Figure 2. Previously published review articles on hybrid PVT system applications [20,44–50].

The present article also includes an analysis of liquid-based spectrum filters in PVT systems, a comparative analysis of PVT technology highlighting its key advantages, challenges, and future potential, and a thorough examination of the impact of PVT system output in industries, as it was noticed that there is a lack of revised and constructive analyses evaluating PVT system industrial applications from both approaches. To fill this gap, the research record on PVT system output was reviewed based on the industrial requirements, and the various factors considered in each analysis were categorized. The use of NEPCM has garnered significant attention due to its positive impact on the thermal regulation of PV units and its ability to improve electricity and thermal energy performance. Thus, this prospect was extensively reviewed based on in-depth articles. Furthermore, various PVT systems for enhancing electricity and thermal energy production were extensively discussed. The literature on experimental applications of hybrid PVT arrangements, with and without PCM, was surveyed, analyzed, and summarized for various industrial applications, including water heating, water desalination, thermal management in buildings, food processing, HVAC, agricultural processes, and thermal power plants. Even though a number of innovative techniques to improve the thermal and electrical performance of PVT systems have been described, it appears that the field of PVT systems that use PCMs and NEPCMs is continuing to grow. According to a search that used the keywords "PVT system, PCM, and NEPCM", it was discovered that the number of publications, which began in the year 2005, grew more than sixty times between the years 2020 and 2023. The specifics of the articles can be obtained from the "Web of Science, Scopus and Science Direct", as shown in Figure 3. From a total of 45,000 publications, approximately 130 articles were chosen for the present review article according to their applicability, suitability and significance. The remaining sections of the article are organized as follows: Section 2 provides a detailed understanding of PVT system behavior and classification, Section 3 focuses on the industrial applications of PVT systems, Section 4 discusses the environmental aspects of PVT systems, Section 5 outlines the technical tasks associated with PVT systems, and Section 6 concludes the article by summarizing the work and suggesting future prospects for this field.

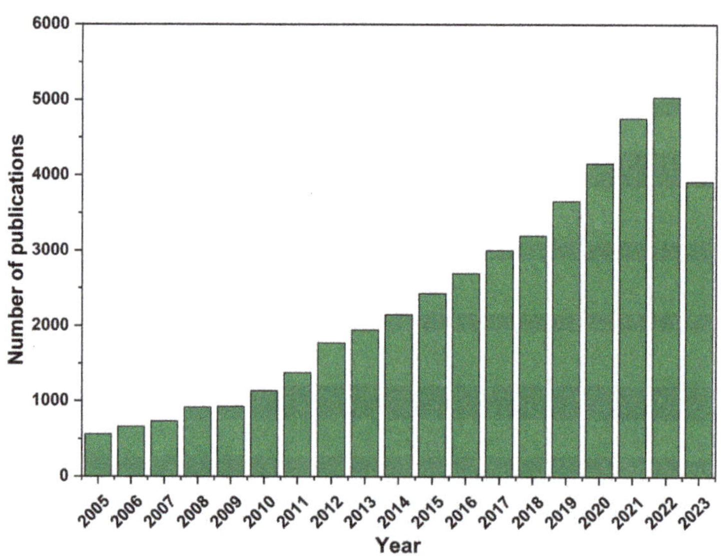

Figure 3. Number of journal publications related to PVT systems, PCMs, and NEPCMs for energy production and storage from 2005 to 2023. Sources: Scopus, Web of Science, and Science Direct.

2. PVT Systems and Their Performances

The PVT system combines photovoltaics with a thermal collector to convert both electrical and thermal energy simultaneously. This integration enhances solar energy utilization, allowing for greater electricity and thermal energy generation per unit area than standalone solar PV or thermal systems.

Various cooling methods are employed to regulate the temperature of PV panels, depending on whether they are passive or active techniques. A comprehensive and up-to-date exploration of cooling technologies is presented in the following sections, categorized according to the type of heat transfer mechanisms involved. The choice of heat transfer fluid (HTF) used in PVT systems determines their performance and implementation. Historically, air and water have been conventionally used as cooling fluids in PVT systems. In the past three to four decades, extensive research has focused on optimizing traditional fluid-based PVT systems, with numerous studies [51,52] conducted in this area. The following section evaluates the findings of these studies, highlighting the use of air and water as HTFs in PVT systems. The segment summarizes the key outcomes, various system configurations, and efficiencies in a tabular format, facilitating a better understanding of the subject matter. Figure 4 illustrates the classification of PVT systems, including air-based PVT systems, water-based PVT systems, bi-fluid PVT systems, heat-pipe-based PVT systems, and PCM-based PVT systems, all falling under the category of PVT.

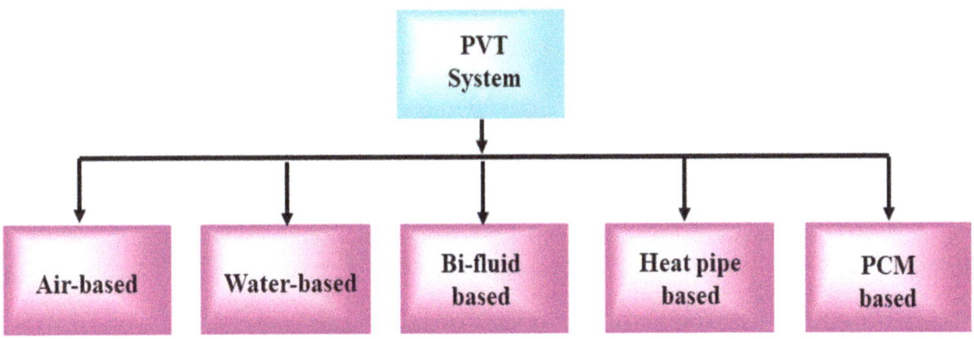

Figure 4. Classification of PVT systems.

2.1. Air-Based PVT Systems

As shown in Figure 5, air-based PVT systems can actively or passively allow air to flow across the PV surface. These systems utilize single- or double-pass configurations and various absorber designs. Numerous researchers have been dedicated to enhancing the performance of conventional air-based PVT systems, including their architecture, operation, and materials [53–56]. Some studies have focused on simulating these systems using computational models to predict their potential performance. Researchers have also explored the energy and exergy approaches of air-based PVT systems.

Ozakin and Kaya conducted an ANSYS Fluent study to analyze the panel temperature and cooling channel air velocity [53]. Their findings were compared to real data, and it was discovered that using small and frequent fins improved the exergy performance of polycrystal panels by 70% and 30%, respectively, compared to empty fins. The thermal efficiencies also increased by approximately 55% and 70%. Furthermore, the PV temperature decreased by around 10% to 15% for all configurations [54].

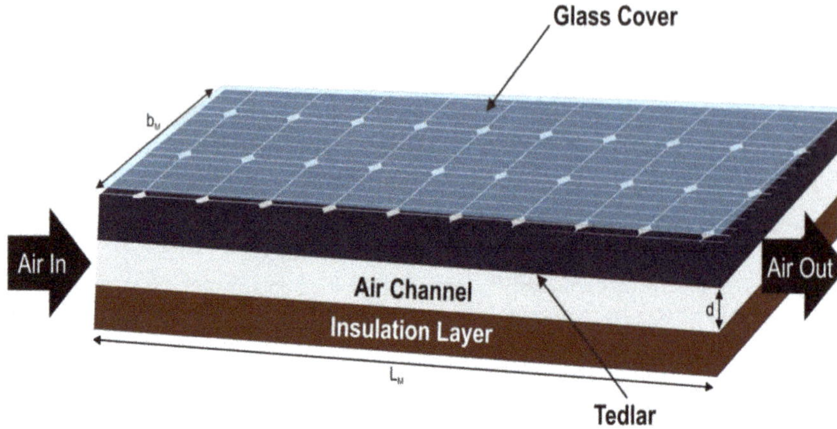

Figure 5. Air-based PVT systems [32].

Agrawal et al. examined a microchannel PV/T module featuring two types of panels: single-channel and multi-channel PV/T [55]. By maintaining a consistent airflow rate, they discovered that the multi-channel configuration exhibited improved thermal energy and exergy efficiency, achieving enhancements of 70.62% and 60.19%, respectively. Another study proposed the integration of a greenhouse solar dryer with a partially enclosed air-based PVT system designed for drying purposes [57]. The researchers evaluated the influence of flow rates on the system's thermal and electrical performance by varying the number of air collectors. As the number of collectors increased from 1 to 5, the equivalent thermal energy and thermal exergy efficiency rose from 3.24 kWh/day to 10.6 kWh/day. Air-based PVT systems require a sufficiently high airflow velocity to absorb the heat from the PV modules, especially during the daytime when temperatures are high [58]. To address this, a rotating airflow device is integrated into the channels to cool the modules. However, the output of air-based PVT systems fluctuates throughout the day due to temperature variations, and the heat absorption potential of air is at its lowest during the hottest hours. Thus, water-based PVT systems are employed to overcome these challenges and improve thermal management.

2.2. *Water-Based PVT Systems*

Air-cooled PVT devices generally have inferior thermal properties and are not very effective in high-temperature applications. Water, on the other hand, has higher thermal properties compared to air. Water-cooled PVT systems can be implemented in active and passive cooling. In both approaches, water tubes are installed behind the PV panel to allow water to flow through them. The water extracts heat from the PV unit, cooling the panel and producing heated water. This heated water can be utilized for various purposes. In one experimental setup, the water-based system consists of several main components: an exposed PVT collector, a storage tank, a forced water circulation circuit, and a pump (see Figure 6). The collector inlet is located at the bottom of the tank, while the heated water flows from the collector into the upper part of the tank. This arrangement creates stratification in the tank, with lower hot water density, improving thermal performance, and reducing the temperature of the photovoltaic cells [59].

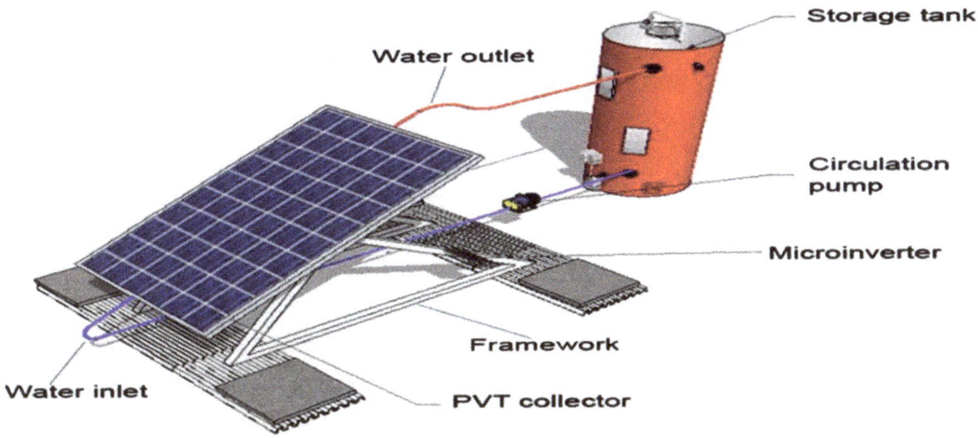

Figure 6. Water-based PVT systems [reproduced with permission from [60], Elsevier, 2016].

Previous studies have shown that water-based PVT systems [61] reach a maximum deflation temperature of 74.2 °C. This temperature level is considered safe in terms of overheating risk and PV efficiency. The PVT systems have also been proven suitable for domestic water preheating, with an average monthly output temperature of the PVT collector exceeding 45 °C for five months. It has also been demonstrated that an 8 m^2 PVT system is sufficient for heating a 300 L storage tank. Simulation results further indicate that the PVT surface area should be roughly double the size of a traditional thermal module to achieve the same thermal capacity. A dual oscillating absorber with a copper pipeline flow was developed and tested using water-based PVT systems. ANSYS 19.2 was used to predict the water outlet and surface temperatures under various irradiation and flow rate conditions [62]. The findings revealed an overall thermal efficiency of 59.6% for the PVT systems. At a mass flow rate of 6 LPM, the highest mean electrical performance values were 10.86% for the PV panel and 11.71% for the water-based PVT system. These results indicate that, as mass transfer improves, the electrical, thermal, and output capacity of PVT systems also increases.

Bevilacqua et al. [63] investigated the electrical performance of a PVT system using back surface spray cooling. They observed a 7.8% increase in electrical power and a 28.2% decrease in cell temperature compared to a non-cooled PV system. Similarly, Bevilacqua et al. [64] analyzed the electrical performance of a PVT system under different climate conditions. They developed a model to analyze panel surface temperature and power generation under various climate conditions. The model's accuracy was confirmed through statistical parameters, demonstrating good agreement between predicted and measured power and temperature. The experimental results were used to validate the simulation findings to evaluate the effectiveness of a novel dual oscillating absorber design for the PVT water system compared to a traditional PV device. The water-based PVT system's electrical, thermal, and overall efficiency were evaluated within an irradiation range of 500 to 1000 W/m^2 and flow rates ranging from 0.03 to 0.1 kg/s. The findings indicated that increasing either the flow rate or the amount of solar radiation can lead to improved thermal and overall efficiency. However, further increases result in a decline in efficiency after reaching an optimum value. Enhancing the cell's electrical efficiency requires lowering its temperature, which can be achieved by increasing the mass flow rate [65]. Bevilacqua et al. [66] compared three different cooling technologies under varying meteorological conditions and found that backside spray cooling provided the best temperature reduction and enhanced electrical performance compared to the other systems.

2.3. Bi-Fluid-Based PVT Systems

The performance of stand-alone air- or water-based PVT systems fell short of expectations, leading researchers to explore alternative approaches. One of the proposed approaches is using two fluids in a PVT unit, known as a bi-fluid-based PVT system (Figure 7). The aim was to overcome the limitations of the separate air-based and water-based PVT systems and simultaneously generate electricity, hot air, and water.

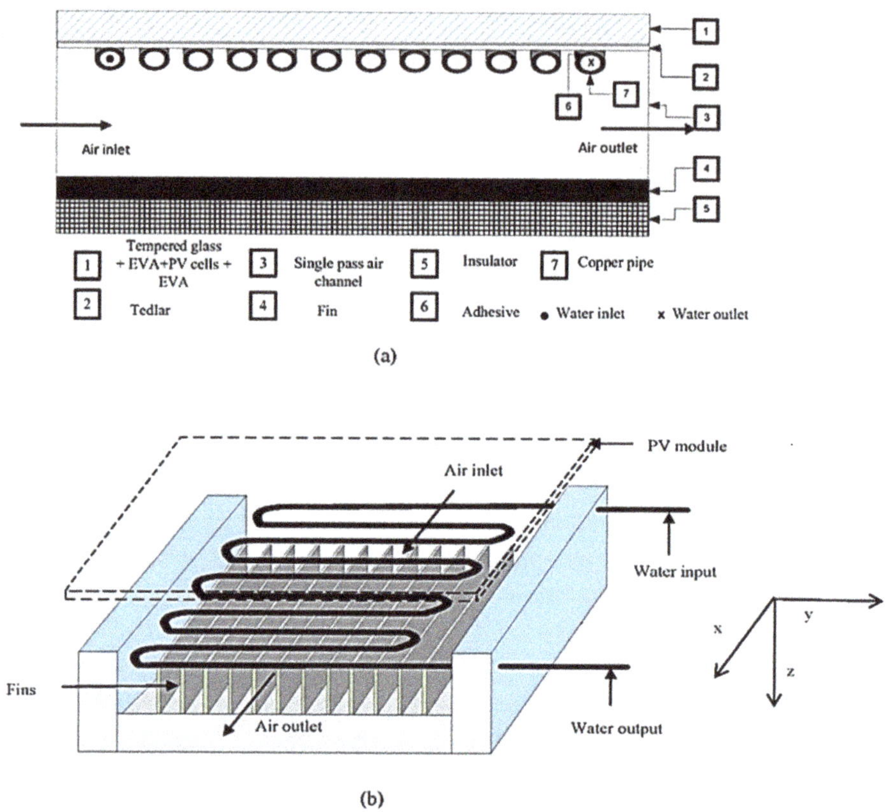

Figure 7. Bi-fluid-based PVT Systems (**a**) schematic illustration bi fluid PVT system (**b**) sectional view of fin and water tubes [reproduced with permission from [67], Elsevier, 2016].

A bi-fluid PVT system was developed, which incorporated a PV laminate, a serpentine copper tube for water heating, and a single-pass air channel for air heating [68]. To evaluate the efficiency of the developed bi-fluid collector, the fluids were analyzed both independently and concurrently using an established model. The simulated thermal and electrical efficiencies were found to be satisfactory when the bi-fluid collector operated independently. However, when the bi-fluid system operated simultaneously, the total performance reached a maximum of 76% efficiency at the specified parameters' maximum flow rate.

Furthermore, experimental analysis was conducted using two different fluids in various geometric channels [69]. Three different geometric forms were utilized as fluid movement channels: circular tube, half tube, and square tube. Pure water and an EG-W (60:40) mixture were used as heat transfer fluids to extract heat from the hybrid systems. The results indicated that the tube configuration was more efficient compared to circular and square tubes. Specifically, using the half-tube form at a mass flow rate of 0.04 kg/s improved the electrical efficiency of the hybrid collector by 0.53% and 1.16% compared to circular and square tubes, respectively. Additionally, the thermal efficiency showed an improvement of 1.17% and 2.6% for the same comparison.

2.4. Heat-Pipe-Based PVT Systems

Figure 8 illustrates a heat-pipe-based system that efficiently transfers heat between two solid surfaces without requiring external assistance. This system incorporates evaporation and condensation principles. Due to pressure variations in the adiabatic components, the working fluid transitions from a liquid phase to a vapor phase using heat from the ambient environment. This passive cooling technique, supported by a heat pipe, is known as a PVT-HP system. The vapor phase travels through the condenser, releasing heat and transitioning back to a liquid phase. It then returns to the evaporator through a wick structure.

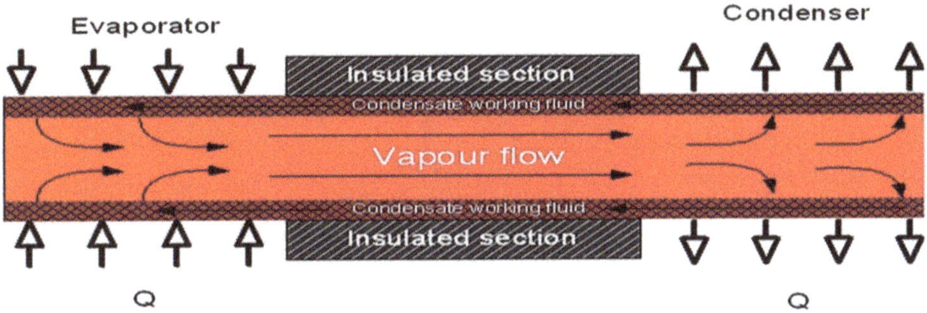

Figure 8. Heat-pipe working cycle [70].

The PVT-LHP system consists of several sub-models and an integrated model, which have been designed and analyzed [71]. The analysis revealed that under the given design conditions, the PV/LHP module achieved electrical, thermal, and overall performance of 12.2%, 55.6%, and 67.8%, respectively. The novel systems exhibited 28% higher efficiency and a 2.2 times higher coefficient of performance (COP) compared to traditional systems.

Another advanced system, the HP-PVT system, was developed to provide energy generation, water preheating, and load reduction in space air conditioning [72]. A dynamic prototype of the HP-PVT system was established and experimentally validated for numerical accuracy. Subsequently, the system's annual efficiency was assessed based on standard weather conditions in Hong Kong. Simulation results indicated that the average cumulative heat transfer through the outside wall could be reduced to less than a quarter of the usual level. The system achieved approximately 35% efficiency in water heating and 10% efficiency in power generation. Additionally, it was reported that the system saved an average of 315 kWh/year per unit surface area.

2.5. PCM-Based PVT Systems

The utilization of thermal energy storage (TES) is crucial to address the demand and availability of electricity at night [73]. Typically, batteries are employed to store electrical energy during the day, which can be used to meet the nighttime demand. Since solar energy is abundant, phase change materials (PCMs) offer an ideal solution for storing thermal energy [74]. PCMs possess desirable characteristics such as rapid charging and discharging capabilities and high performance. Additionally, they are cost-effective, readily available, and pose no toxicity or hazards. Various PCMs with different phase transition temperatures are commercially accessible [75]. These materials effectively reduce the temperature of solar panels during the day while storing the captured heat energy, which can then be released during the night. To enhance the efficiency of PCMs, techniques such as dispersing nanoparticles into the base PCM, encapsulating nano-sized particles with PCM, and creating nano-enhanced PCMs (NEPCMs) have been proposed [26,76–78]. These modifications improve heat transfer and increase thermal conductivity, thereby decreasing the temperature of photovoltaic (PV) panels and extending their lifespan. Figure 9 illustrates the integration of PCMs with a PVT unit.

PVT systems incorporating PCM-based collectors have been primarily used for cooling PV units [79]. Experimental studies have also focused on assessing their electrical performance. The implementation of PCM-based cooling for PV units has contributed to the preservation of their electrical efficiency. The peak efficiency achieved with this approach was reported to be approximately 13.7%, whereas traditional PV systems reached about 7.1% efficiency. The experimental findings also indicate an increase in electricity generation corresponding to the rise in solar irradiance. The cooling process facilitated by PCM integration results in higher power output. In conclusion, utilizing PCM for cooling PVT modules enhances their electrical efficiency [80].

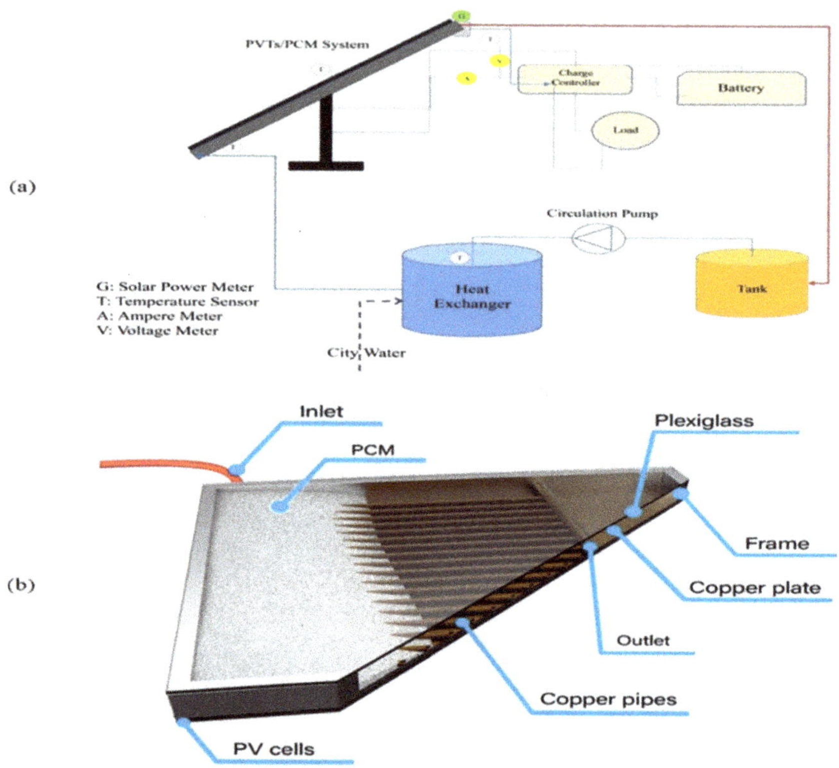

Figure 9. PCM-based PVT systems (**a**) schematic illustration of PCM-based PVT systems, (**b**) cut section of PVT system [reproduced with permission from [81], Elsevier, 2020].

To evaluate the energy performance of PVT systems incorporating PCMs, comparisons were made with standard PV modules based on electricity generation and overall energy performance [81]. Factors such as exergy destruction and external damages were considered in these measurements. The PVT-PCM system demonstrated the greatest relative improvement in energetic and exergetic efficiency. The PV module experienced a temperature drop of 12.6 °C, while the PVT-PCM system achieved a drop of 10.3 °C. The maximum electrical efficiency obtained was 13.72% for PV and 13.8% for PVT. Similarly, the electrical efficiency of PVT-PCM was numerically and experimentally determined to be 13.98% and 13.87%, respectively.

2.6. Summary

A study was conducted to examine the extraction of heat from PV units using various thermal fluids, cooling channel arrangements, and heat transfer enhancement techniques. Water-based PVT systems are commonly employed for dissipating heat from the backside of PV modules. The addition of water-based PVT systems could provide benefits for residential water heating and preheating water for industrial processes that currently use traditional solar collectors. However, the use of water-based systems may be limited in cold regions due to the risk of water freezing. Various methods were extensively discussed, such as glazing, heat pipes, coolant tubes, dual fluids and double channels, and PCMs. Water-based PVT units can also be extended to conventional water heating devices, preheating desalination water, and preheating water for industrial purposes. On the other hand, air-based PVT systems were found to be the least efficient and least effective method due to their low heat transfer properties.

It was observed that using water as a cooling medium for solar panels in PVT systems leads to higher efficiency rates. However, caution must be exercised when using this system in extremely cold environments, and there are other related complications that may cause damage. Both water- and air-based systems demonstrated more significant cooling of PV modules, but their design becomes increasingly complex and complicated when operating in passive mode. Furthermore, it was emphasized that the heat generated from PCMs must be effectively utilized to impact the thermal management system. It is recommended to use highly conductive nanoparticles enhanced nanofluid instead of conventional fluid to enhance the heat transfer rate between the PV panel and the cooling fluid. Moreover, the conductivity and absorption rate of the PCM plays a crucial role in the heat transfer rate and storage performance. Therefore, the use of carbon-based NEPCM (nano-enhanced phase change materials) is suggested to improve electrical performance, absorption rate, heat transfer rate, and storage performance. Implementing nano-enhanced PCMs in PVT systems allows for the extraction and storage of a greater amount of heat energy.

3. Industrial Applications of a Hybrid PVT System

Hybrid PVT systems are designed to generate both electricity and heat energy simultaneously. The heat energy obtained can be utilized for a wide range of applications, including water heating, space heating, drying, solar desalination, textiles, solar power plants, the agricultural sector, preservation of food and pharmaceutical products, automobiles, and medical applications [82].

Similarly, PCM-integrated PVT systems can produce both electricity and heat energy, which can also be stored. PCMs are suitable for energy conversion and heat transfer applications [2]. The stored heat energy can be effectively used for residential and industrial purposes. For instance, PCMs play a crucial role in urgent blood transport, and organ and cold therapies, as well as providing thermal protection for ice creams and food products [83]. Moreover, PCMs find applications in the development of high-performance electronic devices and textiles, cooking, industrial process heating, and regulating temperatures in chemical reactions [84]. This section extensively explores the diverse industrial applications of PVT systems.

3.1. Water Heating

Hot water is essential for a wide range of industrial and household uses. Industrial applications utilize hot water for curing, mixing, washing, cleaning, and sterilizing to facilitate product manufacturing. In a study by Abdallah et al. [85], it was found that adding conductive nanomaterials to the cooling fluid enhances the thermal properties of a PVT system. The cooling fluid reduces the temperature of the PV panel, leading to increased electrical production. Moreover, the nanomaterials in the fluid improve the overall performance of the PVT device by enhancing heat energy absorption for generating hot water for both household and industrial applications.

Researchers at Dalian University conducted a study [86] using a PVT heat pump and a water source heat pump for public hot water supply. They developed an innovative strategy to utilize multiple heat sources and established a calculation model to determine the system's heating characteristics. The study reported an 80% increase in public utility hot water supply and a 49% reduction in hot water production costs, resulting in improved economic performance.

Jafari et al. [87] conducted a thermodynamic investigation on a newly proposed PVT system that incorporated a fuel cell and battery. The excess electrical energy generated by the PVT system was used for hydrogen production. The produced hydrogen was then utilized in the fuel cell to generate electrical energy and heated water. The study revealed an overall electrical energy efficiency of 9% and an electrical energy production cost of USD 0.286/kWh. The PVT system's exergy and energy efficiencies were measured at 16.91% and 14.5%, respectively. Additionally, the combined system demonstrated a total energy efficiency of 41.80%. Kumar et al. [88] developed a model and simulation for industrial solar-assisted process heating (SAPH) applications. The study showed that the maximum heat gain, thermal efficiency, and exergy efficiency were 1420 W, 75%, and 12.72%, respectively. The SAPH system exhibited excellent performance under higher solar irradiation and lower mass flow rates.

A PVT system designed for hostel students, providing hot water and meeting a portion of the electrical requirements, was analyzed by Somasundaram and Tay [89]. The researchers compared two types of PVT systems, unglazed and glazed, as shown in Figure 10. The study indicated that the glazed PVT unit outperformed the unglazed arrangement. The payback period for the system was determined to be 12.5 years, with an electricity cost of USD 0.26 per kW/h.

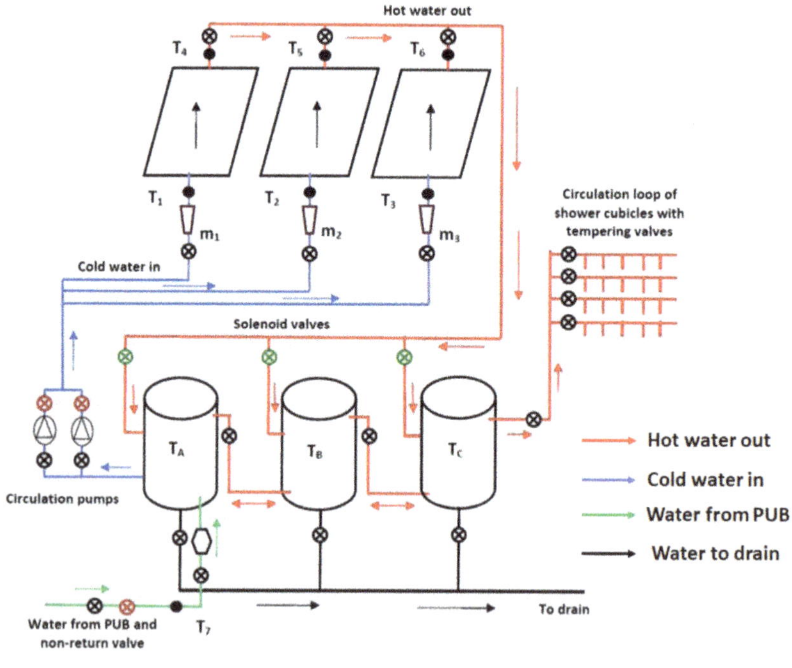

Figure 10. Line diagram of PVT system for water heating [89].

Maatallah et al. [90] conducted an exergoeconomic investigation of a PCM integrated water-based PVT system in Kerala, India, under various weather conditions. The research reported an overall performance of 41.0%. Furthermore, the water-based PVT system exhibited an extended life cycle conversion efficiency compared to conventional systems, with an improvement of 27.0%. The generated hot water was deemed suitable for both household and industrial applications.

Naghdbishi et al. [81] explored the impact of MWCNTs (multi-walled carbon nanotubes)-based nanofluids on a hybrid PVT system incorporating a PCM system. The dispersion of MWCNTs in water was found to enhance the thermal and electrical performance by 23.6% and 4.20%, respectively, compared to conventional water-cooled PVT systems. However, the use of the hybrid system led to higher exergy losses due to heat dissipation, resulting in an overall increase of 76% in exergy destruction.

Rajoria et al. [91] conducted an enviroeconomic and exergetic analysis of hybrid PVT systems, considering two cases based on the flow and design of PVT arrays. Case 2 demonstrated improved performance with lower cell temperature, higher outlet air temperature, and higher electrical efficiency compared to Case 1. The annual extenuation in terms of overall exergy gain and thermal energy gain was higher in Case 2. Additionally, the study evaluated the environmental cost associated with overall exergy and thermal energy for both cases.

Yandri et al. [92] evaluated the thermal performance of a hybrid PVT system, including Joule heating, to enhance thermal energy. The results showed a 13% improvement in thermal performance, suggesting wider applications in areas such as swimming pool water heating, floor heating, and medium-temperature heat pumps. Antony et al. [93] conducted a detailed analysis of a PVT system, considering various parameters such as solar irradiance, operational temperature, heat and electricity demand, thermal storage capacity, rate of flow, and ambient temperature over an entire year, as shown in Figure 11a. The study found that factors like heat removal, flow rate, location (longitude and latitude), tilt angles, and the choice of heat dissipation fluid significantly influenced the system's performance. The analyzed PVT system was recommended for industrial-scale water heating applications.

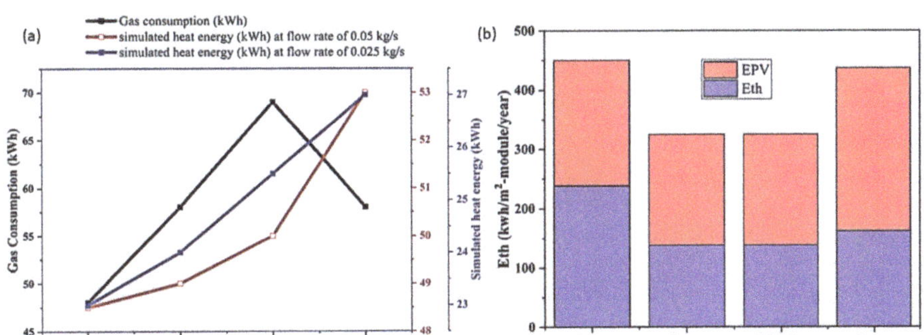

Figure 11. (**a**) Heat demand and supply in winter season [reproduced with permission from [93], Elsevier, 2019], (**b**) yearly thermal (Eth) and photovoltaic (Epv) energies per surface module [reproduced with permission from [61], Elsevier, 2020].

Brottier and Bennacer [61] analyzed the performance of residential water heaters in Western Europe using 28 hybrid non-overglazed PVT systems in France, Portugal, and Switzerland. The study concluded that the PVT systems were suitable for residential water heating, providing an average temperature of 45 °C with five months of autonomy per year. The yearly thermal and electrical energies per surface module are shown in Figure 11b. It was noted that non-overglazed PVT systems exhibited the highest stagnation temperature of 74.2 °C.

Dannemand et al. [94] evaluated the performance of heat-pump-based PVT systems for cold buffer and hot water storage. Amo et al. [84] theoretically analyzed a heat-pump-based PVT system for seasonal storage in educational buildings. Both studies demonstrated high solar coverage of up to 98% for the proposed systems, indicating their economic and technical feasibility.

In summary, various studies have explored different aspects of PVT systems, including the enhancement of thermal properties through the use of nanomaterials, integration with fuel cells and batteries, exergoeconomic analysis, utilization of PCMs, and the evaluation of system performance for different applications. These investigations contribute to the advancement of PVT technology for efficient hot water generation in industrial and household settings.

3.2. Water Desalination

Water salt removal is a process that involves heating saline water to produce water vapor, which is then condensed to obtain pure water. This purification process requires hot water or heat energy. In the context of a PVT system, the hot water outlet can be utilized for the desalination and production of pure water. Mittelman et al. [95] conducted an analysis of a concentrated PVT system for a water desalination setup, as depicted in Figure 12. This hybrid system generates electricity and desalinated water. The designed CPVT solar collectors are capable of operating at high temperatures, enabling the use of more advanced desalination techniques such as vapor-compression or absorption refrigeration systems. Monjezi et al. [96] evaluated a solar-powered osmosis desalination system integrated with a photovoltaic thermal system to produce fresh water. The study found that the specific energy consumption (SEC) for reverse osmosis desalination systems was 4.15 kWh/m^3 with PVT systems and 4.27 kWh/m^3 without PVT systems. Additionally, the novel PVT system with seawater cooling resulted in a 0.12 kWh/m^3 reduction in SEC and a 65% reduction in panel area.

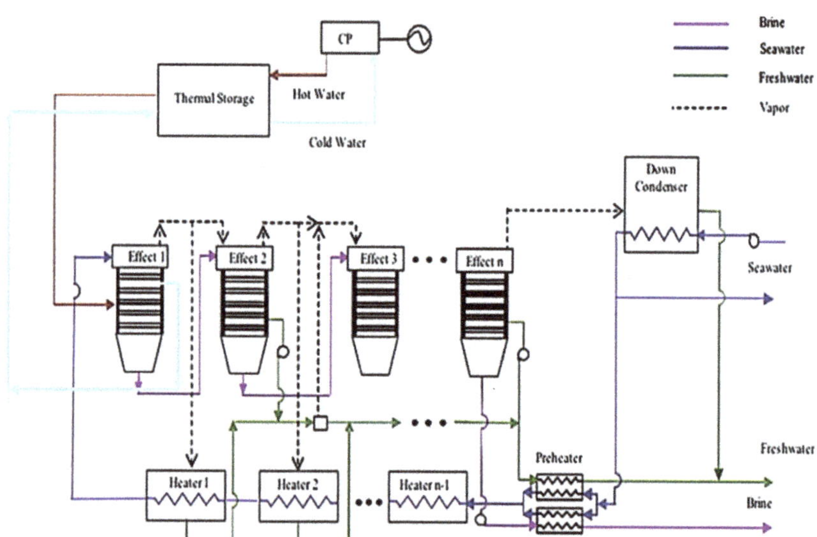

Figure 12. Schematic diagram of CPVT and multiple-effect evaporation plant [reproduced with permission from [95], Elsevier, 2009].

Ong et al. [97] investigated the desalination of water using waste heat obtained from CPVT systems. This hybrid system also generates electrical energy and desalinated water. The results demonstrated that the approach is highly optimized and that effective minimization of heat energy utilization is achievable. Abdelgaied et al. [98] proposed a system that combines PV panels with thermal collectors, a solar dish concentrator, and a reverse osmosis salt removal system with an energy recovery device. They installed a PV unit with a thermal recovery device and a dish concentrator with a solar thermal receiver to power the reverse osmosis desalination process. The dish concentration and PV panels were also used to preheat the purification system. Preheating the input water before pumping it to the reverse osmosis plant resulted in SEC savings ranging from 18.7% to 22.9% for treated seawater and from 24.3% to 35.8% for brackish water, compared to systems without solar preheating. Calise et al. [99] studied a poly-generation system for generating potable water, which includes a PVT system, a single-stage absorber chiller, a multi-effect distillation system, and other components. The PVT system generates electrical and thermal energy, with the electrical energy being sent to the grid. The thermal energy can be utilized to power an air-to-water heat pump that produces chilled water for air conditioning. Finally, the multi-effect distillation system can employ both solar energy and thermal energy from an additional biomass-fired heater to transform seawater into potable water.

Xiao et al. [100] designed a desalination system integrated with a PVT system. The PVT system generates electricity and heat energy, which is used to preheat the saline water in the desalination system, as illustrated in Figure 13a. Potable water production depends on the bottom channel's height, with a deeper channel resulting in decreased water production. As the water quantity increases, the evaporation rate decreases. The use of a PVT system increases the heat transfer rate from the absorber plate to the saltwater by 44%. However, pure water production decreases by 17% when the height of the bottom tank is 0.03 m instead of 0.01 m. Alnaimat et al. [101] designed a solar-diffusion-driving desalination unit, as shown in Figure 13b. This system can generate 100 litres of fresh water daily with an SEC of 3.6 kWh/m^3, using 2 m^2 panels in different weather conditions. Xinxin et al. [102] designed three solar still systems with different glass angles integrated with a concentrated parabolic PVT system, as depicted in Figure 13c. The researchers found that higher solar irradiation increases freshwater generation, electrical efficiency, and thermal efficacy. The temperature difference between the saltwater and the condenser glass is a key factor in freshwater generation. Additionally, the angle of the condenser affects the production of freshwater.

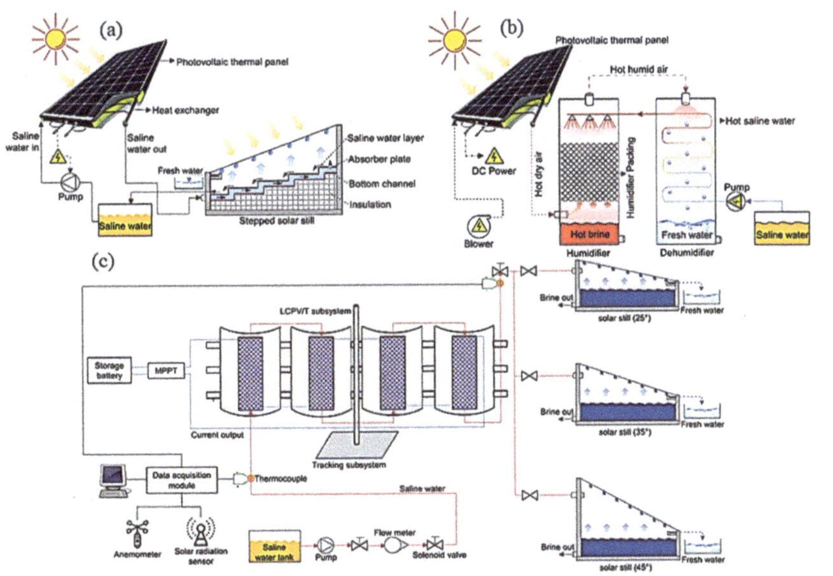

Figure 13. Different configurations of PVT-based solar desalination systems (**a**) stepped solar still with PVT system [reproduced with permission from [100], Elsevier, 2019], (**b**) air assisted PVT integrated desalination [reproduced with permission from [101], Elsevier, 2012], (**c**) low concentration ratio CPC-PVT desalination system [reproduced with permission from [102], Elsevier, 2019].

3.3. Thermal Management in Buildings

The integration of latent heat storage in building materials plays a significant role in building energy management, facilitated by TES technology. Thermal management is crucial in construction to ensure human comfort and the functionality of various structures, from everyday devices to industrial installations [103]. In a study by Zarei et al. [104], a hybrid photovoltaic thermal system was analyzed for eco-friendly solar heating and cooling applications. The system employed two refrigerants, R600 and R290, to reduce global warming potential, replacing R134a. The study found that the outlet water temperature ranged from 31.7 °C to 46.7 °C. Furthermore, using the R290 refrigeration cycle and cooling the panel resulted in an 11.1% improvement in the coefficient of performance (COP), a 9.17 °C increase in the outlet water temperature, and a 60.17% reduction in refrigerant flow rate compared to a system without panel cooling that used R134a refrigerant.

Braun et al. [105] conducted an analysis of a hybrid PVT system for zero-energy office buildings, providing electrical, heating, and cooling functions (trigeneration). The simulation study demonstrated the technical feasibility of the system. Additionally, the proposed PVT system with a heat pump (HP) exhibited higher energy potential for incorporating renewable energies, lower energy production costs, and increased self-consumption of locally generated energy. Hassan et al. [3] investigated the thermal management and regulation of PVT systems integrated with nanofluid-based (PCMs). Three PVT systems were compared: PVT/PCM, PV/PCM, and conventional PV systems. The study revealed that the addition of a 0.1 wt% graphene water nanofluid at a flow rate of 40 LPM improved the overall thermal and electrical efficiency by 14.1%, 20.8%, and 14.5%, respectively, compared to a water-based PV system. An increase in flow rate was found to enhance the electrical and thermal effectiveness of the systems.

Behzadi et al. [83] conducted a techno-economic investigation of PVT systems for building energy management. The study reported a 4.03% increase in exergy, a reduction of EUR 3.64 per Wh in total cost, and a decrease in capital cost achieved by eliminating the battery component. Chemisana et al. [106] analyzed the impact of immersing PVT systems directly in liquids for building applications. The study concluded that immersing the PVT system in a mixture of isopropyl alcohol and deionized water yielded the best results in terms of electrical, thermal, operational, and optical behavior.

Compared to building-integrated photovoltaic (BIPV) systems, building-integrated photovoltaic-thermal (BIPVT) systems offer additional advantages as they can generate both electrical and thermal energy. Debbarma et al. [80] reviewed the thermal performance of BIPVT systems through exergy analysis, highlighting their ability to partially or completely fulfil a building's energy needs.

A PCM-integrated hybrid PVT system was experimentally analyzed to study building thermal management [107]. The system utilized an organic PCM called RT42 for thermal energy storage. The experiment demonstrated a 7.7% improvement in the PCM's electrical energy and a 4.45 V increase in Voc (voltage at open circuit) at an irradiance of 100 W/m^2, validating the effective contribution of PCM in thermal management within buildings.

Balaji et al. [108] discussed the design features of BIPV and BIPVT technologies. They presented two setups, BIPV-wall/facade, where PV panels were vertically mounted on walls, and the BIPV roof, which is suitable for high-rise buildings with increased airflow for efficient PV panel cooling, as shown in Figure 14. The study emphasized several crucial parameters to consider when integrating PV systems into building roofs/facades, including design, buildability, compliance with regulations, safety, maintenance, environmental factors, durability, and performance.

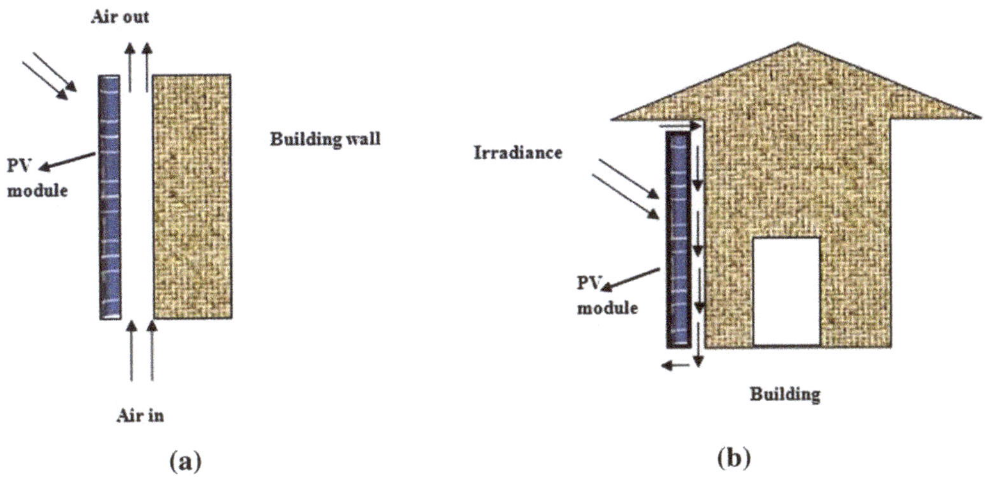

Figure 14. (**a**) Staggered configuration of BIPVT, (**b**) PVT system integrated in building [reproduced with permission from [108], Elsevier, 2016].

3.4. Agricultural Processes Solar Drier

The agricultural industry predominantly utilizes solar dryers to reduce moisture levels and prevent deterioration during storage. A recent experiment [109] involved using a mixed-mode air-based PVT system to dry tomatoes. The initial moisture content of the tomatoes was 91.9%, and they were successfully dried to 23.3% in the mixed-mode solar dryer and to 30.15% in the open-mode solar dryer within a period of 44 h. The maximum thermal and electrical performance achieved was 65% and 12.31%, respectively.

Tiwari et al. explored energy and exergy analysis for solar dryer applications in five different cities: Mumbai, New Delhi, Bangalore, Jodhpur, and Srinagar [110]. They found that the solar dryer's air temperature depends on the air's inlet velocity. This investigation demonstrated self-sustainability and applicability to remote areas, with excellent agreement between numerical and experimental outcomes.

The thermal performance of a PVT arrangement with a desiccant wheel was numerically analyzed, and the produced hot air was utilized for the drying process. By employing the desiccant wheel, the temperature increased from 65 °C to 80 °C, while the humidity ratio of the air decreased from 15 kg water/kg dry air to 8.8 kg water/kg dry air, compared to without the desiccant wheel [111]. Tiwari et al. discussed various types of air-assisted hybrid PVT arrangements with solar greenhouse dryers [57]. They reported an overall thermal output of 26.68% and an electrical output of 11.26% at a flow rate of 0.01 kg/s. Ziaforoughi et al. [112] investigated a novel hybrid PVT system for improving the performance of potato slice drying and agricultural processes in industries. Drying potato slices with thicknesses of 3 mm, 5 mm, and 7 mm at temperatures of 50.0 °C, 60.0 °C, and 70.0 °C, respectively, their system achieved a reduction in electrical consumption of 40% to 69% and a decrease in drying time by 31% to 52%.

Tiwari et al. [113] conducted an experimental study on an optimized drying method using a PVT air collector, which proved beneficial for crop preservation, and is illustrated in Figure 15. The study found that forced convection mode was more efficient for crops with high moisture content, while natural convection mode was better suited for crops with low moisture content. A hybrid PVT system integrated with an indirect solar dryer was analyzed for drying agricultural products [114]. A thermal balance equation was developed and examined using various heat transfer and electric parameters. The numerical analysis demonstrated a 70% improvement in thermal performance, a 10.50% improvement in electrical performance, and an overall performance enhancement of 90.0% at a flow rate of

0.0155 kg/s. Effective crop drying is crucial in rural areas to remove moisture content and prevent quality deterioration.

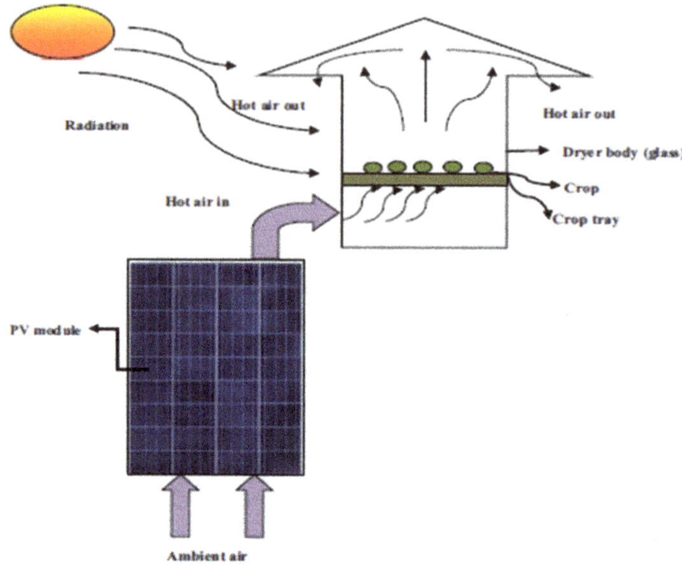

Figure 15. Greenhouse dryer [reproduced with permission from [113], Elsevier, 2018].

The exergoeconomic analysis of hybrid PVT systems, such as that shown in Figure 16 under the forced mode, was conducted, considering parameters such as PV cell temperature, outlet temperature, greenhouse temperature, and crop temperature, at the Indian Institute of Technology in India [115]. Findings revealed the system's efficiency decreased as the PV panel temperature increased. The annual energy outputs were 191.53 kWh for electrical energy, 1182.19 kWh for thermal energy, and 1686.22 kWh for overall energy. The thermal exergy and overall exergy were determined to be 16.52 kWh and 208.05 kWh, respectively. The researchers concluded that the solar dryer was suitable for all climate conditions as it effectively utilized both direct and diffuse radiation.

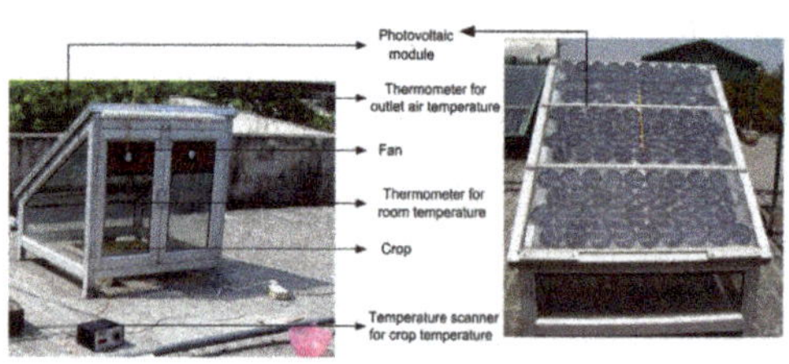

Figure 16. Greenhouse dryer apparatus [reproduced with permission from [115], Elsevier, 2016].

3.5. *Food Processing*

The food processing industry commonly relies on conventional heating sources like hot air and steam. However, the heat energy generated by PVT systems can also be effectively utilized in this industry. Steam and hot air have been the traditional choices for heating in various food processing applications. However, alternative heating methods such as infrared, ohmic, and microwave heating have gained attention due to their high energy efficiency. In recent years, PVT-based heating systems have gained popularity as a more energy-efficient option.

A theoretical analysis conducted by Herrando et al. [116] examined the heating, cooling, and electrical demands of food industries in Spain. The study identified the optimal number (N) of solar collectors that minimizes the payback time for the solar system. In the case of a solar-powered combined cooling, heating and power PVT system, the payback time decreases exponentially with the system's size when N is less than 120. This optimization leads to significant savings in electrical power consumption and increased annual savings overall.

3.6. Heating, Ventilation, and Air Conditioning

The HVAC system, responsible for heating, ventilation, and air conditioning, is crucial for maintaining human comfort and ensuring the protection of goods in residential and commercial buildings. It regulates temperature by heating and cooling the air while aiming to provide high-quality indoor air. During winter, the HVAC system utilizes thermal energy from the PVT system to heat the air. Zarei et al. [104] conducted an analysis of a hybrid PVT system for eco-friendly solar heating and cooling applications. Instead of using R134a, they employed two different refrigerants, R600 and R290, to reduce global warming potential. The study found that the outlet water temperature ranged between 31.7 °C and 46.7 °C. Furthermore, it concluded that the COP of the R290 refrigeration system improved by 11.1%, reducing the refrigerant flow rate by 60.17% and increasing the water temperature by 9.17%.

Braun et al. [105] studied a hybrid PVT system for zero-energy office buildings, focusing on trigeneration (electrical, heating, and cooling) capabilities. Through simulation, they determined that this system was technically feasible. The proposed PVT system with heat pumps exhibited higher energy potential, lower energy production costs, and greater self-consumption of locally generated energy. Zhou et al. [117] enhanced the performance of a PCM-combined hybrid PVT system by incorporating cooling and ventilation. This hybrid system integrated a ventilator, radiative cooling, active PV cooling, and PCM storage. The study showed that the radiant cooling system effectively maintained indoor temperatures while offering significant energy-saving potential. The new hybrid system demonstrated superior energy efficiency compared to conventional systems.

In India, a BIPVT system was analyszed with and without the PVT arrangement [118]. The electrical performance of the BIPVT system with an air duct was reported to be 13.11%, while without the air duct it was 12%. The experimental results aligned well with the theoretical findings. The impact of water flow in a BIPVT system with heat capacity was also investigated in India [119]. The study revealed a significant temperature drop in the PV cell temperature, resulting in a 23.7% improvement in electrical efficiency. Meanwhile, in Portugal, one study focused on the solar-absorption heating and cooling of buildings, considering the energy, economic, and environmental aspects [120]. The research explored three cities—Rome, Lisbon, and Berlin—to investigate solar heating and cooling theoretically. The results indicated that integrated solar systems showed higher economic efficiency in single-family houses and in hotels.

3.7. Thermal Powerplants

Electricity is generated from heat energy in thermal power plants. These power plants typically heat water to produce steam, which then drives steam turbines to generate electrical energy. In a study by Shah et al. [121], the thermal and electrical efficiency of hybrid PVT systems under standard testing conditions (STCs) was examined. These systems generate both electrical and thermal energy, which thermal power plants can utilize. The research findings indicated that employing an active cooling method improved the electrical energy output of the PVT system drastically.

Another study conducted by Wang et al. [122] investigated hybrid PVT systems for solar-powered combined heating and power in power plants and dairy farms (as depicted in Figure 17). The S-CHP hybrid PVT system was found to fulfil 52.0% of the steam demand and 40.0% of the hot water demand. Additionally, the system achieved an electrical efficiency equivalent to 14% of the total electrical demand. Based on the energy consumption of dairy farms, this implementation was projected to reduce 890 tonnes of CO_2 emissions per year. Of this total, 720 tonnes would stem from decreased natural gas consumption, while the remaining 100 would result from displaced electricity.

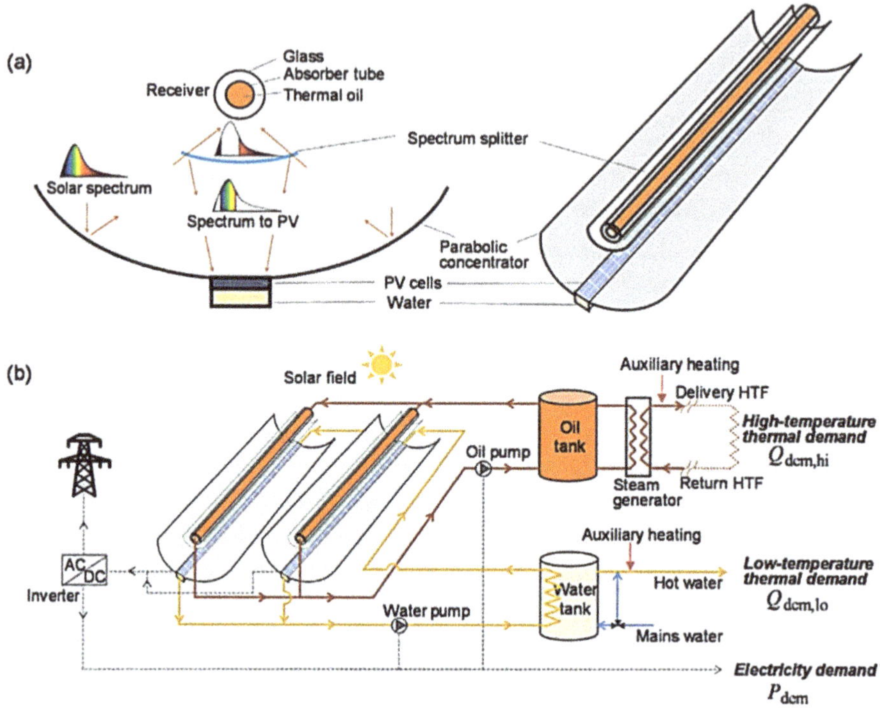

Figure 17. (**a**) Concentrating spectral-splitting hybrid PVT collector (**b**) schematic diagram of S-CHP hybrid PVT system [reproduced with permission from [122], Elsevier, 2020].

3.8. Summary of Hybrid PVT System Applications

Table 1 presents a summary of the performance and industrial applications of various PVT systems across different fields. PVT setups have found utility in water heating, solar salt removal, farming, HVAC, power plants, building heating/cooling, and swimming pool heating. Hybrid PVT systems are particularly suitable for medium-temperature applications like water heating and residential rooftops, as they serve as reliable energy sources for maximizing energy production. Among these systems, the inorganic PCM-integrated PVT system demonstrated superior performance compared to the reference system, achieving a maximum electrical efficiency of 14.99% and a thermal efficiency of 80%. Furthermore, the nanofluid-based PVT system exhibited a remarkable 33.95% improvement in electrical efficiency while effectively reducing the panel's temperature.

Table 1. Hybrid PVT system performances and applications.

Type of System	PCM	Thermal Efficiency	Electrical Efficiency	Overall Efficiency	Place of Study	Observations	Applications	References
Water-cooled PVT system	NA	NA	Increased by 7%	75.8%	Iran	The COP of the system improved by 11%, and the water temperature increased by 9.17 °C.	Solar heating applications	[104]
Nanofluid-based PVT system	NA	NA	Increased by 33.95%	83.26%	Egypt	The PV unit temperature reduced by 12 °C. Water heating applications.	Water heating	[85]
Water-based PVT system	NA	NA	14%	61%	Denmark	Exergy efficiency was increased by 4.03%, and total cost reduced by EUR 3.64/MWh.	Water heating	[83]
Solar combined heating cooling and power-based PVT system	NA	55.1%	8.62%	NA	Italy	Heating, cooling, and power systems for buildings.	Heating and cooling	[123]
Solar combined heating cooling and power-based PVT system	NA	45.4%	6.7–7.9%	-	Spain	Higher efficiency, lower fuel cost and higher reduction of CO_2 emission.	Food processing, beer brewery, pig slaughterhouse, vegetable and fruit processing, and canning plant	[116]
Water-cooled PVT system	NA	NA	9%	41.8%	Iran	The developed system is convenient for all seasons because the produced power satisfies the demand at all times.	Water heating	[87]
Water-based PVT system	NA	30%	10.4%	NA	Singapore	Glazed PVT system had higher performance than unglazed PVT system.	Hot water for student hostel	[89]
Water-based hybrid PVT system	NA	80%	7.9%	NA	Japan	PVT system with PMMA absorber and copper plate provides 80% thermal efficiency, significant cooling effect, and 0.03%/°C reduction in electrical efficiency.	Water heating, swimming pool water heating	[124]
Water-based hybrid PVT system	NA	13% enhanced	NA	NA	Japan	In a moderate wind speed (less than 1 m/s), negligible or no loss in thermal efficiency.	Swimming pool heating, floor heating	[92]
Water-based hybrid PVT system	NA	450 kWh/m^2–module/year	275 kWh/m^2–module/year	NA	Western Europe	The developed non-overglazed PVT system's stagnation temperature is always lower than 74.5 °C, which is lower the chance of overheating the PV panel.	Water heating	[61]

Table 1. *Cont.*

Type of System	PCM	Thermal Efficiency	Electrical Efficiency	Overall Efficiency	Place of Study	Observations	Applications	References
Air-based PVT system	NA	65%	12.31%	NA	Tunisia	The developed prototype system reduces the moisture content from 91.2% to 22.32% for tray 1, 28.9% for tray 2 and 30.15% for using sunlight.	Drying tomatoes	[109]
Air-based PVT system	NA	NA	13.11%	42%	India	The electrical efficiency and temperature with and without air duct is 13.11%, 12% and 10.1 °C, 25 °C, respectively.	Space heating	[118]
Water-based PVT system	NA	NA	15.5%	NA	Egypt	The developed system reduces electricity consumption by 0.12 kWh/m^3.	Solar desalination	[96]
Nanofluid-based PVT system	RT-35HC (35 °C)	Enhanced by 20.8%	Enhanced by 14.5%	Enhanced by 14.1%	Pakistan	PCM-integrated nanofluid-based PVT system increases the thermal and overall energy by 17.50% and 12.0% compared to the water-based PCM-integrated PVT system.	Thermal management in buildings	[3]
Water-based PVT system	Paraffin wax (57 °C)	26.8%	17.33	40.59%	Kottayam, India	Heating water during night time, cooling PV cell during day time.	Water heating	[90]
Nanofluid-based PVT system	Paraffin wax (57–60 °C)	Enhanced by 23.52%	Enhanced by 4.22%	NA	Tehran, Iran	Heating water during night time. Reduces the PV panel temperature by 16–21%.	Water heating	[81]
Water-based hybrid PVT system	$CaCl_2$.6H2O (22–26 °C)	80%	14.99%	NA	China	PCM and active water cooling systems are technically feasible in terms of thermal and electrical energy enhancement.	Heating and ventilation in buildings	[117]
Air-based PCM-integrated PVT system	RT42 (42 °C)	NA	Enhanced by 7.7%	NA	UK	The electrical efficiency improved by 7.7% and the reduction in cell temperature increased by 3.8 °C for PCM embedded PV system.	Building heating/cooling	[107]

CPVT systems were employed for higher-temperature applications such as thermal power plants and solar desalination. The nanofluid PVT arrangement demonstrated higher overall thermal and electrical efficiency among the various PVT configurations. The advantage of PCM-based PVT systems lies in their ability to store thermal energy in the PCM, enabling utilization when solar energy is not available. This stored heat energy can be effectively utilized for water heating, space heating, and crop drying applications. Additionally, grid-connected and building-integrated PVT systems offer a seamless appearance, eliminating the need for separate thermal collectors and PV modules. It is recommended to improve the thermal conductivity of PCMs to enhance thermal energy storage and the heat transfer rate. This can be achieved by incorporating highly conductive nanoparticles into the PCM, further enhancing the heat transfer rate and storage capacity.

3.9. Discussion and Inferences Abount Hybrid PVT Systems

Hybrid photovoltaic thermal (PVT) systems have found widespread use in various industrial applications, with efficiency being the most crucial factor in PVT technologies. The performance of a hybrid PVT system is influenced by several factors, as shown in Figure 18. In evaluating the electrical, thermal, and overall performance of the PVT system, three operational parameters are considered: design, climate conditions, and operating characteristics.

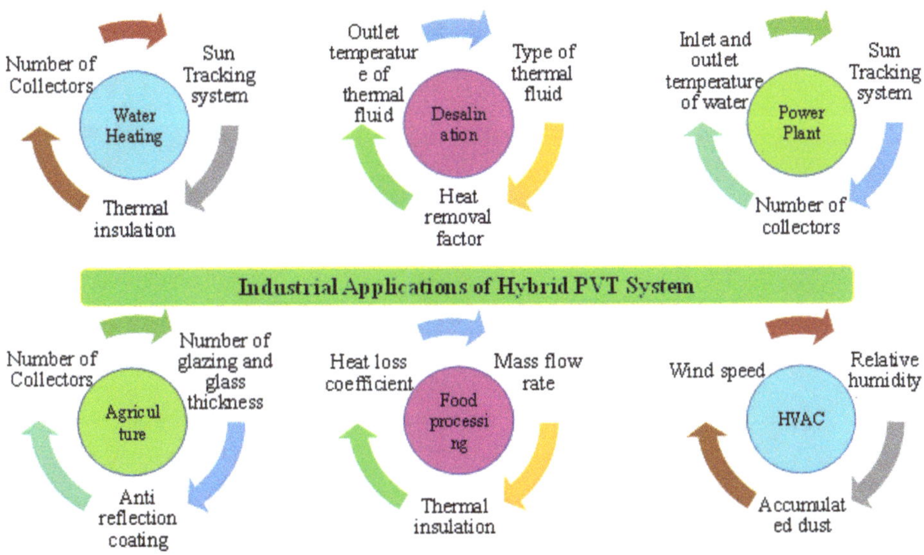

Figure 18. Inferences on hybrid PVT system.

The thermal efficiency of the PVT system decreases as the ambient temperature, inlet water temperature, channel depth, duct length, packing factor, thermal resistance, and flow rate increase. Once the optimal value is exceeded, improving the velocity of the inlet flow rate, tilt angle, wind speed, heat loss coefficient, and thermal conductivity of insulation can enhance the system's performance. Also, the thermal efficiency increases with higher solar irradiation, tilt angles close to the latitude, tracking systems, absorber quality, and supporting metal bars. PV materials play a significant role in enhancing the system's electrical efficiency.

The performance of a water-based PVT system depends on factors such as the type of heat transfer fluid used, solar radiation intensity, and the entry and exit temperatures of the heat transfer fluid. Therefore, using a heat transfer fluid with higher thermal conductivity is advisable. In a solar desalination system, performance relies on solar irradiation, the heat transfer fluid's entry temperature, the absorber plate's quality, and thermal insulation. The use of high-thermal-conductivity fluids like nanofluids can effectively remove heat from the PVT system, utilizing that energy to evaporate salty water in salt removal systems. However, many systems lack proper thermal insulation and higher-thermal conductivity heat transfer fluids. This issue is also relevant in PVT systems used in thermal power stations, where water is heated to generate steam. To improve the performance of thermal power plants, it is necessary to incorporate proper thermal insulation and higher-thermal conductivity heat transfer fluids. The rate of steam production is a critical factor influencing power plant performance. Furthermore, hot air is required for food processing and HVAC systems to enhance their performance. Therefore, optimizing the air flow rate, glazing thickness, and ensuring proper thermal insulation is recommended.

The fluid used in a PVT system can be air, liquid, or a combination of both. Air is a cost-effective and environmentally friendly option that can be used in various locations. However, water is the least expensive and most commonly used fluid due to its purity and usability. Additionally, air-based PVT systems require less maintenance compared to water-based ones. However, certain applications may necessitate the use of different fluids, requiring specific designs or configurations.

4. Environmental Analysis of PVT Systems

Factors such as population growth, industrialization, and technological advancements make it inevitable that greenhouse gas emissions will continue to rise. The primary greenhouse gases of concern include carbon dioxide, methane, sulfur hexafluoride, carbon monoxide, nitrous oxide, and chlorofluorocarbons. To mitigate the release of these gases into the atmosphere, the implementation of PVT systems could be an effective solution. This section explores the significant environmental parameters associated with the impact of PVT systems on the environment.

The introduction of carbon tax and emission trading systems by many nations will impose additional costs on energy usage due to CO_2 emissions [125]. Exposure to greenhouse gases, particularly carbon dioxide, is known to have adverse effects on human health and the environment. Therefore, an analysis was conducted to investigate the reduction of CO_2 emissions from using PVT systems using Equation (1).

$$\bar{\omega}_{ov} = \lambda_{CO_2*} \dot{E}_{ov} \quad (1)$$

$\bar{\omega}_{ov}$ is CO_2 emission ($mtCO_2$), λ_{CO_2*} is electrical equivalent intensity (grams CO_2 equ. per kWh electricity), and $\dot{E}_{ov}$ is overall energy or exergy generation (W).

The overall energy production of the PVT system is estimated using Equation (2).

$$\dot{E}_{ov} = \dot{E}_{el} + \dot{E}_{th} * 0.38 \quad (2)$$

$\dot{E}_{el}$ denotes the electrical power in watts and $\dot{E}_{th}$ is the thermal power in Watts.

Studies have shown that a PV unit can contribute to a reduction of 16.3 kg of CO_2 over a three-month period. Similarly, PCM-embedded PVT systems with water, EG 50%, and EG 100% have demonstrated reductions of 52.56%, 51.38%, and 49.69%, respectively, over the same duration [126]. The implementation of a PVT system with nanofluid in Saudi Arabia could potentially reduce CO_2 emissions by 16,975 tonnes [127]. In another case, a PVT system with air as the working fluid prevented the release of 1.98 kg/h of CO_2 emissions from a solar dryer into the atmosphere [128]. An environmental analysis of a solar-assisted combined cooling, heating, and power (S-CCHP) system revealed that it could prevent the emission of 911 tons of CO_2 per year, which is 16% higher than a similar PV system [123].

The environmental impact assessment of a combined cooling, heating, and power (CCHP) system, coupled with PVT, micro gas turbine, and absorption chiller, was conducted. The analysis divided the proposed CCHP system into three subsystems: power generation mode with MGT, CHP mode combined with PVT cycle, and CCHP mode with MGT combined with absorption systems. The CO_2 emissions from these subsystems were computed using equations (Equations (3)–(5)). The proposed system was found to emit 0.16 (ton/MWh) of CO_2 annually, which is relatively low compared to other methods [129].

$$\varepsilon_{em,GT} = \frac{\dot{m}_{CO_2,emitted}}{\dot{W}_{GT}} \tag{3}$$

$$\varepsilon_{em,CHP} = \frac{\dot{m}_{CO_2,emitted}}{\dot{W}_{GT} + \dot{Q}_{heating}} \tag{4}$$

$$\varepsilon_{em,CCHP} = \frac{\dot{m}_{CO_2,emitted}}{\dot{W}_{GT} + \dot{Q}_{heating} + \dot{Q}_{cooling}} \tag{5}$$

The yearly reduction (ER) in CO_2 emission of the proposed PVT system at the University Sports Center of Bari, Italy, was evaluated based on CO_2 discharge factors of natural gas (f_{ng}) and electricity (f_{el}). Equation (6) was used for the assessment. The yearly reduction in cost of CO_2 emission per unit was found to vary from EUR 120/tCO_2 to a value less than EUR 1/tCO_2.

$$ER = \frac{Q_{COV}}{\eta_{boil}} f_{ng} \tag{6}$$

$$ER_{el} = (E_{COV} + E_{exc}) \cdot f_{el} \tag{7}$$

$$ER_{tot} = ER_{ng} + ER_{el} \tag{8}$$

ER_{tot} denotes the total emission reduction (Equation (7)), ER_{ng} denotes the emission reduction due to displaced natural gas, ER_{el} denotes the emission reduction due to displaced electricity (Equation (8)).

Furthermore, a hybrid PV-ETC system could achieve an annual reduction of 317 tons of CO_2 emissions, while a novel PVT S-CHP system demonstrated a reduction of 445 tons of CO_2 emissions per year. In contrast, an ICE-CHP system only achieved a reduction of 25 tons of CO_2 emissions per year. An ultra-low-cost PVT collector design was found to reduce annual CO_2 emissions by 183.8 kg, and over a 20-year lifespan it could reduce emissions by 3.7 metric tons of CO_2 per square meter of PVT collector installed [130]. Therefore, the implementation of this system can contribute to the European Union's goal of achieving a 40% reduction in carbon emissions by 2030 [8].

The environmental impacts of different configurations of water/carbon nanotube-based PVT systems were examined, revealing the potential to prevent emissions of CO_2, SO_2, NOx, CO, and particulate matter [131]. Additionally, the use of aluminum foam in PCM-integrated PVT systems was found to effectively reduce CO_2 emissions [132].

However, the incorporation of nanomaterials in the cooling fluid can enhance the performance of PVT systems and improve their environmental parameters. It should be noted that while nanofluid-based solar systems can reduce greenhouse gas emissions more effectively and generate higher energy outputs compared to conventional cooling fluids, the production of nanoparticles may also pose contamination risks to other natural sources such as soil and water.

5. Technical Challenges and Limitations

This comprehensive review article delves into the extensive applications of hybrid photovoltaic thermal (PVT) systems. Each type of PVT system holds its own significance and value, depending on its intended use and environmental feasibility. However, for these systems to truly compete with established solar thermal and PVT systems in the commercial market, they require significant upgrades.

One of the challenges associated with conventional working fluids used in PVT systems is their inferior thermal behavior in relation to their expected application. Researchers have been actively searching for improved working fluids with enhanced thermal properties to overcome this. In particular, the freezing of water in cold regions restricts the use of water-based PVT systems, making air-based PVT systems more suitable for heating applications in such areas. Therefore, it is imperative to develop appropriate water-based systems tailored for cold regions.

In terms of system design, active PVT systems necessitate more pumping power compared to passive systems. PVT systems employing natural circulation are more cost-effective than those relying on forced circulation. Water supply availability limits the application of water-based PVT systems, whereas air is readily accessible for air-based PVT systems. The choice of medium depends heavily on the intended application of the PVT system.

The use of conductive adhesives can improve heat transfer and system efficiency. Introducing multiple air inputs into air-based PVT systems is considered beneficial. Moreover, single-laminated PVT units, referred to as TESPI systems, have the potential to reduce thermal resistance provided by tedlers, encapsulant materials, and adhesives, although they have yet to be commercialized. Another approach to enhance thermal performance is to partially cover the absorber layer with PV modules rather than entirely covering it. However, it should be noted that the addition of glazing in PVT systems increases thermal efficiency at the expense of electrical performance.

Despite the promising applications of PVT technology, its commercialization is still limited. To facilitate widespread adoption, focusing on and improving these systems' initial and operating costs is crucial. Additionally, extensive research is required in emerging areas of PVT technology, such as optimizing auxiliary power usage (e.g., pumps or blowers), reducing materials costs, and exploring the utilization of smart materials.

In conclusion, hybrid PVT systems offer diverse applications, but their market penetration requires substantial upgrades and cost improvements. Continued research and development efforts are essential to fully unlock the potential of PVT technology and facilitate its commercial viability.

6. Conclusions and Recommendations

The present research analyzed and summarized the literature on experimental applications of hybrid PVT systems with and without PCMs for various industrial applications and buildings. Industrial applications, such as water heating, water desalination, thermal management in buildings, food processing, HVAC, agricultural processes, and thermal power plants, were taken into consideration. This technology offers benefits such as reduced electrical consumption, shifting of cooling and heating peak loads to off-peak periods, minimized temperature fluctuations, and improved thermal comfort. Additionally, the incorporation of nanoparticles into PCMs has been shown to enhance the performance of PVT systems.

The present study discussed the various types of cooling systems of PVT systems. The thermal management system bi-fluid cooling system performs better than other systems. Also, the PCM-integrated PVT system has the advantages of cooling the PVT system and storing excess thermal energy. The stored energy can be used when solar energy is unavailable. Studies have demonstrated that PVT systems incorporating NEPCMs exhibit higher thermal performance. During charging, these systems stored 8.3% more heat energy; during discharging, they released 25.1% more heat energy than traditional PVT systems. Moreover, employing a solar-powered combined cooling, heating, and power (S-CCHP) system instead of a conventional PV system can prevent the emission of 911 tons of CO_2 per year. Enhancing the overall efficiency of hybrid PVT systems leads to reduced payback periods and CO_2 emissions.

Based on the analysis conducted in this article, the following recommendations for future research in the field of hybrid PVT systems are presented:

- Investigate the impact of dust and dirt on the efficiency of PVT systems under various weather conditions, as most previous research has focused on clean conditions with low occurrence and intensity of rain.
- Conduct experiments that simultaneously explore the use of nanofluid as the heat transfer fluid (HTF) and PCM/NEPCM as the cooling method, as this area has received limited attention compared to studies on nanofluid- and water-based PVT systems.
- Perform long-term tests to assess the performance of PVT devices over extended periods, including the examination of the effects of prolonged solar exposure on adhesives, glazes, selective coating materials, and other components.
- Explore alternative heat transfer techniques, such as PCMs and NEPCMs, as cooling media in PVT systems, as the majority of the existing literature primarily focuses on water and air as the cooling mediums. Investigate various PCM and NEPCM options for PVT systems.
- It is necessary to work on flame retardant and leakage-preventable PCMs to avoid leakage during the phase transition of the PCM-integrated PVT system.
- Consider studying nano-fluid-based jet impingement systems, as they have the potential to generate a greater amount of heat energy and electricity.

By addressing these recommendations, future research can further advance the understanding and effectiveness of hybrid PVT systems in industrial applications.

Funding: This research was funded by Universiti Malaysia Pahang (RDU192208).

Data Availability Statement: Data sharing is not applicable to this article.

Acknowledgments: The author would like to thank Universiti Malaysia Pahang (UMP) for the financial support under grant RDU192208.

Conflicts of Interest: The author declares no conflict of interest.

Nomenclature

ALCC	Annualized life cycle cost
ALCS	Annualized life cycle savings
COP	Coefficient of performance
CPVT	Concentrated photovoltaic thermal
ECS	Electricity savings
EG	Ethylene glycol
Epv	Photovoltaic energy
Et	Thermal energy
HTF	Heat transfer fluid
LCC	Life cycle cost
LPM	Liters per minute
MC	Maintenance cost
N	Optimum number
NCOT	Nominal cell operating temperature
NEPCM	Nano-enhanced phase change material
PCM	Phase change materials
PMMA	Polymethyl methacrylate
PV	Photovoltaic
PVT	Photovoltaic thermal
PWF	Present worth factor
SAH	Solar air heater
STC	Standard temperature conditions
tCO_2	Tonne of carbon dioxide
Tmean	Mean temperature
V	Volt
VAR	Vapor absorption refrigeration
Voc	Open circuit voltage
W	Watts

References

1. Zhan, S.; Dong, B.; Chong, A. Improving energy flexibility and PV self-consumption for a tropical net zero energy office building. *Energy Build.* **2023**, *278*, 112606. [CrossRef]
2. Laghari, I.A.; Samykano, M.; Pandey, A.K.; Kadirgama, K.; Tyagi, V.V. Advancements in PV-thermal systems with and without phase change materials as a sustainable energy solution: Energy, exergy and exergoeconomic (3E) analytic approach. *Sustain. Energy Fuels* **2020**, *10*, 4956–4987. [CrossRef]
3. Hassan, A.; Wahab, A.; Qasim, M.A.; Janjua, M.M.; Ali, M.A.; Ali, H.M.; Javaid, N. Thermal management and uniform temperature regulation of photovoltaic modules using hybrid phase change materials-nanofluids system. *Renew. Energy* **2020**, *145*, 282–293. [CrossRef]
4. Verma, S.; Verma, A.; Kumar, V.; Gangil, B. Materials Today: Proceedings Concentrated photovoltaic thermal systems using Fresnel lenses—A review. *Mater. Today Proc.* **2021**, *44*, 4256–4260. [CrossRef]
5. Sadeghi, G.; Pisello, A.L.; Safarzadeh, H.; Poorhossein, M.; Jowzi, M. On the effect of storage tank type on the performance of evacuated tube solar collectors: Solar radiation prediction analysis and case study. *Energy* **2020**, *198*, 117331. [CrossRef]
6. IEA. International Energy Agency (IEA) World Energy Outlook 2022. 2022. Available online: https://Www.Iea.Org/Reports/World-Energy-Outlook-2022/Executive-Summary (accessed on 20 June 2023).
7. Jaganmohan, M. Global Cumulative Installed Solar PV Capacity 2000–2019. 2021. Available online: https://www.statista.com/statistics/280220/global-cumulative-installed-solar-pv-capacity/ (accessed on 20 June 2023).
8. Diwania, S.; Agrawal, S.; Siddiqui, A.S.; Singh, S. Photovoltaic–thermal (PV/T) technology: A comprehensive review on applications and its advancement. *Int. J. Energy Environ. Eng.* **2020**, *11*, 33–54. [CrossRef]
9. Raja, A.A.; Huang, Y. Novel parabolic trough solar collector and solar photovoltaic/thermal hybrid system for multi-generational systems. *Energy Convers. Manag.* **2020**, *211*, 112750. [CrossRef]
10. Yang, L.; Huang, J.N.; Zhou, F. Thermophysical properties and applications of nano-enhanced PCMs: An update review. *Energy Convers. Manag.* **2020**, *214*, 112876. [CrossRef]
11. Mandal, S.; Ghosh, S.K. Experimental investigation of the performance of a double pass solar water heater with reflector. *Renew. Energy* **2020**, *149*, 631–640. [CrossRef]
12. Chopra, K.; Tyagi, V.V.; Pandey, A.K.; Sharma, R.; Sari, A. PCM integrated glass in glass tube solar collector for low and medium temperature applications: Thermodynamic & techno-economic approach. *Energy* **2020**, *198*, 117238. [CrossRef]
13. Khanmohammadi, S.; Baseri, M.M.; Ahmadi, P.; Al-Rashed, A.A.A.; Afrand, M. Proposal of a novel integrated ocean thermal energy conversion system with flat plate solar collectors and thermoelectric generators: Energy, exergy and environmental analyses. *J. Clean. Prod.* **2020**, *256*, 120600. [CrossRef]
14. Ma, T.; Li, M.; Kazemian, A. Photovoltaic thermal module and solar thermal collector connected in series to produce electricity and high-grade heat simultaneously. *Appl. Energy* **2019**, *261*, 114380. [CrossRef]
15. Liang, R.; Wang, C.; Wang, P.; Zhao, L. Performance modeling and analysis of a PVT-HP system with the roll-bond plate as the evaporator during winter conditions. *Appl. Therm. Eng.* **2023**, *224*, 120102. [CrossRef]
16. Qiu, L.; Ouyang, Y.; Feng, Y.; Zhang, X. Review on micro/nano phase change materials for solar thermal applications. *Renew. Energy* **2019**, *140*, 513–538. [CrossRef]
17. Pathak, S.K.; Tyagi, V.V.; Chopra, K.; Rejikumar, R.; Pandey, A.K. Integration of emerging PCMs and nano-enhanced PCMs with different solar water heating systems for sustainable energy future: A systematic review. *Sol. Energy Mater. Sol. Cells* **2023**, *254*, 112237. [CrossRef]
18. Nejadhasan, S.; Zaheri, F.; Abiri, E.; Salehi, M.R. PVT-compensated low voltage LNA based on variable current source for low power applications. *AEU Int. J. Electron. Commun.* **2022**, *143*, 154042. [CrossRef]
19. Liu, W.; Yao, J.; Jia, T.; Zhao, Y.; Dai, Y.; Zhu, J.; Novakovic, V. The performance optimization of DX-PVT heat pump system for residential heating. *Renew. Energy* **2023**, *206*, 1106–1119. [CrossRef]
20. Abdelrazik, A.S.; Al-Sulaiman, F.A.; Saidur, R.; Ben-Mansour, R. A review on recent development for the design and packaging of hybrid photovoltaic/thermal (PV/T) solar systems. *Renew. Sustain. Energy Rev.* **2018**, *95*, 110–129. [CrossRef]
21. Kumar, R.R.; Samykano, M.; Pandey, A.K.; Kadirgama, K.; Tyagi, V.V. Phase change materials and nano-enhanced phase change materials for thermal energy storage in photovoltaic thermal systems: A futuristic approach and its technical challenges. *Renew. Sustain. Energy Rev.* **2020**, *133*, 110341. [CrossRef]
22. Santosh, M.P.R.; Kumaresan, M.R.S.G. Study of effect of Al and Cu microparticles dispersed in D-Mannitol PCM for effective solar thermal energy storage. *J. Therm. Anal. Calorim.* **2020**, *139*, 895–904. [CrossRef]
23. Kumar, R.; Pandey, A.K.; Samykano, M.; Aljafari, B.; Ma, Z.; Bhattacharyya, S.; Tyagi, V.V. Phase change materials integrated solar desalination system: An innovative approach for sustainable and clean water production and storage. *Renew. Sustain. Energy Rev.* **2022**, *165*, 112611. [CrossRef]
24. Sharma, A.; Tyagi, V.V.; Chen, C.R.; Buddhi, D. Review on thermal energy storage with phase change materials and applications. *Renew. Sustain. Energy Rev.* **2009**, *2*, 318–345. [CrossRef]
25. Kumar, R.; Pandey, A.K.; Samykano, M.; Mishra, Y.N.; Mohan, R.V.; Sharma, K.; Tyagi, V.V. Effect of surfactant on functionalized multi-walled carbon nano tubes enhanced salt hydrate phase change material. *J. Energy Storage* **2022**, *55*, 105654. [CrossRef]
26. Alsaqoor, S.; Alqatamin, A.; Alahmer, A.; Nan, Z.; Al-Husban, Y.; Jouhara, H. The impact of phase change material on photovoltaic thermal (PVT) systems: A numerical stud. *Int. J. Thermofluids* **2023**, *18*, 100365. [CrossRef]

27. Kabeel, A.E.; Sathyamurthy, R.; Manokar, A.M.; Sharshir, S.W.; Essa, F.A.; Elshiekh, A.H. Experimental study on tubular solar still using Graphene Oxide Nano particles in Phase Change Material (NPCM's) for fresh water production. *J. Energy Storage* **2020**, *28*, 101204. [CrossRef]
28. Kumar, R.R. Experimental Investigations on Thermal Properties of Copper (II) Oxide Nanoparticles Enhanced Inorganic Phase Change Materials for Solar Thermal Energy Storage Applications. In Proceedings of the 2022 Advances in Science and Engineering Technology International Conferences (ASET), Dubai, United Arab Emirates, 21–24 February 2022; pp. 1–6. [CrossRef]
29. Kumar, R.; Samykano, M.; Pandey, A.K.; Kadirgama, K.; Tyagi, V.V. A comparative study on thermophysical properties of functionalized and non-functionalized Multi-Walled Carbon Nano Tubes (MWCNTs) enhanced salt hydrate phase change material. *Sol. Energy Mater. Sol. Cells* **2022**, *240*, 111697. [CrossRef]
30. Kumar, R.; Samykano, M.; Ngui, W.K.; Pandey, A.K.; Kalidasan, B.; Kadirgama, K.; Tyagi, V.V. Investigation of thermal performance and chemical stability of graphene enhanced phase change material for thermal energy storage. *Phys. Chem. Earth* **2022**, *128*, 103250. [CrossRef]
31. Samykano, M. Role of phase change materials in thermal energy storage: Potential, recent progress and technical challenges. *Sustain. Energy Technol. Assess.* **2022**, *52*, 102234. [CrossRef]
32. Dwivedi, P.; Sudhakar, K.; Soni, A.; Solomin, E.; Kirpichnikova, I. Advanced cooling techniques of P.V. modules: A state of art. *Case Stud. Therm. Eng.* **2020**, *21*, 100674. [CrossRef]
33. Şirin, C.; Goggins, J.; Hajdukiewicz, M. A review on building-integrated photovoltaic/thermal systems for green buildings. *Appl. Therm. Eng.* **2023**, *229*, 120607. [CrossRef]
34. Sathe, T.M.; Dhoble, A.S. A review on recent advancements in photovoltaic thermal techniques. *Renew. Sustain. Energy Rev.* **2017**, *76*, 645–672. [CrossRef]
35. Bevilacqua, P.; Morabito, A.; Bruno, R.; Ferraro, V.; Arcuri, N. Seasonal performances of photovoltaic cooling systems in different weather conditions. *J. Clean. Prod.* **2020**, *272*, 122459. [CrossRef]
36. Jia, Y.; Alva, G.; Fang, G. Development and applications of photovoltaic–thermal systems: A review. *Renew. Sustain. Energy Rev.* **2019**, *102*, 249–265. [CrossRef]
37. Pandey, A.; Laghari, I.A.; Kumar, R.R.; Chopra, K.; Samykano, M.; Abusorrah, A.M.; Sharma, K.; Tyagi, V. Energy, exergy, exergoeconomic and enviroeconomic (4-E) assessment of solar water heater with/without phase change material for building and other applications: A comprehensive review. *Sustain. Energy Technol. Assess.* **2021**, *45*, 101139. [CrossRef]
38. Kalidasan, B.; Pandey, A.K.; Shahabuddin, S.; Samykano, M.; Thirugnanasambandam, M. Phase change materials integrated solar thermal energy systems: Global trends and current practices in experimental approaches. *J. Energy Storage* **2019**, *27*, 101118. [CrossRef]
39. Ahmadi, A.; Ehyaei, M.; Doustgani, A.; Assad, M.E.H.; Hmida, A.; Jamali, D.; Kumar, R.; Li, Z.; Razmjoo, A. Recent residential applications of low-temperature solar collector. *J. Clean. Prod.* **2020**, *279*, 123549. [CrossRef]
40. Al-Waeli, A.H.A.; Kazem, H.A.; Chaichan, M.T.; Sopian, K. A review of photovoltaic thermal systems: Achievements and applications. *Int. J. Energy Res.* **2021**, *45*, 1269–1308. [CrossRef]
41. Al-waeli, A.H.A.; Kazem, H.A. Advances in Nano-Materials Used in Photovoltaic/Thermal Systems. In *Advances in Energy Materials, Advances in Material Research and Technology*; Springer: Berlin/Heidelberg, Germany, 2020; pp. 105–133. [CrossRef]
42. Alessandro, M.; Niccolao, A.; Claudio, D.P.; Fabrizio, L. Photovoltaic-thermal solar-assisted heat pump systems for building applications: Integration and design methods. *Energy Built Environ.* **2023**, *4*, 39–56. [CrossRef]
43. Prasetyo, S.D.; Prabowo, A.R.; Arifin, Z. The use of a hybrid photovoltaic/thermal (PV/T) collector system as a sustainable energy-harvest instrument in urban technology. *Heliyon* **2023**, *9*, e13390. [CrossRef]
44. You, T.; Wu, W.; Yang, H.; Liu, J.; Li, X. Hybrid photovoltaic/thermal and ground source heat pump: Review and perspective. *Renew. Sustain. Energy Rev.* **2021**, *151*, 111569. [CrossRef]
45. Das, D.; Kalita, P.; Roy, O. Flat plate hybrid photovoltaic- thermal (PV/T) system: A review on design and development. *Renew. Sustain. Energy Rev.* **2018**, *84*, 111–130. [CrossRef]
46. Good, C. Environmental impact assessments of hybrid photovoltaic-thermal (PV/T) systems—A review. *Renew. Sustain. Energy Rev.* **2016**, *55*, 234–239. [CrossRef]
47. Chaibi, Y.; El Rhafiki, T.; Simón-Allué, R.; Guedea, I.; Luaces, S.C.; Gajate, O.C.; Kousksou, T.; Zeraouli, Y. Air-based hybrid photovoltaic/thermal systems: A review. *J. Clean. Prod.* **2021**, *295*, 126211. [CrossRef]
48. Tirupati, V.; Sekhar, Y.R. Hybrid Photovoltaic/Thermal (PVT) Collector Systems With Different Absorber Configurations For Thermal Management—A Review. *Energy Environ.* **2023**, *34*, 690–735. [CrossRef]
49. Abdelrazik, A.S.; Shboul, B.; Elwardany, M.; Zohny, R.N.; Osama, A. The recent advancements in the building integrated photovoltaic/thermal (BIPV/T) systems: An updated review. *Renew. Sustain. Energy Rev.* **2022**, *170*, 112988. [CrossRef]
50. Ekka, J.P.; Kumar, D. A review of industrial food processing using solar dryers with heat storage systems. *J. Stored Prod. Res.* **2023**, *101*, 102090. [CrossRef]
51. Fikri, M.A.; Samykano, M.; Pandey, A.; Kadirgama, K.; Kumar, R.R.; Selvaraj, J.; Rahim, N.A.; Tyagi, V.; Sharma, K.; Saidur, R. Recent progresses and challenges in cooling techniques of concentrated photovoltaic thermal system: A review with special treatment on phase change materials (PCMs) based cooling. *Sol. Energy Mater. Sol. Cells* **2022**, *241*. [CrossRef]
52. Anand, R.N.B.; Shankar, R.; Murugavelh, S.; Rivera, W.; Prasad, K.M. A review on solar photovoltaic thermal integrated desalination technologies. *Renew. Sustain. Energy Rev.* **2021**, *141*, 110787. [CrossRef]

53. Özakin, A.N.; Kaya, F. Effect on the exergy of the PVT system of fins added to an air-cooled channel: A study on temperature and air velocity with ANSYS Fluent. *Sol. Energy* **2019**, *184*, 561–569. [CrossRef]
54. Li, J.; Zhang, W.; Xie, L.; Li, Z.; Wu, X.; Zhao, O.; Zhong, J.; Zeng, X. A hybrid photovoltaic and water/air based thermal(PVT) solar energy collector with integrated PCM for building application. *Renew. Energy* **2022**, *199*, 662–671. [CrossRef]
55. Agrawal, S.; Tiwari, G.N. Energy and exergy analysis of hybrid micro-channel photovoltaic thermal module. *Sol. Energy* **2011**, *85*, 356–370. [CrossRef]
56. Touti, E.; Masmali, M.; Fterich, M.; Chouikhi, H. Experimental and numerical study of the PVT design impact on the electrical and thermal performances. *Case Stud. Therm. Eng.* **2023**, *43*, 102732. [CrossRef]
57. Tiwari, S.; Tiwari, G.N. Energy and exergy analysis of a mixed-mode greenhouse-type solar dryer, integrated with partially covered N-PVT air collector. *Energy* **2017**, *128*, 183–195. [CrossRef]
58. Ali, H.M. Recent advancements in PV cooling and efficiency enhancement integrating phase change materials based systems—A comprehensive review. *Sol. Energy* **2019**, *197*, 163–198. [CrossRef]
59. Bevilacqua, P.; Perrella, S.; Cirone, D.; Bruno, R.; Arcuri, N. Efficiency improvement of photovoltaic modules via back surface cooling. *Energies* **2021**, *14*, 4. [CrossRef]
60. Aste, N.; Del Pero, C.; Leonforte, F.; Manfren, M. Performance monitoring and modeling of an uncovered photovoltaic-thermal (PVT) water collector. *Sol. Energy* **2016**, *135*, 551–568. [CrossRef]
61. Brottier, L.; Bennacer, R. Thermal performance analysis of 28 PVT solar domestic hot water installations in Western Europe. *Renew. Energy* **2020**, *160*, 196–210. [CrossRef]
62. Misha, S.; Abdullah, A.L.; Tamaldin, N.; Rosli, M.A.M.; Sachit, F.A. Simulation CFD and experimental investigation of PVT water system under natural Malaysian weather conditions. *Energy Rep.* **2019**, *6*, 28–44. [CrossRef]
63. Bevilacqua, P.; Bruno, R.; Rollo, A.; Ferraro, V. A novel thermal model for PV panels with back surface spray cooling. *Energy* **2022**, *255*, 124401. [CrossRef]
64. Bevilacqua, P.; Perrella, S.; Bruno, R.; Arcuri, N. An accurate thermal model for the PV electric generation prediction: Long-term validation in different climatic conditions. *Renew. Energy* **2021**, *163*, 1092–1112. [CrossRef]
65. Indartono, Y.S.; Suwono, A.; Pratama, F.Y. Improving photovoltaics performance by using yellow petroleum jelly as phase change material. *Int. J. Low-Carbon Technol.* **2016**, *11*, 333–337. [CrossRef]
66. Bevilacqua, P.; Bruno, R.; Arcuri, N. Comparing the performances of different cooling strategies to increase photovoltaic electric performance in different meteorological conditions. *Energy* **2020**, *195*, 116950. [CrossRef]
67. Jarimi, H.; Bakar, M.N.A.; Othman, M.; Din, M.H. Bi-fluid photovoltaic/thermal (PV/T) solar collector: Experimental validation of a 2-D theoretical model. *Renew. Energy* **2016**, *85*, 1052–1067. [CrossRef]
68. Bakar, M.N.A.; Othman, M.; Din, M.H.; Manaf, N.A.; Jarimi, H. Design concept and mathematical model of a bi-fluid photovoltaic/thermal (PV/T) solar collector. *Renew. Energy* **2014**, *67*, 153–164. [CrossRef]
69. Hissouf, M.; Feddaoui, M.; Najim, M.; Charef, A. Performance of a photovoltaic-thermal solar collector using two types of working fluids at different fluid channels geometry. *Renew. Energy* **2020**, *162*, 1723–1734. [CrossRef]
70. Jouhara, H.; Chauhan, A.; Nannou, T.; Almahmoud, S.; Delpech, B.; Wrobel, L.C. Heat pipe based systems—Advances and applications. *Energy* **2017**, *128*, 729–754. [CrossRef]
71. Diallo, T.M.; Yu, M.; Zhou, J.; Zhao, X.; Shittu, S.; Li, G.; Ji, J.; Hardy, D. Energy performance analysis of a novel solar PVT loop heat pipe employing a microchannel heat pipe evaporator and a PCM triple heat exchanger. *Energy* **2018**, *167*, 866–888. [CrossRef]
72. Long, H.; Chow, T.T.; Ji, J. Building-integrated heat pipe photovoltaic/thermal system for use in Hong Kong. *Sol. Energy* **2017**, *155*, 1084–1091. [CrossRef]
73. Nazir, H.; Batool, M.; Osorio, F.J.B.; Isaza-Ruiz, M.; Xu, X.; Vignarooban, K.; Phelan, P.; Inamuddin; Kannan, A.M. Recent developments in phase change materials for energy storage applications: A review. *Int. J. Heat Mass Transf.* **2019**, *129*, 491–523. [CrossRef]
74. Asefi, G.; Ma, T.; Wang, R. Techno-economic evaluation of photovoltaic thermal system integrated with porous phase change materials: Case studies in China. *Energy Convers. Manag.* **2022**, *290*, 117227. [CrossRef]
75. Zhao, J.; Li, Z.; Ma, T. Performance analysis of a photovoltaic panel integrated with phase change material. *Energy Procedia* **2019**, *158*, 1093–1098. [CrossRef]
76. Khan, M.M.A.; Ibrahim, N.I.; Mahbubul, I.M.; Ali, H.M.; Saidur, R.; Al-Sulaiman, F.A. Evaluation of solar collector designs with integrated latent heat thermal energy storage: A review. *Sol. Energy* **2018**, *166*, 334–350. [CrossRef]
77. Hosouli, S.; Gomes, J.; Loris, A.; Pazmiño, I.-A.; Naidoo, A.; Lennermo, G.; Mohammadi, H. Evaluation of a solar photovoltaic thermal (PVT) system in a dairy farm in Germany. *Sol. Energy Adv.* **2023**, *3*. [CrossRef]
78. Hossain, M.S.; Pandey, A.K.; Selvaraj, J.; Rahim, N.A.; Islam, M.M.; Tyagi, V.V. Two side serpentine flow based photovoltaic-thermal-phase change materials (PVT-PCM) system: Energy, exergy and economic analysis. *Renew. Energy* **2019**, *136*, 1320–1336. [CrossRef]
79. Al-Waeli, A.H.A.; Kazem, H.A.; Yousif, J.H.; Chaichan, M.T.; Sopian, K. Mathematical and neural network modeling for predicting and analyzing of nanofluid-nano PCM photovoltaic thermal systems performance. *Renew. Energy* **2020**, *145*, 963–980. [CrossRef]
80. Bhakre, S.S.; Sawarkar, P.D.; Kalamkar, V.R. Numerical study on photovoltaic thermal phase change material system in hot climatic conditions. *Appl. Therm. Eng.* **2023**, *227*, 120423. [CrossRef]

81. Naghdbishi, A.; Yazdi, M.E.; Akbari, G. Experimental investigation of the effect of multi-wall carbon nanotube—Water/glycol based nanofluids on a PVT system integrated with PCM-covered collector. *Appl. Therm. Eng.* **2020**, *178*, 115556. [CrossRef]
82. Dhoke, A.; Sharma, R.; Saha, T.K. An approach for fault detection and location in solar PV systems. *Sol. Energy* **2019**, *194*, 197–208. [CrossRef]
83. Behzadi, A.; Arabkoohsar, A.; Yang, Y. Optimization and dynamic techno-economic analysis of a novel PVT-based smart building energy system. *Appl. Therm. Eng.* **2020**, *181*, 115926. [CrossRef]
84. Ma, Z.; Lin, W.; Sohel, M.I. Nano-enhanced phase change materials for improved building performance. *Renew. Sustain. Energy Rev.* **2016**, *58*, 1256–1268. [CrossRef]
85. Abdallah, S.R.; Saidani-Scott, H.; Abdellatif, O.E. Performance analysis for hybrid PV/T system using low concentration MWCNT (water-based) nanofluid. *Sol. Energy* **2019**, *181*, 108–115. [CrossRef]
86. Mi, P.; Ma, L.; Zhang, J. Integrated optimization study of hot water supply system with multi-heat-source for the public bath based on PVT heat pump and water source heat pum. *Appl. Therm. Eng.* **2019**, *176*, 115146. [CrossRef]
87. Jafari, M.; Armaghan, D.; Mahmoudi, S.M.S.; Chitsaz, A. Thermoeconomic analysis of a standalone solar hydrogen system with hybrid energy storage. *Int. J. Hydrogen Energy* **2019**, *44*, 19614–19627. [CrossRef]
88. Kumar, L.; Hasanuzzaman, M.; Rahim, N.A.; Islam, M.M. Modeling, simulation and outdoor experimental performance analysis of a solar-assisted process heating system for industrial process heat. *Renew. Energy* **2021**, *164*, 656–673. [CrossRef]
89. Somasundaram, S.; Tay, A.A.O. Performance study and economic analysis of photo-voltaic thermal system under real-life thermal loads in tropical climate. *Sustain. Environ. Res.* **2019**, *1*, s42834. [CrossRef]
90. Maatallah, T.; Zachariah, R.; Al-Amri, F.G. Exergo-economic analysis of a serpentine flow type water based photovoltaic thermal system with phase change material (PVT-PCM/water). *Sol. Energy* **2019**, *193*, 195–204. [CrossRef]
91. Rajoria, C.S.; Agrawal, S.; Tiwari, G.N. Exergetic and enviroeconomic analysis of novel hybrid PVT array. *Sol. Energy* **2013**, *88*, 110–119. [CrossRef]
92. Yandri, E. The effect of Joule heating to thermal performance of hybrid PVT collector during electricity generation. *Renew. Energy* **2017**, *111*, 344–352. [CrossRef]
93. Antony, A.; Wang, Y.D.; Roskilly, A.P. A detailed optimisation of solar photovoltaic/thermal systems and its application. *Energy Procedia* **2018**, *158*, 1141–1148. [CrossRef]
94. Dannemand, M.; Perers, B.; Furbo, S. Performance of a demonstration solar PVT assisted heat pump system with cold buffer storage and domestic hot water storage tanks. *Energy Build.* **2019**, *188–189*, 46–57. [CrossRef]
95. Mittelman, G.; Kribus, A.; Mouchtar, O.; Dayan, A. Water desalination with concentrating photovoltaic/thermal (CPVT) systems. *Sol. Energy* **2009**, *83*, 1322–1334. [CrossRef]
96. Monjezi, A.A.; Chen, Y.; Vepa, R.; Kashyout, A.E.-H.B.; Hassan, G.; Fath, H.E.-B.; Kassem, A.E.-W.; Shaheed, M.H. Development of an off-grid solar energy powered reverse osmosis desalination system for continuous production of freshwater with integrated photovoltaic thermal (PVT) cooling. *Desalination* **2020**, *495*, 114679. [CrossRef]
97. Ong, C.L.; Escher, W.; Paredes, S.; Khalil, A.S.G.; Michel, B. A novel concept of energy reuse from high concentration photovoltaic thermal (HCPVT) system for desalination. *Desalination* **2012**, *295*, 70–81. [CrossRef]
98. Abdelgaied, M.; Abdullah, A.S.; Kabeel, A.E.; Abosheiasha, H.F. Assessment of an innovative hybrid system of PVT-driven RO desalination unit integrated with solar dish concentrator as preheating unit. *Energy Convers. Manag.* **2022**, *258*, 115558. [CrossRef]
99. Calise, F.; Dentice, M.; Piacentino, A. A novel solar trigeneration system integrating PVT (photovoltaic/thermal collectors) and SW (seawater) desalination: Dynamic simulation and economic assessment. *Energy* **2014**, *67*, 129–148. [CrossRef]
100. Xiao, L.; Shi, R.; Wu, S.; Chen, Z. Performance study on a photovoltaic thermal (PV/T) stepped solar still with a bottom channel. *Desalination* **2019**, *471*, 114129. [CrossRef]
101. Alnaimat, F.; Klausner, J.F. Solar diffusion driven desalination for decentralized water production. *DES* **2012**, *289*, 35–44. [CrossRef]
102. Xinxin, G.; Heng, Z.; Haiping, C.; Kai, L.; Jiguang, H.; Haowen, L. Experimental and theoretical investigation on a hybrid LCPV/T solar still system. *Desalination* **2019**, *468*, 114063. [CrossRef]
103. Elaouzy, Y.; El Fadar, A. Investigation of building-integrated photovoltaic, photovoltaic thermal, ground source heat pump and green roof systems. *Energy Convers. Manag.* **2023**, *283*, 116926. [CrossRef]
104. Zarei, A.; Liravi, M.; Rabiee, M.B.; Ghodrat, M. A Novel, eco-friendly combined solar cooling and heating system, powered by hybrid Photovoltaic thermal (PVT) collector for domestic application. *Energy Convers. Manag.* **2020**, *222*, 113198. [CrossRef]
105. Braun, R.; Haag, M.; Stave, J.; Abdelnour, N.; Eicker, U. System design and feasibility of trigeneration systems with hybrid photovoltaic-thermal (PVT) collectors for zero energy office buildings in different climates. *Sol. Energy* **2020**, *196*, 39–48. [CrossRef]
106. Chemisana, D.; Fernandez, E.F.; Riverola, A.; Moreno, A. Fluid-based spectrally selective filters for direct immersed PVT solar systems in building applications. *Renew. Energy* **2018**, *123*, 263–272. [CrossRef]
107. Sharma, S.; Tahir, A.; Reddy, K.S.; Mallick, T.K. Performance enhancement of a Building-Integrated Concentrating Photovoltaic system using phase change material. *Sol. Energy Mater. Sol. Cells* **2016**, *149*, 29–39. [CrossRef]
108. Baljit, S.S.S.; Chan, H.Y.; Sopian, K. Review of building integrated applications of photovoltaic and solar thermal systems. *J. Clean. Prod.* **2016**, *137*, 677–689. [CrossRef]
109. Fterich, M.; Chouikhi, H.; Bentaher, H.; Maalej, A. Experimental parametric study of a mixed-mode forced convection solar dryer equipped with a PV/T air collector. *Sol. Energy* **2018**, *171*, 751–760. [CrossRef]

110. Tiwari, G.N.; Nayak, S.; Dubey, S.; Solanki, S.C.; Singh, R.D. Performance analysis of a conventional PV/T mixed mode dryer under no load condition. *Int. J. Energy Res.* **2009**, *33*, 919–930. [CrossRef]
111. Kabeel, A.E.; Abdelgaied, M. Performance of novel solar dryer. *Process Saf. Environ. Prot.* **2016**, *102*, 183–189. [CrossRef]
112. Ziaforoughi, A.; Esfahani, J.A. A salient reduction of energy consumption and drying time in a novel PV-solar collector-assisted intermittent infrared dryer. *Sol. Energy* **2016**, *136*, 428–436. [CrossRef]
113. Tiwari, S.; Agrawal, S.; Tiwari, G.N. PVT air collector integrated greenhouse dryers. *Renew. Sustain. Energy Rev.* **2018**, *90*, 142–159. [CrossRef]
114. El, M.; Slimani, A.; Amirat, M.; Bahria, S.; Kurucz, I.; Aouli, M. Study and modeling of energy performance of a hybrid photovoltaic/thermal solar collector: Configuration suitable for an indirect solar dryer. *Energy Convers. Manag.* **2016**, *125*, 209–221. [CrossRef]
115. Tiwari, S.; Tiwari, G.N. Exergoeconomic analysis of photovoltaic-thermal (PVT) mixed mode greenhouse solar dryer. *Energy* **2016**, *114*, 155–164. [CrossRef]
116. Herrando, M.; Simón, R.; Guedea, I.; Fueyo, N. The challenges of solar hybrid PVT systems in the food processing industry. *Appl. Therm. Eng.* **2021**, *184*, 116235. [CrossRef]
117. Zhou, Y.; Zheng, S.; Zhang, G. Study on the energy performance enhancement of a new PCMs integrated hybrid system with the active cooling and hybrid ventilations. *Energy* **2019**, *179*, 111–128. [CrossRef]
118. Gaur, A.; Tiwari, G.N.; Ménézo, C.; Al-Helal, I.M. Numerical and experimental studies on a Building integrated Semi-transparent Photovoltaic Thermal (BiSPVT) system: Model validation with a prototype test setu. *Energy Convers. Manag.* **2016**, *129*, 329–343. [CrossRef]
119. Gupta, N.; Tiwari, G.N. Effect of water flow on building integrated semitransparent photovoltaic thermal system with heat capacity. *Sustain. Cities Soc.* **2018**, *39*, 708–718. [CrossRef]
120. Mateus, T.; Oliveira, A.C. Energy and economic analysis of an integrated solar absorption cooling and heating system in different building types and climates. *Appl. Energy* **2009**, *86*, 949–957. [CrossRef]
121. Shah, R.; Srinivasan, P. Hybrid Photovoltaic and Solar Thermal Systems (PVT): Performance Simulation and Experimental Validation. *Mater. Today Proc.* **2018**, *11*, 22998–23006. [CrossRef]
122. Wang, K.; Pantaleo, A.; Herrando, M.; Faccia, M.; Pesmazoglou, I.; Franchetti, B.M.; Markides, C. Spectral-splitting hybrid PV-thermal (PVT) systems for combined heat and power provision to dairy farms. *Renew. Energy* **2020**, *159*, 1047–1065. [CrossRef]
123. Herrando, M.; Pantaleo, A.M.; Wang, K.; Markides, C.N. Solar combined cooling, heating and power systems based on hybrid PVT, PV or solar-thermal collectors for building applications. *Renew. Energy* **2019**, *143*, 637–647. [CrossRef]
124. Yandri, E. Development and experiment on the performance of polymeric hybrid Photovoltaic Thermal (PVT) collector with halogen solar simulator. *Sol. Energy Mater. Sol. Cells* **2019**, *201*, 110066. [CrossRef]
125. Pfahler, M.; Branner, S.; Refior, H. Die komplette Rotatorenmanschettenruptur—Differenzierte Op-Techniken und mittelfristige Ergebnisse. *Z. Orthop. Und Ihre Grenzgeb.* **2008**, *137*, 1037045. [CrossRef] [PubMed]
126. Kazemian, A.; Taheri, A.; Sardarabadi, A.; Ma, T.; Passandideh-Fard, M.; Peng, J. Energy, exergy and environmental analysis of glazed and unglazed PVT system integrated with phase change material: An experimental approach. *Sol. Energy* **2020**, *201*, 178–189. [CrossRef]
127. Lari, M.O.; Sahin, A.Z. Design, performance and economic analysis of a nanofluid-based photovoltaic/thermal system for residential applications. *Energy Convers. Manag.* **2017**, *149*, 467–484. [CrossRef]
128. Arslan, E.; Aktaş, M. 4E analysis of infrared-convective dryer powered solar photovoltaic thermal collector. *Sol. Energy* **2020**, *208*, 46–57. [CrossRef]
129. Gholamian, E.; Hanafizadeh, P.; Ahmadi, P.; Mazzarella, L. A transient optimization and techno-economic assessment of a building integrated combined cooling, heating and power system in Tehran. *Energy Convers. Manag.* **2020**, *217*, 112962. [CrossRef]
130. Widyolar, B.; Jiang, L.; Brinkley, J.; Hota, S.K.; Ferry, J.; Diaz, G.; Winston, R. Experimental performance of an ultra-low-cost solar photovoltaic-thermal (PVT) collector using aluminum minichannels and nonimaging optics. *Appl. Energy* **2020**, *268*. [CrossRef]
131. Hassani, S.; Saidur, R.; Mekhilef, S.; Taylor, R.A. Environmental and exergy benefit of nanofluid-based hybrid PV/T systems. *Energy Convers. Manag.* **2016**, *123*, 431–444. [CrossRef]
132. Yousef, M.S.; Sharaf, M.; Huzayyin, A.S. Energy, exergy, economic, and enviroeconomic assessment of a photovoltaic module incorporated with a paraffin-metal foam composite: An experimental study. *Energy* **2022**, *238*, 121807. [CrossRef]

buildings

MDPI

Review

Climate Zoning for Buildings: From Basic to Advanced Methods—A Review of the Scientific Literature

Alexey Remizov, Shazim Ali Memon * and Jong R. Kim

Department of Civil and Environmental Engineering, School of Engineering and Digital Sciences, Nazarbayev University, Astana 010000, Kazakhstan
* Correspondence: shazim.memon@nu.edu.kz

Abstract: Understanding the link between the energy-efficiency of buildings and climatic conditions can improve the design of energy-efficient housing. Due to global climate change and growing requirements for building energy-efficiency, the number of publications on climate zoning for buildings has grown over the last 20 years. This review attempted to give the reader an up-to-date assessment of the scientific literature in the field of climate mapping for buildings on a global and national scale, filling in the gaps of previous works and focusing on details that were not presented before. There were 105 scientific sources examined. The most dominant climate zoning variables were thoroughly analyzed. A clear categorization of climate zoning methods with specific criteria was shown. The most used methods were evaluated, emphasizing their similarities and differences, as well as their essential components and advantages. The main literature review was supported with bibliometric and bibliographic analysis. The existence of many climate zoning methods can be an indicator of the lack of agreement on the most effective strategy. A tendency has been established for the popularization among scientists of methods based on machine learning and building energy simulations, which are relatively easy to use and have proven to be the most reliable climate zoning methods. A transformation is emerging by shifting from a climate-based to a building performance-based climate zoning approach.

Keywords: building energy-efficiency; building energy simulation; climate zoning; climatic variables; cluster analysis; degree-days; machine learning

check for updates

Citation: Remizov, A.; Memon, S.A.; Kim, J.R. Climate Zoning for Buildings: From Basic to Advanced Methods—A Review of the Scientific Literature. *Buildings* **2023**, *13*, 694. https://doi.org/10.3390/buildings13030694

Academic Editor: Adrian Pitts

Received: 29 January 2023
Revised: 22 February 2023
Accepted: 28 February 2023
Published: 6 March 2023

1. Introduction

People are becoming more conscious about the link between energy use and environmental impacts as global warming and climate change progress more significantly [1,2]. The present energy-related greenhouse gas (GHG) emissions are around 39 Gt CO2 equivalent, according to the International Energy Agency. The building industry was directly or indirectly responsible for nearly 50% of global energy consumption and 39% of total GHG emissions in 2018 [3]. While developed countries have taken significant progress to reduce their energy consumption, the energy demand for buildings rose by over 20% between 2000 and 2017 due to factors including the rapidly expanding floor area of dwellings, the relatively small reduction in energy intensity, and the rising energy requirements of the energy services [4]. Existing and future buildings will be largely responsible for determining global energy consumption [5–10]. Future growth in energy use and accompanying emissions is prominent. The increased access of billions of people in developing countries to decent housing, electricity, and improved cooking facilities is a significant trend. By 2040, buildings are expected to be the most significant source of GHG emissions [11]. In addition to the issue of climate change, there are important economic reasons why energy-efficient buildings are becoming increasingly attractive. There are between 100 and 150 million people in developed countries that are unable to afford the cost of energy due to low incomes [12]. In 2018, nearly 13% of Europeans said they live in homes that are too cold, and 20% said they live in homes that are not properly protected from the heat [13]. In 2022, the situation worsened, and wholesale electricity prices rose significantly in many countries, especially in Europe. Power prices in the first half of 2022 were more than four times as high as the average in the first half of 2016 to 2021, primarily due to gas prices climbing to more than five times the value of the reference period [14]. Therefore, during seasons of extremely high or low ambient temperatures, low-income households might confront significant heating, cooling, and health difficulties. Indeed, buildings represent a critical piece of a low-carbon future and a global challenge for integration with sustainable development. Therefore, reducing the energy and GHG footprint in both existing and new buildings represents a key challenge and an opportunity to tackle global warming and energy safety.

Building energy consumption, in turn, is influenced by several elements, where environmental or climatic factors are one of the most important [15]. With other factors (socioeconomic conditions, occupant behavior, energy management, and building design) being equal, changes in climate characteristics affect building energy consumption [16–19]. A wide range of climate variables influence buildings' thermal performance [20,21]. The impact of climate variables is different in different geographic regions. Increasing energy-efficiency is a key goal for the building sector, and the use of climatic zoning for buildings (CZB) as a tool in the establishment of design guidelines that address lower energy consumption is an important factor to consider. However, climatic zoning (CZ) methods are diverse and there is no "standard" technique for CZB, although some are widely acknowledged and implemented [22–24]. It is also known that not all existing CZB approaches are directly related to building energy consumption [25–27].

Recognizing the relationship between the power consumption of buildings and climate conditions can help with the engineering of climate-appropriate dwellings for various geographical locations [28]. The relevance of precise CZ for building energy consumption is demonstrated by the fact that discrepancies in CZB led to a significant increase in heating and cooling energy needs [29,30]. The design of the buildings should be maximized to take into account regional priorities. Defining climate zones makes it feasible to identify and prevent the negative effects of the environment on buildings by identifying basic zonal construction criteria [31,32]; additionally, this makes it possible to support the efficient use of resources [33]. Over the past 20 years, the number of publications on CZB and the interest of scientists has increased significantly due to global climate change and higher requirements for the energy-efficiency of buildings [22–25,27,34–54]. Recently, numerous measures of energy-efficiency and sustainability, specifically LEED, BREAM, VERDE, and Passivhaus certificates, have demonstrated an inclination toward integrating environmental criteria into buildings [55,56].

Two review articles on CZB were identified in the analyzed literature [16,24]. Walsh et al. [38] reviewed the domestic and international standards, laws, scientific journals, and other documents about the climatic categorization of buildings and energy-saving measures implemented by 54 countries. Mainly using data from national and international building codes (90% of the cases were related to normative documents), methodologies for CZB were explored country by country. In addition to a more modern set of sources (51% of the publications we reviewed were published between 2017 and 2022), this review is different since it is focused on studying only scientific publications in the narrow field of CZB. The purpose of this study is to review academic publications in the CZB field to quantify the research output and current progress supported with bibliometric and bibliographic analysis. Additionally, a revised criterion for determining CZ methods, which was re-established with two new techniques, was used. We attempted to give the reader an up-to-date assessment of the scientific literature, filling in the gaps of previous works and focusing on details such as primary sources of climate data, its form, and the period of observation of the climate, which was not presented in Walsh's review. Verichev et al. [16] investigated the most-cited climate-related studies in building from 1979 to 2019. One hundred twenty-eight publications were used in this paper, all with more than

35 citations. The studies were investigated by employing both manual and bibliographic analysis. This paper covers a wide range of topics, both directly and indirectly linked to climate, climate zones, and buildings. However, the author deals with the issue of the climate component for buildings only in a small part of the article. Conversely, this review is purely focused on the CZB, its methods, variables, and their impact on the building's energy usage. To summarize, this study is unique in the following ways:

1. The information on the CZB from scientific publications in 37 countries and 95 affiliations was collected and reviewed. The Scopus database was selected as a primary source of publications. Research articles represent 84% of the materials we analyzed, while conference papers account for 10%;
2. This study is state-of-the-art since 51% of the publications reviewed were published between 2017 and 2022;
3. The study essentially differentiates buildings' CZ variables and buildings' CZ methods, which were typically bundled in previously published works. Each of the categories was extensively reviewed and analyzed;
4. An organized categorization of the most commonly used building CZ variables and building CZ methods (with criteria used in determining each method) is presented. The most commonly used CZB methods were evaluated emphasizing their similarities and differences, as well as their essential components and advantages. The current development of this field was explored and traced;
5. Several additional machine learning (ML) methods for CZB have been revealed. In light of this, the category of conventional clustering techniques was expanded and given a new term, "Machine Learning Methods" (MLM). Additionally, a previously rare term, "The Interval Judgment Method" (IJM), has been put into use;
6. Covering the gaps of prior works and concentrating on information that was not previously published, the primary sources of climate data and the form in which climate data are commonly used were recognized. The data on climate observation periods for CZB methods were also collected and analyzed. Other details such as the most commonly used software for energy simulations and the number of archetypes were mentioned;
7. All collected data are shown in the condensed table with the following extracted features: sources, publication years, authors, publication type, country or region of study, CZ methods used, their number and combinations, number of climate zones, etc.
8. Several promising studies regarding future climate scenarios in CZB were identified. In this review, 12% of publications dealt with future CZ, and their main principles are given;
9. Using bibliometric and bibliographic analysis for evaluating and analyzing the performance of research activities, this paper indicates substantially contributing authors, nations, the co-citation and bibliographic coupling networks, the direct citation network, etc.

2. Brief Historical Background of Climatic Zoning and Its Purpose

People have attempted the climatic classification of the earth since ancient times for different purposes. Several attempts by Greek philosophers (Pythagoras, Aristotle, Plutarch, and Ptolemy) are known to map and classify the climate [57,58]. The 19th century can be considered the beginning of the modern climate classification era, with the first published maps based on temperature and, later, precipitation parameters [59]. Vegetation-based climate classifications were started by Köppen; his first scheme was published in 1900. Still, the Köppen map remains the most widely used climate classification map, which was presented in its most recent edition in 1961 by Rudolf Geiger. It is still constantly updated and refined [38]. Because the Köppen–Geiger (KG) classification is primarily concerned with vegetation growth, it is limited in formulating the link between outdoor and indoor climates, as well as how climate influences building energy consumption. In the first half of the 20th century, building codes and standards, which contained requirements for climatic protection and durability in conformity with defined climate zones, were introduced in some countries [60,61]. Over time, the requirements of building codes and standards gradually expanded. In addition to the requirements for proper weather protection and interior comfort, the guidelines for the energy-efficiency of buildings, which sufficiently depend on accurate climate classification, were introduced [62,63].

3. Methodology

3.1. Literature Review Framework

The framework of the literature review is shown in Figure 1 and is composed of the following steps:

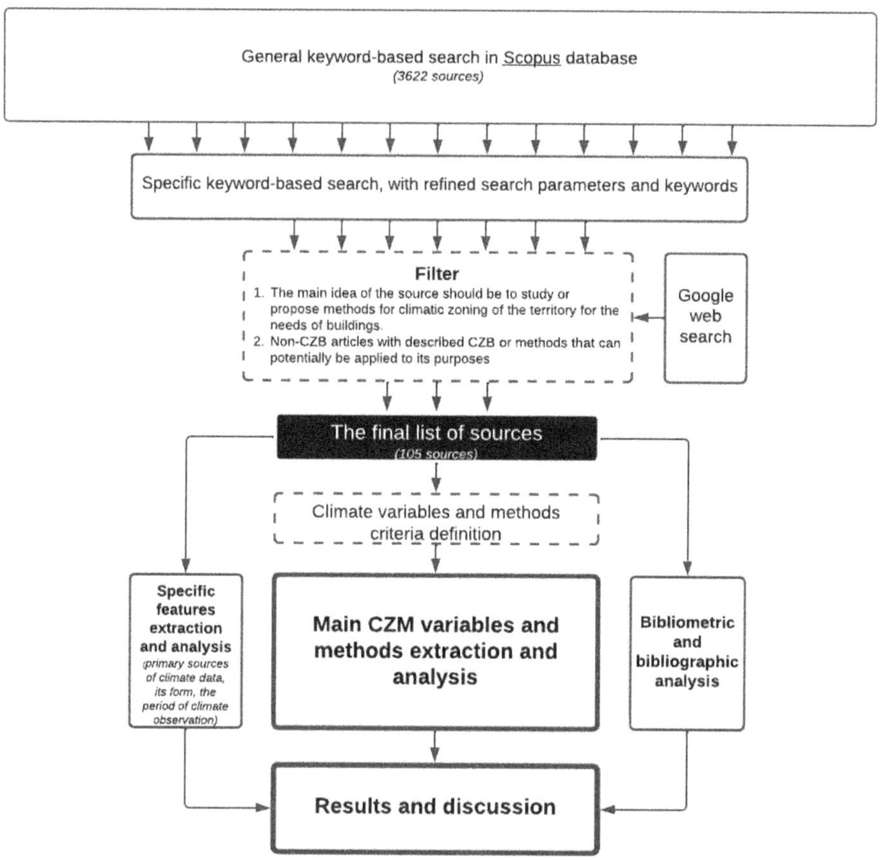

Figure 1. The review methodology flowchart.

1. General keyword-based search in Scopus database;
2. Specific keyword-based search, with refined search parameters and keywords to find the most relevant sources;
3. The composition of the final list of sources using the following criteria. The main idea of the source should be to study or propose methods for climatic zoning of the territory for the needs of energy-efficient buildings or a non-CZB article with described methods which influence CZB or can potentially be applied to its purposes;
4. Identifying and screening additional articles. The sources that were cited by an article from the shortlist became additional candidate sources. Relevant sources outside of

Scopus were also identified by Google web search. Further, the candidate sources were checked following the established criteria, and the selected ones were added to the final list;
5. Criteria were established to distinguish between climate variables and CZB methods;
6. The review of each source and extraction of information on climate variables and CZB methods. Specific features (more details are included in the Data Collection section) were also extracted from the sources at this stage;
7. All data were subjected to in-depth quantitative analysis (descriptive analysis);
8. Sources cited in Scopus were subjected to basic bibliometric and bibliographic analysis to identify bibliometric networks;
9. Discussion (interpretations of the findings, directions for future research, recommendations).

3.2. Adopted Bibliometric and Bibliographic Analysis

Bibliometrics is a valuable technique for evaluating and analyzing the performance of research activities. It corresponds to scientific progress in a variety of ways, including evaluating progress, recognizing the most authoritative sources, developing the academic basis for analyzing novelties, identifying significant scientific performers, constructing bibliometric measures to evaluate academic output, and so on [64].

In this work, the bibliometric analysis supported the mail literature review with the citation analysis, an indication of research performance and collaboration identification. Techniques such as co-citation, bibliographic coupling, co-authorship, citation, and keyword co-occurrence networks were implemented; for more information about adopted bibliometric analysis principles please refer to [65–68]. For bibliometric network visualization, the VOSviewer software was used [64,68–70]. The results of the bibliometric analysis contributed to the following data, which were explored and incorporated into the findings:

1. A map of affiliations or public organizations which publish more articles than others in a CZB research field;
2. The top 10 most cited articles;
3. The most contributing authors in the CZB area;
4. The most popular journals for CZB;
5. Citation over time analysis;
6. The co-citation networks of researchers;
7. The bibliographic coupling network of the top 100 authors;
8. A direct citation network;
9. The network of co-occurrences of keywords;
10. The bibliometric coupling network of countries.

4. Data Collection

The Scopus database was selected as a primary source of publications. The data collection was carried out in a general-to-specific order. The search included all languages and documents for all years of operation of Scopus (until 2022). The result of a general keyword-based search was a long-list of 3622 articles. Local climate zoning (LCZ) articles were excluded from the scope of this review. After final filtering, the specific list of publications from Scopus was formed with 93 publications. During this stage, two review articles [16,24] were found, which served as additional sources of publications. Additionally, this review included a few non-CZB-related articles [71–78]. However, the methods presented there still influence CZB or can potentially be applied to its purposes. Google search was used to find possibly valuable publications outside the Scopus database; additionally, five more articles were found.

After the filtering and selection of candidate sources, the final review list comprised 105 documents. The academic literature published from 1990 onwards was reviewed to capture the most recent published findings. More than half of the publications in the final list were published after 2017. Figure 2 depicts the timeline of all papers considered in this study. The timeline has an increasing trend, with the highest number of documents published in recent years (2017, 2021, and 2022). The typology of the final list of sources is shown in Figure 3.

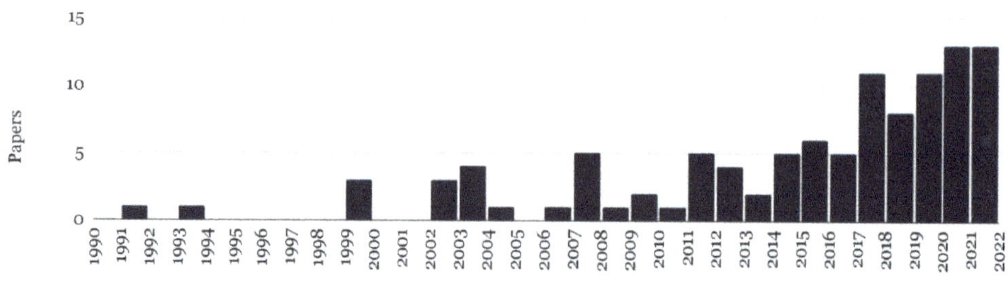

Figure 2. Histogram of publication years.

Figure 3. The sources of the academic papers (**a**) and types of documents in the review lists (**b**).

During the literature review, the information was gathered, which formed the basis for further quantitative analysis. All the data are shown in the condensed table (Table 1). In the following sections, the reader will be provided with information on the climate variables used for CZB, quickly highlighting their impact on building energy consumption; then, potential climate data sources and the period for climate observations will be discussed. In the final part, the essence and characteristics of each CZB method will be addressed in the order of their popularity, starting with the most frequent. In addition, information regarding combinations of methods, as well as a discussion of the benefits and drawbacks associated with each technique, will be provided.

Table 1. Data collected during the review.

Number	Reference	Year of the Source Publication	Entire Territory of or Part of the Country	Country/Region	Main Variables	Number of Climate Variables Used for Climate Zoning	Climate Data Source	Climate Data Source Name	Initial Data Form	Data Information Observation Period (Years)	Methods Used for Climate Zoning	Number of Methods Used	Methods Details	Number of Zones Defined
1	[9]	2020	Entire territory of	Belgium	DDs	1	Web database	Agri4Cast dataset	Daily mean values	1976–2004 (30)	DDM	1	Base temperatures: HDD 18 °C; CDD 18 °C	7
2	[27]	2019	Part of	China (hot summer and cold winter (HSCW) zone)	DDs, RH, SR, W, TMY	5	National meteorological service	National meteorological service	Daily mean/maximum/minimum values, Hourly values TMY	2006–2015 (10)	DDM, MLM, BES	3	Two-tier classification with hierarchical agglomerative clustering (HAC). EnergyPlus simulations with 1 archetype	7
3	[23]	2017	Entire territory of	Nicaragua	AT, RH, SR, W	4	Software	Autodesk Green Building Studio (GBS)	Hourly values		DDM, MLM, Administrative division	3	K-nearest neighbors algorithm	3
4	[79]	2019	Entire territory of	Italy	DDs, AI, TMY	3	National meteorological service	Italian Military Air Force weather stations.	Daily mean values, Hourly values TMY	2000–2009 (10)	DDM, BES	2	Base temperatures: HDD 12 °C; CDD 12 °C, TRNSYS simulations with 13 archetypes	6
5	[80]	2017	Entire territory of	Iran	AT, RH, DDs	3	National meteorological service	Iran Meteorological Organization	Daily mean values	1995–2014 (20)	DDM, BCM	2	Milne-Givoni chart	8
6	[81]	2012		Europe	DDs	1			Monthly mean values		DDM, BES	2	Base temperatures: HDD 18 °C; CDD 18 °C	5
7	[74]	2019	Entire territory of	Madagascar	RH, GHI, Pr	3					MLM	1	Hierarchical k-means clustering on principal components (HCPC)	3
8	[82]	2009	Entire territory of	Madagascar	AT, SR, W	3	National meteorological service	Meteorological forecast utilities of Antananarivo.	Monthly mean values	(20)	BCM	1		6
9	[25]	2018	Entire territory of	Nicaragua	TMY	1	Software	Autodesk Green Building Studio (GBS)	Hourly values TMY		DDM, MLM, Administrative division, BES	4	EnergyPlus simulations with 4 archetypes	3
10	[10]	2012	Entire territory of	Iran	DDs	1			Daily mean values	1961–1990 (40)	DDM	1	Base temperatures: HDD 18 °C; CDD 24 °C	

Table 1. *Cont.*

Number	Reference	Year of the Source Publication	Entire Territory of or Part of the Country	Country/Region	Main Variables	Number of Climate Variables Used for Climate Zoning	Climate Data Source	Climate Data Source Name	Initial Data Form	Data Information Observation Period (Years)	Methods Used for Climate Zoning	Number of Methods Used	Methods Details	Number of Zones Defined
11	[32]	2019	Part of	United States (States of Florida, Georgia, and Tennessee)	TMY	1	National meteorological service	The U.S. Department of Energy (DOE)	Hourly values TMY		DDM, BES	2	EnergyPlus simulations with 13 archetypes	4
12	[83]	2019	Entire territory of	Chile	TMY, DDs, SR, Pr, RH, W	6	Software	Autodesk Green Building Studio (GBS) Mesoscale Meteorological Model, Version 5 (MM5)	Hourly values TMY	2007–2017 (11)	DDM, MLM, BCM, BES	4	Base temperatures: HDD 18 °C; CDD 10 °C	5
13	[84]	2010	Entire territory of	China	AT, RH	2	Web database	CRU TS 2.1 data set from the University of East Anglia	1224 records of monthly minimum temperature, maximum temperature and vapor pressure, annual cumulative heat and cold stresses	1901–2002 (102)	MLM, HCI	2	Hierarchical cluster tree of comfort index and heat/cold stresses,	8
14	[46]	2020	Entire territory of	Brazil	DDs, AT, RH, Pr	4	National meteorological service	INMET database	Hourly values TMY	(10)	DDM, KGM, BES, enhanced degree-day method, MLM, etc.	7	Base temperatures: HDD 18 °C; CDD 10 °C	8
15	[85]	2016	Entire territory of	Turkey	DDs	1			Hourly values TMY	1989–2009 (20)	DDM	1	Base temperatures: HDD 18 °C; CDD 18 °C	4
16	[86]	2011	Entire territory of	United States	DDs	1			Daily mean values	(5)	DDM	1		5
17	[41]	2015	Part of	Spain (Andalusia)	DDs, SR, AT, Al	4	National meteorological service	Agencia Andaluza de la Energía (Andalusian Energy Agency)			CSIM, BES	2	Approximation and interpolation method (AIM), CERMA software simulations with 1 archetype	3
18	[87]	2007	Entire territory of	China	SR	1			Monthly mean values	1957–2000 (10–44)	MLM	1		5

Table 1. *Cont.*

Number	Reference	Year of the Source Publication	Entire Territory of or Part of the Country	Country/Region	Main Variables	Number of Climate Variables Used for Climate Zoning	Climate Data Source	Climate Data Source Name	Initial Data Form	Data Information Observation Period (Years)	Methods Used for Climate Zoning	Number of Methods Used	Methods Details	Number of Zones Defined
19	[88]	2002	Entire territory of	Thailand	AT, RH	2	National meteorological service	Meteorological Department of Thailand	3 h values	1981–1998 (18)	FDV	2	Frequency distribution of occurrence of maximum and minimum values	4
20	[89]	2014	Entire territory of	United States	TMY	1	Research institution	National Renewable Energy Laboratory.	Monthly mean values	1991–2005 (15)	BES	1	EnergyPlus simulations with 9 archetypes	7
21	[43]	2008	Part of	Spain (Andalusia)	DDs, SR, Al	3	National meteorological service	the Andalusian Regional Government	Monthly mean values	1970–2006 (37)	DDM	1	Al correction and approximation and interpolation method (AIM)	12
22	[73]	2017	Entire territory of	Chile	AT, Pr	2	Web database	Global Historical Climate Network Dataset (GHCN), FAOClim 2.0	Annual and monthly mean values	1950–2000 (50)	KGM	1		25
23	[90]	2018	Part of	Chile (southern part)	DDs, SR	2	National meteorological service	the Ministry of Agriculture of Chile (Agromet), the Ministry of Environment (MMA) and the Di-rectorate General of Civil Aviation (DGAC)	Hourly values	2008–2018 (10)	DDM, CSIM	2	Base temperature: HDD 15 °C	5
24	[91]	2007	Part of	India (northeast region)	AT, RH, Pr, W	4	National meteorological service	Regional Meteorological Centre, Guwahati, India	Monthly mean values	(30)	BCM	1	Milne, Givoni charts	4
25	[76]	2011	Entire territory of	Egypt	AT, RH	2	National meteorological service	General Meteorological Authority, Cairo, Egypt	Monthly mean values	(30)	BCM	1	ASHRAE charts	8
26	[47]	2016	Part of	Spain (Extremadura)	DDs, SR, Al	3	National meteorological service	the National Meteorological Agency and the Regional Government of Extremadura.	Monthly mean values	1976–2011 (10)	CSIM	1	Approximation and interpolation method (AIM)	5
27	[92]	1999	Entire territory of	Brazil	AT, RH	2			Monthly mean values		BCM	1		8

Table 1. *Cont.*

Number	Reference	Year of the Source Publication	Entire Territory of or Part of the Country	Country/Region	Main Variables	Number of Climate Variables Used for Climate Zoning	Climate Data Source	Climate Data Source Name	Initial Data Form	Data Information Observation Period (Years)	Methods Used for Climate Zoning	Number of Methods Used	Methods Details	Number of Zones Defined
28	[93]	2003	Entire territory of	Israel	AT, RH, SR, W	4			Daily mean values		MLM	1	Hierarchical clustering	7
29	[94]	2003	Entire territory of	Turkey	AT, Pr	2	National meteorological service	National Weather Service of Turkey	Monthly mean values	1951–1998 (47)	MLM	1	Hierarchical clustering	7
30	[95]	1993	Entire territory of	United States	AT, Pr	2	National meteorological service	National climatic data center	Monthly mean values	1931–1980 (50)	MLM	1	Principal component analysis, hierarchical clustering	8, 14, 25
31	[48]	2018	Entire territory of	India	AT, RH, SR, TMY	4	National meteorological service	Indian Society of Heating Refrigeration and Air-Conditioning Engineers (ISHRAE)	Hourly values TMY		BCM, BES	2	Ecotect analysis program simulation with 1 archetype (duplex house)	5
32	[96] *	2015	Entire territory of	Saudi Arabia	AT, RH, W, SR	4	National meteorological service	MEPA. The Meteorology and Environmental Protection Administration (MEPA) weather tapes in Jeddah, Saudi Arabia		1960–2010 (20)	KGM, TCCM, MLM, The World Health Organization classification method, etc.	16		6
33	[72] *	2012		World	AT, Pr, SR	3	Web database	CRU and GPCC data sets, ERA-Interim, MODIS	Monthly mean values	2001–2007 (8)	KGM, MLM	2	Principal component analysis, k-means clustering	12
34	[97]	2017		World	AT, Pr	2	Climate model	GCM ensemble	Monthly mean values	1981–2000 (20)	MLM	1	k-means clustering, hierarchical clustering	60
35	[98]	2021	Entire territory of	China	AT, RH, SR, W	4	National meteorological service	National Meteorological Information Center	Hourly values	1961–2010 (50)	BES	1	Transient system simulation program software (TRNSYS) with 1 archetype (office)	
36	[44]	2021	Entire territory of	China	TMY	1	Software	Medpha database of China	Hourly values TMY		BES, MLM	2	K-Means and agglomerative hierarchical clustering. DeST (designer's simulation toolkits) with 1 archetype (20-story office)	5
37	[99]	2018		Europe	AT, RH, SR	3	Web database	EnergyPlus website	Hourly values TMY	1982–1999 (18)	KGM, MLM	2	k-means clustering, k-medoids clustering	5

Table 1. *Cont.*

Number	Reference	Year of the Source Publication	Entire Territory of or Part of the Country	Country/Region	Main Variables	Number of Climate Variables Used for Climate Zoning	Climate Data Source	Climate Data Source Name	Initial Data Form	Data Information Observation Period (Years)	Methods Used for Climate Zoning	Number of Methods Used	Methods Details	Number of Zones Defined
38	[100] *	2020		Europe	AT, RH, SR	3	Web database	EnergyPlus website	Hourly values TMY	1982–2000 (18)	KGM, MLM	2	hierarchical clustering with Euclidean distances	7
39	[71]	2020	Entire territory of	China	AT, RH, SR, AP	4	National meteorological service	National Climate Center of China	10 days mean values	2004–2013 (10)	MLM	1	K-nearest-neighbor and sparse subspace representation.	5
40	[77]	2018		World	AT, Pr	2	Climate model	WorldClim V1 and V2, and CHELSA V1.2, CHPclim V1; Table 1	Monthly mean values	1980–2016 (37)	KGM	1		30
41	[69]	2018	Entire territory of	Turkey	AT, Pr	2	National meteorological service	Turkish State Meteorological Service		1950–2010 (60)	TCCM	1		9
42	[101]	2016	Part of	Brazil (State of Paraná)	AT, rainfall, evapotranspi-ration	3	Climate model	European Center for Medium-Range Weather Forecast (ECMWF) models	Monthly mean values	1989–2014 (25)	KGM, TCCM, Camargo Climatic Classification	3		3
43	[102]	2007		World	AT and Pr	2	Web database	Global Historical Climatology Network (GHCN) version 2.0 dataset	Monthly mean values	1909–1993 (70)	KGM	1		30
44	[103]	2019		World	AT, Pr, solar irradiation	3	Web database	"GPCCv2018" "CRU TS4.01"	Monthly mean values	1950–2016 (68)	KGM	1	Köppen–Geiger-Photovoltaic (KGPV)	12
45	[42] *	2011		World	AT, RH, TMY	3	Software	Ecotect climate classification tool (Autodesk Incorporated 2011)	TMY, monthly mean values		BCM, BES	2	Standard psychometric charts with each location's actual temperature and RH, "SUNREL" software (National Renewable Energy Laboratory 2010) with 3 archetypes	
46	[35]	2020	Entire territory of	China	AT, DDs, Pr, SR, PW	5	National meteorological service	National Climate Center (NCC)	Daily mean value	1997–2013 (17)	DDM, MLM	2	Base temperatures: HDD 18 °C; CDD 10 °C, CD 26 °C; hierarchical clustering;	17

Table 1. *Cont.*

Number	Reference	Year of the Source Publication	Entire Territory of or Part of the Country	Country/Region	Main Variables	Number of Climate Variables Used for Climate Zoning	Climate Data Source	Climate Data Source Name	Initial Data Form	Data Information Observation Period (Years)	Methods Used for Climate Zoning	Number of Methods Used	Methods Details	Number of Zones Defined
47	[104]	2019	Entire territory of	China	AT, DDs	2	National meteorological service	National Climate Center (NCC)	Daily mean value	1997–2013 (17)	IJM, BES	2	Simulations with 1 archetype	5
48	[75]	2016		World	AT, Pr	2	Web database	WorldClim global climate dataset	Monthly mean values	1950–2000 (50)	MLM	1	32 clustering methods, hierarchical clustering, partitioning around medoids	5
49	[105]	2013		Europe	TMY	1	Software	Meteonorm	Hourly values TMY		BES	1	HAMBase and mathlab software with 1 archetype	
50	[106]	2020	Entire territory of	Algeria	TMY	1	Web database	United States Department of Energy	Hourly values TMY	2004–2018 (15)	BES	1	EnergyPlus simulations V8.9.0, energy demand and indoor-discomfort hours with typical multifamily social residential building	
51	[107] *	2019	Entire territory of	Brazil	AT, RH, SR, TMY	4	Software	EnergyPlus database	Hourly values TMY	2005–2018 (13)	BES, MLM	2	EnergyPlus simulations V8.9.0, annual building cooling thermal loads as indicators. 1 archetype. k-means clustering with the sum of squares (within SS) and Hubert index	5
52	[108]	2015	Entire territory of	Italy	DDs	1	National meteorological service	Italian meteorological database	Monthly mean values	1978–2013 (35)	DDM	1	Base temperatures: HDD 18 °C, HDD 20 °C, HDD 22 °C; CDD 22 °C, CD 24 °C, CDD 26 °C	
53	[109]	2021	Entire territory of	Spain	AT, DDs	2	National meteorological service	State Meteorological Agency (AEMET)		2015–2018 (4)	CSIM, BES	2	HULC tool for simulation	19
54	[110]	2015	Entire territory of	Algeria	DDs	1	National meteorological service	CNERIB, Réglementation thermique du bâtiment			DDM	1	Base temperatures: HDD 18 °C; CDD 26 °C; territory is classified into climatic zones according to the annual cost of energy consumption	7

Table 1. *Cont.*

Number	Reference	Year of the Source Publication	Entire Territory of or Part of the Country	Country/Region	Main Variables	Number of Climate Variables Used for Climate Zoning	Climate Data Source	Climate Data Source Name	Initial Data Form	Data Information Observation Period (Years)	Methods Used for Climate Zoning	Number of Methods Used	Methods Details	Number of Zones Defined
55	[111]	2021	Entire territory of	Belarus	AT, Pr, W	3	National meteorological service	State Climate Cadastre of the Republic of Belarus (Belhy- dromet 2019)	Daily values	1971–2000 (30)	BES	1	A historical simulation and an evaluation simulation	
56	[112]	2013	Entire territory of	Czech Republic		14				1961–2000 (40)	Quitt's Climate Classification	1		23
57	[7]	2017		Europe (Cyprus, France, Greece)										
58	[113]	2017	Part of	Spain (Galicia)	AT, RH	2	National meteorological service		Monthly mean values	(15)	BCM	2	Givoni bioclimatic charts	5
59	[114]	2021	Entire territory of	Ireland	DDs	1			Daily values	2003–2017 (15)	DDM	1	Base temperature: HDD 15, 5 °C	
60	[115]	2015	Entire territory of	South Korea	DDs	1	National meteorological service	Korea Meteorological Administration	Daily values, 3 h values	1981–2010 (30)	DDM	1		4
61	[116] *	2017	Entire territory of	Philippines	Pr	1	National meteorological service	Philippine Atmospheric, Geophysical, and Astronomical Services Administration (PAGASA)	Monthly mean values	1961–2015 (55)	MLM	1	K-means clustering, hierarchical clustering	6
62		2002	Entire territory of	Nigeria	AT, RH	2				(20)	MM	1		9
63	[117]	2020	Entire territory of	Nigeria	AT, RH	2	National meteorological service	Meteorological center of Nigeria	Monthly mean values	(5)	BCM	1	Olgyay charts	5
64	[118]	2002	Entire territory of	Venezuela	AT, RH, Al	3					IJM	1	Al correction	5
65	[119]	1999	Part of	United States (Puerto Rico)	AT, Pr	2			Seasonal mean values	1960–1990 (31)	MLM	1	Principal component analysis, artificial neural networks (ANN)	4

Table 1. *Cont.*

Number	Reference	Year of the Source Publication	Entire Territory of or Part of the Country	Country/Region	Main Variables	Number of Climate Variables Used for Climate Zoning	Climate Data Source	Climate Data Source Name	Initial Data Form	Data Information Observation Period (Years)	Methods Used for Climate Zoning	Number of Methods Used	Methods Details	Number of Zones Defined
66	[120]	2010	Entire territory of	Dominican Republic	AT, Pr	2	National meteorological service	ONAMET network, National Institute for Hydrologic Resources (Instituto Nacio- nal de Recursos Hidráulicos, or INDRHI)	Monthly mean values	1971–2000 (30)	TCCM	1		9
67	[121]	2018	Entire territory of	Dominican Republic	AT, RH	2	National meteorological service	ONAMET network, National Institute for Hydrologic Resources (Instituto Nacio- nal de Recursos Hidráulicos, or INDRHI)	30 min values	1998–2016 (18)	FDV	1	Frequency distribution of maximum and minimum values of AT and RH	8
68	[34]	2021	Part of	China (cold climate zone)	TMY	1			Hourly values TMY		BES, MLM	2	EnergyPlus simulation with 1 archetype, k-means clustering	4
69	[53]	2022	Entire territory of	Ethiopia	AT, RH, SR, TMY	4	Web database, climate model, software	WorldClim repository, CRU CL v. 2.0: (A high-resolution data set of surface climate over global land areas), Meteonorm software	Monthly mean values		BES, MLM	2	K-means clustering, principal component analysis, Mahoney method, EnergyPlus simulation with 2 archetypes	10
70	[51]	2020	Part of	Chile	TMY	1	Climate model	the Mesoscale Meteorological Model ver.5 (MM5)	Hourly values TMY		BES	1	Simulation with 1 archetype	
71	[52]	2020	Entire territory of	China	AT, RH	2	National meteorological service	National Meteorological Information Center		1984–2013 (30)	MLM	1	Mahalanobis distance as an indicator for evaluating the distances between samples	7
72	[39]	2022	Part of	Brazil (semiarid region)	AT, Pr, SR, W, PW	5	Web database	World-Clim 2 Data Portal	Monthly mean values	1970–2000 (30)	MLM	1	Principal component analysis, hierarchical clustering	3

Table 1. *Cont.*

Number	Reference	Year of the Source Publication	Entire Territory of or Part of the Country	Country/Region	Main Variables	Number of Climate Variables Used for Climate Zoning	Climate Data Source	Climate Data Source Name	Initial Data Form	Data Information Observation Period (Years)	Methods Used for Climate Zoning	Number of Methods Used	Methods Details	Number of Zones Defined
73	[122]	2018	Part of	Chile (south part (La Araucanía, Los Ríos, and Los Lagos))	DDs, SR	2	National meteorological service	Ministry of Agriculture-National Agroclimatic Network	Hourly values	2011–2015 (4)	CSIM	1	Base temperatures: HDD 20 °C; CDD 20 °C	3
74	[123]	2017	Entire territory of	Sweden	Primary energy consumption	1					EBSM	1	Primary energy consumption (measured in kWhp/m2),	3
75	[22]	2021	Entire territory of	Spain	DDs	1	National meteorological service	Spanish State Meteorological Agency (AEMET) (State Meteorological Agency-AEMET-Spanish Government, 2020)	Hourly values	2015–2018 (4)	CSIM	1	Base temperatures: HDD 20 °C; CDD 20 °C	12
76	[124]	2009	Entire territory of	Thailand	AT	1	National meteorological service	Department of Meteorological	Hourly values	1981–1999 (18)				3
77	[6]	2012	Entire territory of	Greece	AT, SR	2			Hourly values	1961–1990 (30)	BES	1	TRNSYS software simulations with 3 archetypes	13
78	[40]	2021	Part of	Spain (Andalusia)	TMY (EPW)	1	Software	METEONORM	EPW		BES, MLM	2	Simulations with 8 archetypes, k-means clustering	12
79	[125]	2016	Part of (Catalonia)	Spain	Primary energy consumption	1		Catalan Institute of Energy	Primary energy consumption	1980–2008 (30)	EBSM	1	Primary energy consumption (measured in kWhp/m^2),	3
80	[126] *	2021	Entire territory of	Mexico	TMY (EPW)	1	Software	EnergyPlus	EPW		BES	1	Open Studio with 3 archetypes	10
81	[127]	2021	Entire territory of	Libya	DDs	1			Daily mean values		DDM	1	Base temperatures: HDD 24 °C; CDD 18 °C	conti nu-ous
82	[128]	2014	Entire territory of	Turkey	AT	1	National meteorological service	State Meteorology General Directorate		(42)	FDV	1	The outdoor temperature distributions	8

Table 1. *Cont.*

Number	Reference	Year of the Source Publication	Entire Territory of or Part of the Country	Country/Region	Main Variables	Number of Climate Variables Used for Climate Zoning	Climate Data Source	Climate Data Source Name	Initial Data Form	Data Information Observation Period (Years)	Methods Used for Climate Zoning	Number of Methods Used	Methods Details	Number of Zones Defined
83	[50]	2021	Entire territory of	Turkey	Wall insulation, fuel types, DDs	3					MLM	1	Fuzzy c-mean clustering	5
84	[129] *	2014	Entire territory of	Nepal	AT, RH, W	3	National meteorological service	Department of Hydrology and Meteorology	TMY		BCM	1	Givoni charts	4
85	[130]	2017	Entire territory of	South Korea	AT	1	National meteorological service	Korea Meteorological Administration (KMA)		2001–2010 (10)	FDV	1	Graph pattern of the cumulative temperature density	4
86	[131]	2011	Entire territory of	Iran	AT, RH, AP	3					BCM	1		4
87	[132]	2017	Part of (Pampas)	Argentina	AT, RH, W, Pr, Al	5	National meteorological service	National Meteorological Service (SMN, Argentina)	Monthly mean values	1960–2010 (50)	MLM	1	Ward agglomerative hierarchical clustering method	8
88	[49]	2021	Entire territory of	Iran	AT, RH	2				1995–2014 (20)	MLM, BCM	2	Givoni's bioclimatic chart modified by Brown and Dekay	19
89	[133]	2007	Part of (San Luis Potosí, central–northeastern region of México)	Mexico	AT, Pr	2	National meteorological service	México's Comisión Nacional del Agua (CNA)	Monthly mean values	1940–1997 (28)	KGM, MLM	2	Principal component analysis	6
90	[134]	2019	Entire territory of	India	AT, RH, DDs	3	National meteorological service	Indian Society of Heating Refrigerating and Air-conditioning Engineers (ISHRAE)	Hourly weather data		MLM	1	Hierarchical clustering	8
91	[135]	2015	Entire territory of	India	AT, RH, Pr	3	Web database	SEWRA and UNEP NASA's Surface meteorology	Raster and shape file datasets	1983–2005 (25)	MM	1		62

Table 1. *Cont.*

Number	Reference	Year of the Source Publication	Entire Territory of or Part of the Country	Country/Region	Main Variables	Number of Climate Variables Used for Climate Zoning	Climate Data Source	Climate Data Source Name	Initial Data Form	Data Information Observation Period (Years)	Methods Used for Climate Zoning	Number of Methods Used	Methods Details	Number of Zones Defined
92	[136]	2019	Part of (Kerala)	India	AT, DHs, Al	3	Software	Meteonorm	TMY		DDM, BES	2	Al correction	3
93	[137]	2004	Entire territory of	Thailand	AT, RH, Pr, SR, Al	5	National meteorological service	Thai meteorological service		1990–2000 (10)			Al correction	
94	[138]	2021	Entire territory of	Morocco	AT, RH, Pr, SR	4	Web database	The Prediction of Worldwide Energy Resource (POWER)		1984–2017 (32)	MLM	1	Hierarchical clustering, k-means clustering, support vector machine classifier (SVM-C)	8
95	[45]	2022	Entire territory of	India										
96	[139] *	2003												
97	[140] *	2003	Entire territory of	United States	AT, RH, DDs	3					IJM, DDM	2		17
98	[141]	2022	Entire territory of	Ethiopia	AT, DDs	2	National meteorological service	National Center for Environmental Prediction (NCEP)		1979–2013 (34)	DDM	1	Heating degree-days (HDD) and cooling degree-days (CDD) with base temperature of 18.3 °C	5
99	[142]	2006	Entire territory of	China	AT, RH	2	National meteorological service	National Meteorological Centre of China	Daily values	1971–2000 (30)	BCM	1		9
100	[143] **	2014	Territory of	Eastern Africa (Tanzania, Kenya, Uganda)	AT, RH	2					BCM, MM	2	Givoni charts	6
101	[144] **	1991	Entire territory of	Ethiopia	AT, Pr, Al	3			Annual mean values		IJM	1	Al correction	6
102	[145]	2020												
103	[24] ***	2017												

Table 1. *Cont.*

Number	Reference	Year of the Source Publication	Entire Territory of or Part of the Country	Country/Region	Main Variables	Number of Climate Variables Used for Climate Zoning	Climate Data Source	Climate Data Source Name	Initial Data Form	Data Information Observation Period (Years)	Methods Used for Climate Zoning	Number of Methods Used	Methods Details	Number of Zones Defined
104	[146] ****	2007	Entire territory of	Bolivia	AT, RH	2	National meteorological service	National meteorological service (SENAMHI)		1970–2004 (35)	BCM	1	Givoni bioclimatic charts, ABC software	8
105	[147] ****	2020	Entire territory of	Ethiopia	AT, RH, DDs	3	National meteorological service	National Center for Environmental Prediction NMAE, National Meteorological Agency of Ethiopia		1974–2013 (30)	DDM	1	Heating DDs (base 18.3 °C)	5

* Conference paper. ** Book. *** Review paper. **** Thesis.

5. Climate Variables Used for CZB

Several factors can influence building energy consumption: building location and design, occupant behavior, energy management of the building, climate, and socioeconomic and legal-related characteristics [17,18]. The issue of the energy demand of buildings is quite complex; however, speaking of the impact of climate on buildings, it can confidently be said that with other factors being equal, changes in climate characteristics significantly influence building energy consumption. A wide range of climate variables, including outdoor air temperature (AT), solar radiation (SR), relative humidity (RH), degree-days (DDs), wind (W), etc., impact buildings' thermal performance [20]. However, specific types of buildings are influenced by climate variables differently. Climate, for example, does not affect large venue buildings' extreme daily heating or cooling energy usage, but for commercial buildings, daily excessive heating energy consumption is closely connected to maximum and minimum AT, dry-bulb temperature (DBT), and SR [21]. In contrast, for residential buildings, only DBT has an impact. The wet-bulb temperature (WBT) is the primary impacting climatic parameter for the extreme cooling energy consumption of commercial buildings, which indicates that the combination of AT and RH influences the cooling energy consumption [21]. For many building types, the DBT significantly impacts hourly severe heating energy consumption, followed by the WBT. In contrast, cooling energy consumption is closely connected to the WBT for commercial buildings or has no apparent relationship to climate for large venue buildings [11,21]. However, the exact mechanism by which climate influences the amount of energy that buildings consume is a complicated, contentious, and highly studied topic in the scientific community [11,17,21,36,148–150].

This review found 99 documents with information about the variables used. The list of the main variables that were identified consists of AT, RH, DDs, SR, precipitation (Pr), W, typical meteorological year (TMY), altitude (Al), atmospheric pressure (AP), and the pressure of water vapor (PW). Building energy simulation cases were explicitly indicated as a TMY since a compilation of meteorological elements was used during the simulation, not exact variables. In the analysis, TMY was considered as a separate independent variable. Additionally, one climatic derivative (DDs) and one geographical feature (Al) were identified as separate variables. To avoid confusion, the variables indicated in the analysis are, in fact, a set of related (close) variables grouped to simplify the further analysis; see Table 2.

Ten main variables in CZB processes, and the number of variables used simultaneously, were revealed (Figure 4). As expected, AT was the most common variable, which was alone or in combination with other variables used for the CZ in 63 cases out of 99 cases (63.6%). AT and RH were the most common combination of variables among the examined documents. To simplify, DBT, WBT, temperature ranges, diurnal, seasonal, and extreme temperatures were referred to the AT category. RH was the second most popular variable. Out of 99 cases with available data on climate variables in this literature review, 40 cases using RH for climate classification were found (40.4%). DDs are also common in climate classification and were found in 30 cases (30.3%) when implemented alone or in combination with other variables. DD and DH are essentially a derivative of AT; however, these variables were mentioned separately as independent variables in cases where climate classification was based on DDs or DHs. Despite the importance of SR for CZB, its use is not so significant; it was revealed in 25.3% of cases (25 out of 99). In this review, the SR category also combines global horizontal irradiation (GHI), sunshine duration, and daily clearness index. The use of Pr as one of the variables was encountered in 25 cases out of 99 (25.3%). This is due to the use of this variable in the popular KG method [102], which uses monthly AT and Pr for classification, and due to the frequent use of this variable in hot climate countries, such as Brazil [101], the Dominican Republic [120], Chile [73], Colombia [151], Mexico [133], and the Philippines [152]. Wind component as a climate variable is not common in CZB and was found in 12 cases (12.1%) in combination with other variables. We combined wind speed, wind frequency, and wind direction under this category. TMY was found in 15.2% of cases, and Al correction in 13%.

Table 2. CZB variables definitions.

#	Variable Group Name	Abbreviations	Variables Included
1	AT	AT	Dry-bulb temperature
			Wet-bulb temperature
			Temperature ranges
			Seasonal and extreme temperatures
2	Relative humidity	RH	Relative humidity
			Moisture content
3	Degree-days	DDs	Degree-days
			Degree-hours
4	Solar radiation	SR	Direct solar radiation
			Diffuse solar radiation
			Global horizontal irradiation (GHI)
			Sunshine duration
			Daily clearness index
5	Precipitation	Pr	Rainfall
			Snowfall
6	Wind	W	Wind speed
			Wind frequency
			Wind direction
7	Typical meteorological year	TMY	Set of meteorological data (TMY, EPW, RMY, etc.)
8	Altitude	Al	Altitude
9	Atmospheric pressure	AP	Atmospheric pressure
10	The pressure of water vapor	PW	The pressure of water vapor

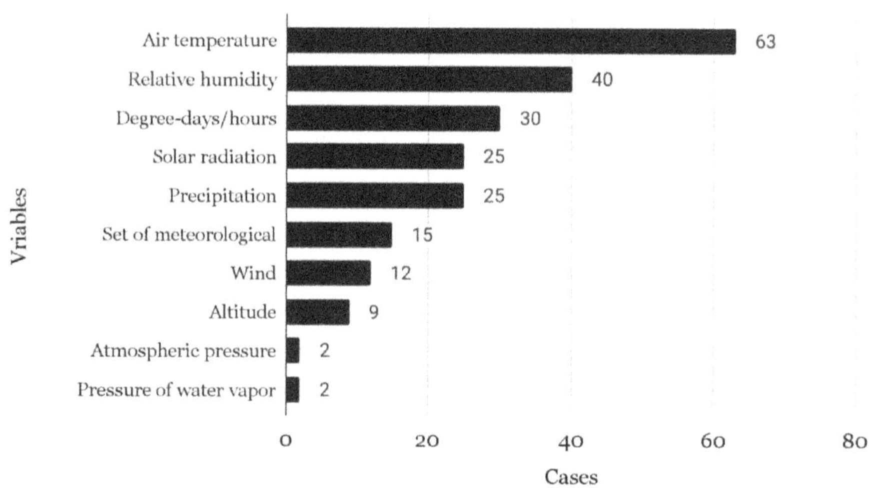

Figure 4. Main variables for CZB.

Most often, climate classification used two variables at the same time (29.3%). The usage of only one variable or a combination of three variables was slightly less frequent, accounting for 28.3% and 22.2%, respectively. It was infrequent to use more than three variables simultaneously for CZB (Figure 5). Furthermore, when only one variable is utilized for zoning, as seen in the first bar of Figure 4, DDs and TMY are the most popular choices. AT and RH were the most common combination of two variables among the examined documents, with a significant gap between AT and SR, which was the second most prevalent combination (Figure 6).

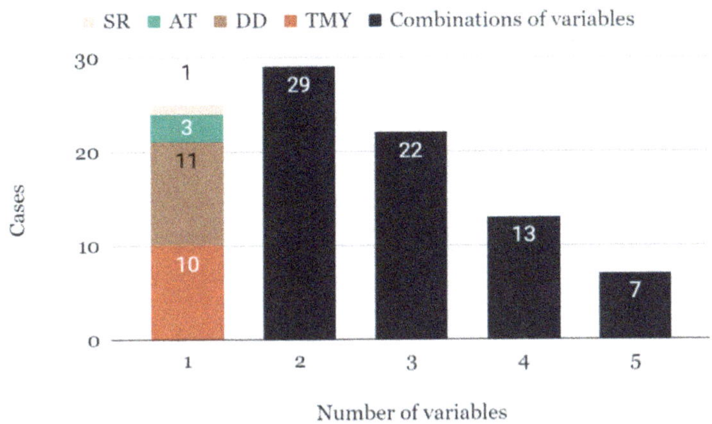

Figure 5. The number of climate variables used for CZB.

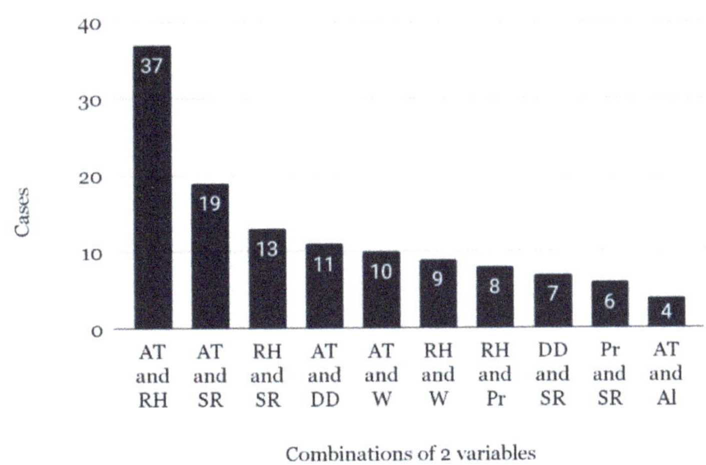

Figure 6. The most popular combinations of two climate variables used for CZB.

6. Climate Data Sources

During the course of the review, several climatic data sources became evident. Most publications use data obtained from national meteorological services (60.0%). Authors also frequently use web data sources such as the WorldClim repository, EnergyPlus website, or Global Historical Climate Network Dataset (18.7%). Cases of using databases of various software were revealed in 13.3%, with the EnergyPlus database, Autodesk Green Building Studio, and Meteonorm being the most common. Climate models were used in 6.7% of cases; basically, these are publications related to future climate scenarios [51,77,97,101] (Figure 7).

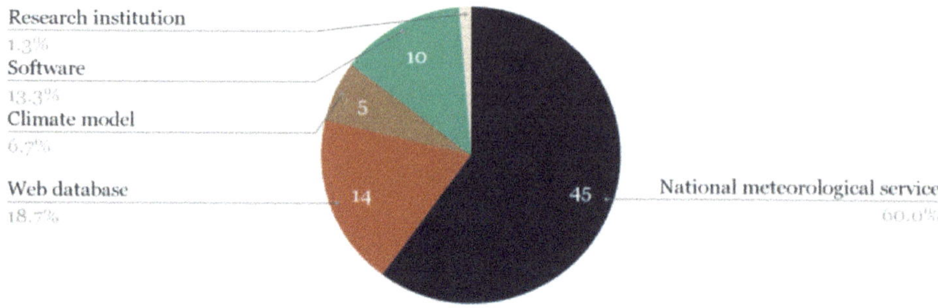

Figure 7. The most common climate data sources.

The forms in which climate data are used for classification have also been identified. Monthly mean values, TMY, daily mean and hourly values are four primary data forms. Often, the data undergo some processing and can be significantly changed from the data source to the classification itself. However, the forms that provided the core for the classifications are depicted in Figure 8.

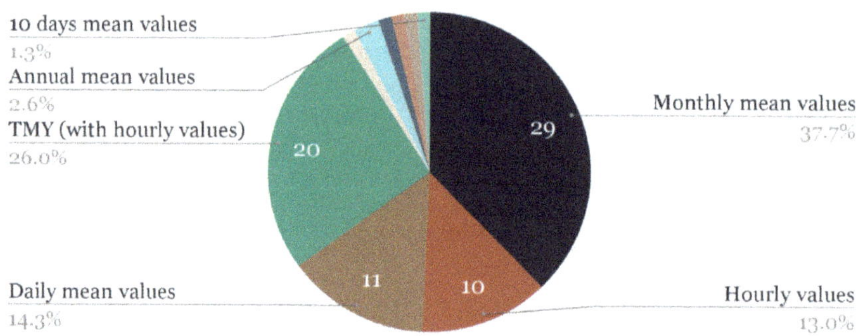

Figure 8. The most prevalent forms of climatic data used to develop the classifications.

7. Period of Climate Observation

Of interest is the period of climate observations, namely the monitoring duration and the time elapsed from the end of observation to the use of data in the climate classification. Climate change makes it preferable to use the data of recent years for accurate CZ [104]. There were 71 sources identified with an indication of the period of observations. The longest monitoring time of 102 years used for climate classification was found in the article by Wan [84]. Several authors used short observation periods of up to 5 years [22,109,122]. The average climate monitoring period was 28 years. Figure 9 shows the distribution of the observation periods. Figure 10 shows a histogram of the number of years that have passed since the last observation date until the document was published. An average of 8.5 years elapses from the end of monitoring to publication, with a period from 2 to 8 years being the most common. Given the quickly changing climatic conditions, which need the use of the most appropriate, up-to-the-moment data, it is reasonable that, on average, it takes from two to eight years between the end of observations and their utilization.

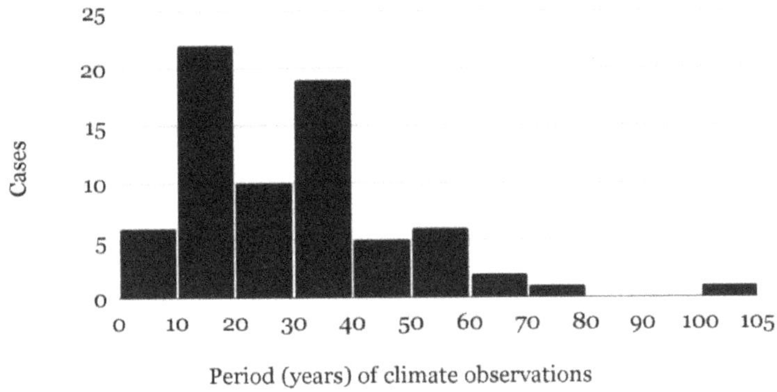

Figure 9. Histogram of climate observation periods.

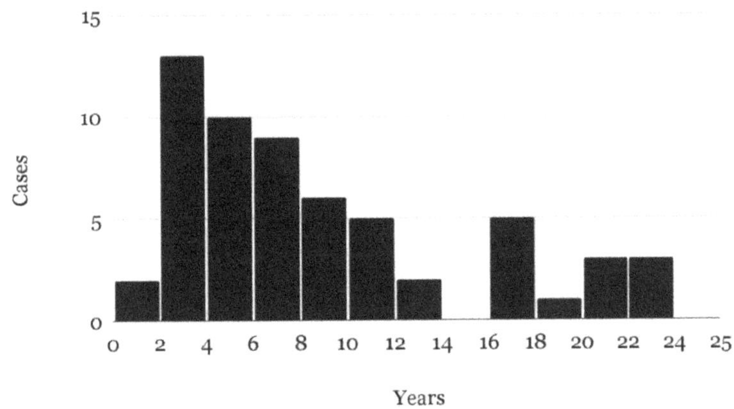

Figure 10. Histogram of years passed from last observation date to publication.

8. Methods Used for CZB

Indeed, CZB is characterized by a wide variety of applied techniques. Here, a detailed analysis of the most often used methods, the number of methods utilized simultaneously, and information about authors working with specific methods will be given.

A contextual framework or a consolidated and logical system based on opinions, ideas, and values that drive the activities taken by policymakers, researchers, or other users for CZ is viewed here as a CZ method. Along with "method", other terms such as "approach," "technique," "strategy," and "mechanism" will be used to prevent lexical repetition. Based on previous studies, we classified CZ approaches with specific criteria. The criteria used in determining each method of CZB in this review are indicated in Table 3. The conventional CZ techniques list was expanded by including the IJM. The category for cluster analysis (CA) has been enhanced with additional ML techniques; accordingly, we used the new MLM term for this domain.

There were 98 sources found with information on applied CZ methods. Figure 11 shows the diversity of CZ approaches, where the twelve most commonly used methods were identified. MLM, degree-day method (DDM), and buildings energy simulations (BES) were the three most popular methods among researchers. Table 4 shows different CZB methods and the researchers working with them. It can be seen that most of the authors work with several methods. The number of methods used simultaneously is shown in Figure 12 Often, only one method is used for CZB. Of 98 cases, one approach was used in 64 cases (65.3%), a combination of two methods was used in 27 cases (27.5%), and the simultaneous use of three or more approaches was identified in 7.2% of cases (Figure 12).

In addition, when only one method was used for CZB, as shown in the first bar of Figure 12, the most prevalent choices were MLM, DDM, BES, and BCM. Next, a quick summary of the main principles and key aspects of each method will be provided. A more detailed review of each CZ method will be presented in a separate article on which the authors are already working.

Table 3. CZ methods definitions.

#	Name of a Method	Abbreviations	Criteria
1	Machine learning methods	MLM	Classification is based on clustering techniques, neural networks, support vector machines, sensitivity analysis, or principal component analysis
2	Degree-days/-hours methods	DDM	Classification is based DDs values only. OR If several variables are used in the classification, then DDs should be the primary variable
3	Building energy simulation	BES	Classification is based on BES results
4	Bioclimatic charts method	BCM	Classification is based on Givoni, Lamberts, Milne, and Olgay charts with the combination of temperature and humidity as the main variables
5	Köppen–Geiger method	KGM	Classification is followed by the Köppen–Geiger system and is based on seasonal precipitation and temperature pattern
6	Climate severity index method	CSIM	Classification is based on the climate severity index (a site-specific value that defines the severity index of a specific climate) according to Formulas 4 and 5
7	Interval judgment method (the complex combination of climate variables based on the repeatability of their elements)	IJM	The classification is based on a combination of different variables with established limits (threshold) of variables for each zone
8	Frequency distribution of climate variable	FDV	Classification is based on the different types of probability distributions of variable(s)
9	Mahoney method	MM	Classification is based on Mahoney tables
10	Thornthwaite climate classification method	TCCM	Classification is based on Thornthwaite climate classification
11	Existing building stock performance method	EBSM	Classification is based on actual data of building stock performance
12	Heating or cooling index	HCI	Classification is based on heating and cooling indexes. This index is commonly used to determine how ambient temperature, relative humidity, and radiation affect human comfort
13	Other		Quitt's climate classification Roriz method The World Health Organization (WHO) classification method Administrative division Approximation and interpolation method (AIM) Camargo climatic classification

Table 4. Authors working with different CZB methods.

Abbreviations	Cases	%	First Authors	References
MLM	34	34.7%	Aliaga, Alrashed, Anas, Bai, Benevides, Bhatnagar, Bienvenido-Huertas, Deng, Erell, Falquina, Fovell, Lau, Malmgren, Mazzaferro, Netzel, Pernigotto, Pineda-Martínez, Praene, Roshan, Shi, Tükel, Unal, Walsh, Wan, Wang, Xiong, Yang, Zeleke, Zscheischler	[10,23,25,27,34, 35,39,40,44,46,49, 50,52,53,71,72,74, 75,80,84,87,93–97,99,100,107, 119,132–134,138]
DDM	25	25.5%	Abebe, Asfaw, Bai, Bawaneh, Briggs, D'Amico, De Rosa, Elmzughi, Ghedamsi, Mazzaferro, Muddu, Nair, Noh, Pusat, Rakshit, Ramon, Sánchez de la Flor, Tsikaloudaki, Verichev, Walsh, Xiong	[9,23,25,27,32,35, 43,46,79,81,85,86, 90,108,110,114, 115,127,136,140, 141,147]
BES	25	25.5%	Asimakopoulos, Bai, Bienvenido-Huertas, Carpio, Cory, D'Amico, Danilovich, Deng, Díaz-López, Kishore, Mazzaferro, Meng, Nair, Semahi, Tsikaloudaki, van Schijndeln, Verichev, Walsh, Wang, Xiong, Zeleke	[6,25,27,32,34, 40–42,44,46,48,51,53, 79,81,89,104–107, 109,111,136,153]
BCM	16	16.3%	Bodach, Cory, da Casa Martín, Kishore, Lam, Mahmoud, Mobolade, Moradchelleh, Navarro, Rakoto-Joseph, Roriz, Roshan, Singh	[42,48,49,76,80, 82,91,92,113,117, 131,142,146,154]
KGM	11	11.2%	Alrashed, Aparecido, Ascencio-Vásquez, Beck, Mazzaferro, Peel, Pernigotto, Pineda-Martínez, Sarricolea, Zscheischler	[46,72,73,77,96, 99–103,133]
CSIM	6	6.1%	Carpio, Diaz-Lopez, Moral, Verichev	[22,41,47,90,109, 122]
IJM	4	4.0%	Bai, Briggs, Ferstl, Hobaica	[104,118,140,144]
FDV	4	4.0%	Coskun, Felix, Khedari, Kim	[88,121,128,130]
MM	3	3.0%	Ogunsote, Pawar, Zeleke	[53,116,135]
TCCM	2	2.0%	Aparecido	[101]
EBSM	2	2.0%	Gangolells, Hjortling	[123,125]
HCI	1	1.0%	Wan	[84]

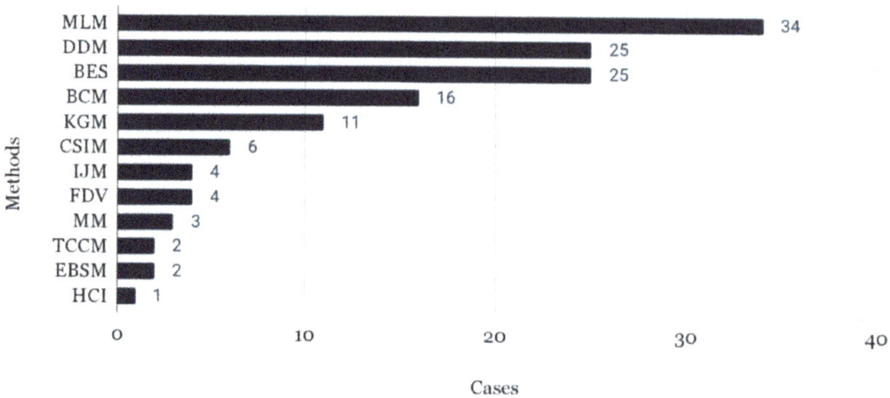

Figure 11. Different methods in CZB.

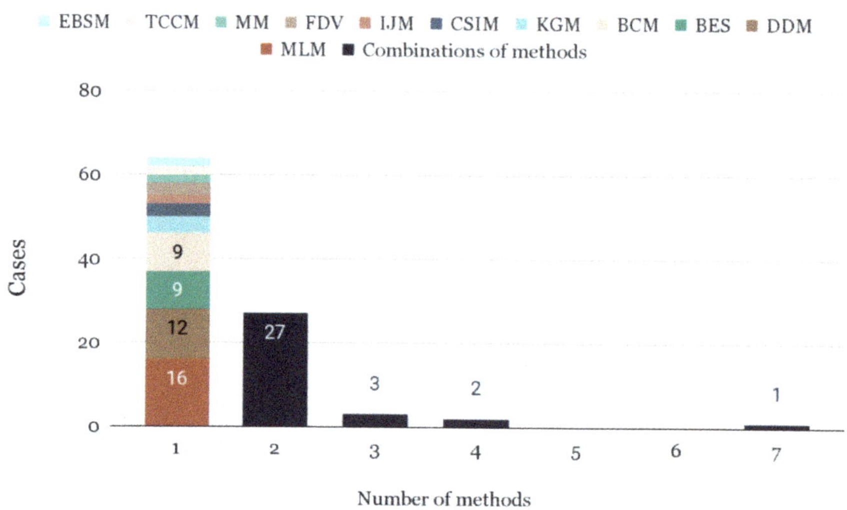

Figure 12. Frequency of the number of methods used for CZB.

Machine learning methods (MLM) denote different types of multivariate data classification and segmentation algorithms that can be successfully applied for the CZ of territories for building energy-efficiency. It can simultaneously involve many climatic and geographic variables or even be combined with some building properties, avoiding oversimplification and obtaining more meaningful results [84]. Although the most common type of ML classification in the development of climate maps is CA, other various ML approaches for CZB were revealed during this review (neural networks, principal component analysis, sensitivity analysis, etc.) [39,45,52,53,72,119,133]. Accordingly, we extended this category and used the new MLM term. Still, among all possible ML methods, CA was the most frequent technique for climate classification with hierarchical, k-nearest-neighbor, and k-means clustering.

The study by Fovell et al. [95] is one of the early attempts to adopt MLM for climate zoning. The U.S. territory was divided into climate zones using principal component analysis (PCA) and hierarchical clustering (HC) with AT and Pr as the main climate variables. In total, three different candidate clustering levels (8, 14, and 25) were tested. As the most effective with a satisfactory level of detail, an option with 14 clusters was chosen.

Conflicting buffer zones and cluster overlap were found as drawbacks. According to the authors, it is important to have data points outside of the research area's boundaries that were not included in the analysis for a categorization near the study area's borders to make the final climate classification more accurate. Stepinski and Netzel [75] attempted to classify the climate of the entire globe based on a comprehensive clustering approach. This study included 32 different clustering-based classifications. All methods were then compared with the KG classification. The authors concluded that using three climate variables (AT, Pr, and temperature range) provides the best results. About half of the climatic zones detected by clustering were accurately matched with the KG classification classes; the rest, however, differed in their climatic characteristics and geolocation. Additionally, the authors pointed out that the k-means algorithm should be preferred over the HC algorithm in the process of climate classification. Praene et al. [74] applied HC on principal components and spatial interpolation using GIS to propose a new CZB of Madagascar. As main climate variables, RH, daily GHI, and Pr were chosen. Three climate zones that correspond to dry, humid, and highland environments were derived. The authors also investigated the relationships between established CZ and the thermal comfort levels of conventional building typologies. It was demonstrated that some building types can ensure a higher yearly comfort rate and be the most effective when erected in a specific climate zone. Walsh et al. [23] performed a survey on building energy performance maps of Nicaragua. Three CZ techniques (DDs, CA, and administrative division) were compared. It was concluded that CA can provide a more in-depth understanding of CZ than other methods, but for better results, it needs to take BES into account during the development of climate maps. It was emphasized that the appropriate selection of the method plays a central role in CZ for building energy-efficiency purposes.

Given its primary advantages, MLM provides significant potential to go deeper into the CZB and acquire more reliable, previously unavailable results. One of the most important issues is data dimensionality reduction and the proper number of clusters. PCA can be applied for data dimensionality reduction and elbow method or Hubert index for solving the number of clusters problem. However, often, MLM is supplemented with other approaches, such as BES, to generate more accurate results or to evaluate them [23,34,40,44]. It should be noted that although the problem of studying climate and its classification is location-oriented [155,156], none of the found sources used the principles of spatial analysis for CZB. Spatial analysis has become a standard in many research areas (such as epidemiology, sociology, ecology, and tourism) [157–159], but this has not happened in the field of CZB yet. The core concepts of geographical dependency and spatial autocorrelation are founded on The First Law of Geography, which states "everything is related to everything else, but closer things are more related than distant things" [160]. Therefore, spatial objects and phenomena should be analyzed based mainly on their locations and relationships. In this way, the role of space in CZB is emphasized, and the understanding of the working and representation of space, spatial patterns, and processes is enhanced. In CZ, the recognition of the spatial dimension is expected to yield different and more meaningful results and helps to avoid erroneous conclusions [161].

Degree-days methods (DDM) are well-known techniques that have been used for decades [162]. They indicate the severity of the climate in various locations by documenting when the external AT falls below or rises above a specified temperature during a given year, necessitating heating or cooling. This method is generally defined as the sum of the temperature differences between the outdoor mean temperature over 24 h and a base temperature daily (with 18 °C as the most frequent base temperature value). The base temperature is arbitrary, but it is commonly described as the outside temperature at which heating, ventilation, and air-conditioning (HVAC) systems do not have to operate to keep the building's internal climate comfortable. According to ASHRAE standard [163], cooling and heating degree-days (CDD, HDD) are calculated according to Equations:

$$HDD_t = \sum_{d=1}^{Dt} (\ T_b - \overline{T})_d^+ \quad (1)$$

$$CDD_t = \sum_{d=1}^{Dt} (\ \overline{T} - T_b)_d^+ \quad (2)$$

where

Tb—base temperature;

+—only positive values are considered;

$\overline{T}$—the mean value of the maximum and minimum temperatures in a given day, as shown in Equation (3):

$$\overline{T} = \frac{(T_{max} + T_{min})}{2} \quad (3)$$

Other methods for the calculation of HDD and CDD are the mean daily degree-hours method [164], UKMO approach [162], Hitchin method [165], and Schoenau and Kehrig technique [166]. It should be emphasized that except for the mean daily degree-hours method, all other DDs calculation methods are aimed to approximate the true sum of the daily outside temperature variations. Due to approximate estimations using only the maximum and minimum values of the day or monthly mean values, those methods create considerable inaccuracies [167]. Thus, if more detailed weather data, covering the hourly outdoor dry-bulb temperature of the selected region, is available for more accurate climate classification, the mean daily degree-hours method [164] is preferred over all others.

Quayle and Diaz [168] and Le Comte and Warren [169] have found that DDs are significantly connected to electricity, natural gas, and heating consumption. HDD has also been proven to be a reliable indicator of residential energy use [170,171]. Briggs et al. [140] were some the first to use DDs for climate classification. The authors proposed a new climate classification for use in characterizing the performance of energy-efficiency measures for buildings in the USA. First, according to a table composed by Strahler [172], humid/dry/marine zones were determined, and further division was carried out based on DDs, with established intervals for each zone. The authors also compared the new classification with the existing IECC 90.1-2001 system. According to the authors, the proposed classification provided a better perspective of climate, with more uniform climate zones that better characterize U.S. climates. This classification was later included in ASHRAE Standard 169-2006. Pusat and Ekmekci [85] applied DDM to the CZB of Turkey. Unlike the Turkish official CZ code, which only considers HDDs, the authors used the HDD/CDD combination (18 °C base temperature). Six main climatic zones were identified instead of the four of the national code. The authors proposed a reclassification of the country's climate from both a heating and cooling point of view. The findings emphasized the need for the cooling loads consideration in DD climate zone classification. Katerina Tsikaloudaki et al. [81] presented an approach for defining climatic zones in Europe based on the number of DDs. The article also concludes that the most realistic classification can be obtained with the simultaneous use of HDD and CDD. Tükel et al. [50] performed the climate classification of Turkey in terms of the thermo-economic perspective utilizing the combination of DDM and MLM. Turkey's 80 provinces were divided into five zones by fuzzy c-means clustering technique based on DDs, thermal insulation, main wall component, and fuel type. The results showed that 16 out of 80 provinces moved to a new climatic zone when the suggested CZ was compared with the existing national thermal zones. The new classification proved that Turkey's present national CZB is insufficient, particularly in mild regions with significant cooling demands. Verichev et al. [90] used a combination of HDD with a climate severity index (CSI) to update the boundaries of climate zones in the southern regions of Chile, where three climate zones were found. After, the relative energy consumption of houses was examined for cooling and heating in the summer and winter seasons. The investigation showed that the energy expenditures for cooling the same house during the summer may vary by 50% within the boundaries of a single thermal zone. The question of whether using only the HDD method is adequate was brought up by these substantial differences in the theoretical estimation of a building's energy consumption. Mazzaferro et al. [46] proposed a new climate zoning of Brazil. The total number of methods used in this work was seven, the largest in this review. Among other methods, the technique of enhanced DDs was proposed, where a DBT setpoint of 10 °C was used to determine DDs values. The DDs data were then divided into groups using k-means clustering. There was a total of eight fixed clusters. The authors concluded that the enhanced DDM presented considerable performance, even though it did not involve building performance data in its application. However, the authors concluded that the quality of CZ could be significantly improved by using a preliminary assessment of zoning by building performance data.

The popularity of the DDM for CZB is partly related to its ease of understanding and close connection with energy use, especially in cold regions [162]. DDM can provide quality CZ for a range of applications due to its documented relationship to building energy consumption. However, DDM operates primarily with outside AT alone, ignoring other key climatic factors that affect a building's energy consumption. Choosing the correct base temperature is also crucial, as the incorrect base temperature will lead to inaccurate DDs [173]. Bioclimatic comfort zones, as described in [80], can be used to determine the base temperature. The DDM is the second most common method we found in CZB.

Building energy simulation (BES) has shown great potential when applied to CZB [24,25]. BES is a method of determining how a building and its components will behave in real-world scenarios. This is accomplished by using a mathematical model that simulates situations in a virtual environment. Within the framework of this review, the term is understood as the energy consumption of a building in a given climate based on the heating, cooling, and lighting loads. The climate classification procedure based on BES consists of generating performance maps showing how a set of chosen indicators, such as energy consumption or thermal comfort, vary throughout the territory (country or region) for given archetype buildings, for a typical year of climatic data. These maps are produced using building performance simulation results. The building performance is then linked to each climatic zone under study [25]. The idea of using BES for CZB is that the performance of a building model inside a single climate zone should stay essentially the same. BES is less common than DDM, but becomes more widely used in CZB applications.

Shaan Cory et al. [42] proved that for better climate classification, data on the weather and climate of the region are not enough. The authors noticed that sometimes when using external climate data for classifying climatic zones, buildings of different types could be assigned to different climatic zones during the thermal simulation within the same location. An adjustment to the traditional approach to climate classification for buildings was proposed. The authors used the climate indicator determined from the results of simulations, which was simply a three-level definition of climate challenges of building: heating-dominated, cooling-dominated, or mixed heating- and cooling-dominated. Thus, the refined classification was made based on the external climate characteristics together with the reference thermal data of the building. The authors also noted that buildings in cold regions could not be classified solely by using external climatic conditions. Walsh et al. [25,32] used BES to validate the CZB produced by other methods and reduce misclassification. The concept of the Percentage Misclassified Areas (PMA) was introduced, which is based on the idea that each climatic zone should have its unique climate conditions, which leads to a unique performance (cooling and heating energy demand) for a particular type of building. By this, there should be no overlap observed in the building performance of an identical building type placed in different zones. Results also showed the challenges of developing CZ to address multiple building types, as each archetype showed particular sensitivity to climate. Mazzaferro et al. [107] developed a data-driven CZ methodology to increase the robustness of CZB. In this study, climatic zone validation was performed by thermal loads of three different-sized office buildings as building performance indicators. An assessment was performed by comparing building thermal load results within each climatic zone obtained by clustering climatic variables (DBT, RH, and GHI). It has been proven that the utilization of BES results, data analysis, and CA methods can greatly contribute to the development of CZ methodology. In 2020, in the development of their previous study, the same team carried out work to determine the climatic zones of Brazil for building energy-efficiency regulations [46]. Climate classification was performed with known CZ methodologies (ASHRAE 169, Koppen-Geiger, Brazilian regulation method NBR 15220, Roriz method) and alternative methods (data-driven, enhanced DD, and decision tree method) supported by building performance results. The authors concluded that the quality of CZ could be greatly improved by using a preliminary assessment of zoning by building performance data.

BES is currently regarded as the most accurate method for predicting thermal building performance and has demonstrated significant promise as a policy tool [174] when applied to CZ, mainly through parametric analysis. Detailed climatic data and BES, according to several sources, could aid in the construction of a more robust climatic categorization [23,46,83,174]. However, there are certain limitations to its use in CZB, such as the necessity to pre-define a design hypothesis that differs depending on building type, occupational patterns, and HVAC systems. Furthermore, thorough meteorological data are required, which are not always available for the locations of interest [41].

Many authors use the BES method to work on the topic of the impact of climate change on buildings and CZB [89,98,111]. This is partly because simulations can be easily used to gain insight into the future energy consumption of buildings and, based on that, analyze the possible future climate zones. The data from future climate models can be easily represented as future weather files. Different methodologies and software tools can be applied [175,176]. These files are then used to simulate buildings under future scenarios.

The bioclimatic chart method (BCM) can also be used to change or create new CZ classifications. The BC analysis usually leads to identifying the possibility of passive design methods to preserve thermal comfort in outdoor spaces while also contributing to a more energy-efficient built environment. BC usually depicts the combination of AT and RH at any given time, so it becomes easier to analyze the climate features of a given location. This paper includes Givoni [177,178], Lamberts [92], Milne [179], Olgyay [180], and other psychometric chart methods in the BC category. Early attempts to develop concepts for bioclimatic building design were made by Olgyay [180]. He conducted a study on the influence of climate on building design concepts around the world and identified four major climate groups: cool, temperate, hot and arid, and hot and humid. Human tolerance ranges were also established from a BC using a combination of RH and DBT. In addition to the average radiant temperature, wind speed and SR were taken into consideration. A zone in the middle of a psychrometric chart established the range of conditions people find comfortable in different situations. Milne and Givoni [179] developed BC based on typical psychrometric charts frequently used to assess the characteristics of moist air. Additionally, four main climate categories—hot, warm–temperate, cool–temperate, and cold—were identified, along with eleven additional sub-climatic categories. The study of the effects of the environment on occupant comfort and thermal adaptability showed promising implications for the design of HVAC systems and the enhancement of building energy-efficiency. Bodach [129] proposed a foundation for creating BC zones for building design in Nepal. The psychrometric chart was used to identify passive design ideas for each location using a BC approach. Finally, a summary of building design tactics that are suited for the summer and winter in each zone was created. According to the authors, to create more climate-responsive and energy-efficient buildings, planners and architects can use proposed climate classification to make general judgments at an early design stage. Additionally, it might help with the creation of proper building energy regulations. The BCM, however, has some limitations because it only takes two climate variables into account (AT and RH). Da Casa Martin [113] offered a way for creating a CZB based on Givoni's design principles for the Autonomous Community of Galicia (Spain), creating general zoning between five geographical regions (CZ) with comparable behavior. New climate classification could give the designers knowledge of the techniques that can be employed to establish direct recommendations and different bioclimatic strategies that can be adopted to obtain more energy-efficient buildings. New climate classification based on BCM and DB-HE-2013 code were compared to identify their similarities and differences, finding that each should have its own specific applications.

Even though the primary goal of BCs is to provide comfortable indoor conditions under thermal comfort requirements, a consequence of adopting BCM is an increase in the building's energy-efficiency due to reduced energy demand. Additionally, a significant correlation is demonstrated between bioclimatic potential and annual building energy consumption [48,83]. Nonetheless, several resources recommend utilizing BCM in conjunction with BES [42,48,83,181] to produce secure results in terms of climatic zoning.

Köppen–Geiger method (KGM), as was mentioned, is one of the oldest and most well-established methods. A monthly record of average AT and Pr data is the only need for determining KG climatic zones. The elevational component can be taken into account in some circumstances. The KG climate map of the world is constantly updated and refined [78,102]. Based on a global data set of long-term monthly AT and Pr time series data, Peel et al. [102] used the KG method to produce a new global climate map. Climate variables were interpolated for each continent on a 0.1 × 0.1-degree grid. With each variable being interpolated separately, the updated KG world climate map was produced utilizing statistics from stations during their complete record period from 1951 to 2000. Beck et al. [77] published updated global KG climate maps with a 1 km resolution for current conditions (1980–2016) and predicted future climate change scenarios (2071–2100). The current map was made using an ensemble of four high-resolution, spatially adjusted climate maps. Particularly in regions with sharp spatial or elevation gradients, the new maps offer far more detail and classification accuracy than prior ones. Not having an unambiguous connection with CZB or energy-efficient CZ, the KGM is still used for these purposes [99,100,103]. Pernigotto et al. [99,100] discussed the application of the KGM with CA (HC) to address the issues of climate classification and representative climate identification. Results were found to be similar to KG, but the proposed climatic categories were fewer and more homogenous. Overall, KG improved by CA is intended to characterize building performance under various climatic conditions, support the creation of national or international energy policies, and give an analytical reference for CZ and the selection of representative climates.

KGM can hardly be used for precisely characterizing the performance of energy-efficiency measures for buildings. KGM alone does not allow for the accumulation of accurate data needed to address the issue of CZB. Multiple sources compared the precision of a KGM classification to that of a CA (k-means clustering) and BES, revealing that CA and BES exceed traditional KG classification quality [72,74,75].

The climate severity index method (CSIM) is capable of characterizing climate severities; accordingly, there is a wide variety of approaches to the calculation of such indices [182–184]. The core concept of CSI is that climate variables can be combined into a single site-specific value that defines the severity of a specific climate. The data can be examined like any other meteorological parameter to identify trends, develop sector-specific applications, and examine climate patterns or individual seasons to put their severity into context. Speaking of the CSI in this article, we mean the index used in the practice of Spain [43] and Chile [122], which uses the HDD, CDD, hours of sunshine duration, and regression coefficients for calculation. Accordingly, the summer climate severity (SCS) is determined using Equation (4), and winter climate severity (WCS) is defined by Equation (5).

$$SCS = a{\cdot}CDD20jun - sep + b{\cdot}CDD20jun - sep^2 + c \quad (4)$$

$$WCS = a{\cdot}HDD20oct - may + b\frac{n}{N} + c{\cdot}HDD20oct - may^2 + d{\cdot}\left(\frac{n}{N}\right)^2 + e \quad (5)$$

where

$CDD20_{(jun\text{-}sep)}$—the hourly sum of CDD (20 °C base temperature) from June to September;

$HDD20_{(oct\text{-}may)}$—the sum of HDD (20 °C base temperature) from October to May;
n—the sum of hours of sunshine duration from October to May;
N—the sum of maximum hours of sunshine duration for October through May;
a, b, c, d, e—the regression coefficients.

In this case, SCS and WCS were produced using absolute data on heating and cooling energy consumption. The index of relative energy use for cooling and heating buildings was then connected with the climate-related values of meteorological indicators. Since CSIM is based on building energy consumption translated into conditional indices, it has a certain potential and can be applied to CZB. It is easy to establish climatic severity indices and classify the territory into climate zones using the simple formula, DD, and sunshine duration. In general, the idea of utilizing a regression model based on a number of the most crucial CZB characteristics, such as building energy consumption, DD, and SR, appears to have the potential to be successful. However, it is difficult to adapt the strategy to other regions outside of Spain because the calculation of the regression coefficients is still not entirely clear.

Other methods. The climate classification methods found during the literature review are not limited to the six methods described previously. Less popular techniques are the IJM [118,144], the frequency distribution of climate variables (FDV) [88,121] the Mahoney method (MM) [116,135], Thornthwaite climate classification (TCCM) [101], heating or cooling index (HCIM) [84], Quitt's climate classification [112], and existing building stock performance (EBSM) [123,125].

As was previously stated, the issue of CZ has long been studied, initially predominantly for agricultural purposes and later for building and construction applications. In the first half of the 20th century, building climatology was more responsible for weather protection and interior comfort provision rather than energy-efficiency. KGM and IJM, which used a limited number of basic climate variables (AT and Pr), were then adequate [185]. Indeed, these fundamental methods were limited in their ability to incorporate a large number of climatic variables into the analysis and were far from establishing an obvious and understandable link between the variables and the energy usage of buildings. Later, during the 1970–1980s, a clear correlation was established between the main climate variables (mostly AT and its various derivatives) and the energy consumption of buildings [168–171,186–188]. DDs, used alongside or sometimes replacing other climate variables, started to occupy one of the leading places in CZB. At the same time, work began in the field of bioclimatic architecture, and BC were introduced [177,179]. In the early 1990s, CA was implemented in CZ with PCA, hierarchical clustering, and artificial neural networks, which can be considered as the beginning of a new stage in the development of CZB [95,119]. Over time, CA has become one of the main methods in the study of CZ along with DDM [84,87,94,96,139]. Later, experiments started on the use of BES for CZB needs, and more MLM were incorporated into this field [42,174,189]. Since the 2010s, scientists' enthusiasm for MLM and BES has grown considerably. The possibilities of traditional and contemporary methods were extensively compared. In this way, more recent MLM and BES were interpreted with older KG, BCM, and DDM [23,25,46,100,101,103,107,122]. Recently, BES and MLM have shown great potential when applied to CZB. In addition, the significance of using BES to validate CZB was proved by several publications [23,25,46,79,83,104]. Overall, BES and MLM methods are simple to implement and have shown to be reliable in defining CZ by transitioning from a climate-based to a performance-based approach. Additionally, a combination of approaches yields much better and more robust zoning classification results. Any combination of DDM, BES, and MLM techniques is likely to be the most powerful, efficient, and promising, delivering the most consistent results. It is reasonable to claim that there is now a solid scientific basis for widely applying BES and MLM in official national CZ codes and regulations. However, the presence of so many strategies indicates that there is still no consensus about the optimal CZB strategy, emphasizing the importance of further research work in this field.

In the next section, we will support the findings of this review with bibliometrics and bibliographic analysis, which aims to indicate substantially contributing nations, affiliations, journals, authors, and connections among them. A network of keywords, articles and overall link strength among publications will be discussed. We will present the data step by step, most clearly reflecting the performance of research activities and academic output in the CZB field.

9. Bibliometric Analysis

In this work, 98 (Scopus cited) articles were subjected to bibliometric analysis. To identify the most significant affiliations or public organizations in a CZB research field and their geographical location, a map (Figure 13) was formed which indicates the affiliations in which the researchers are registered. The larger the mark on the map, the greater the contribution (number of papers published). In addition, a list of affiliations that have three or more publications in a CZB is shown; those 10 affiliations are marked on the map with signatures. Among 98 publications, the largest number of publications (six each) belong to authors from the University of Granada (Spain), Xi'an University of Architecture and Technology (China), and the Austral University of Chile.

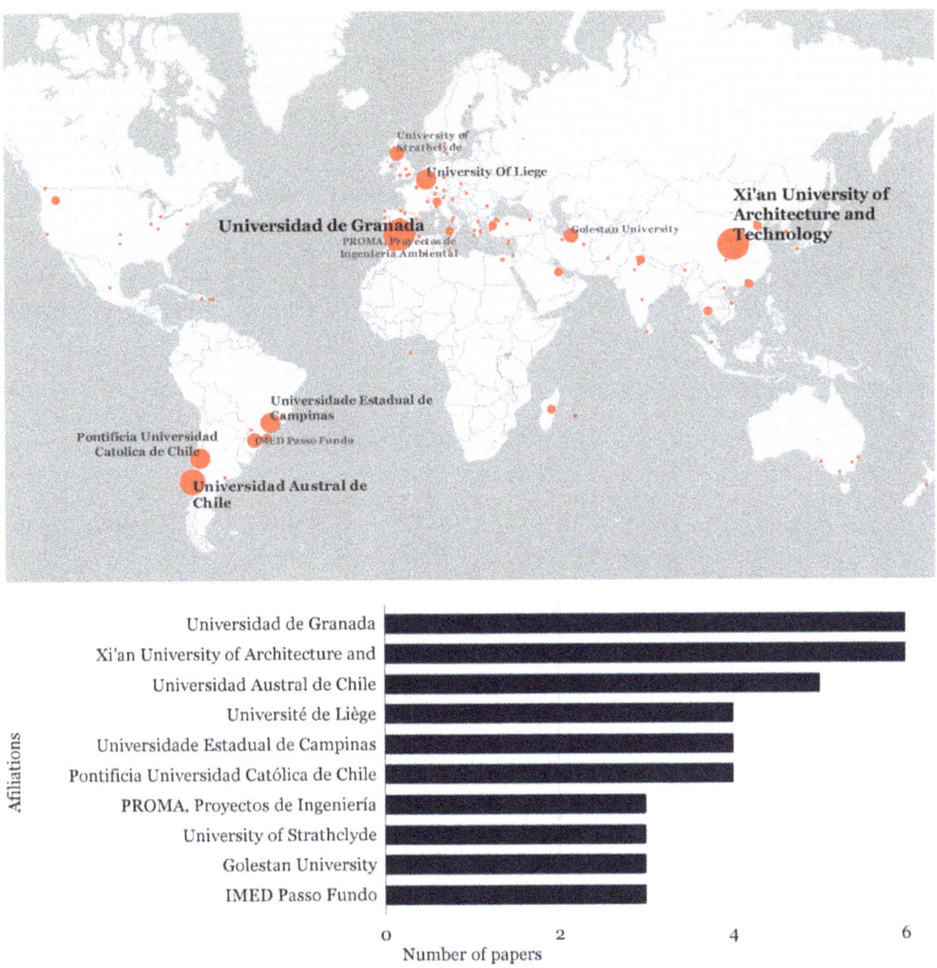

Figure 13. A map of affiliations or public organizations which publish more articles than others in CZB.

The publications considered in this analysis had a total of 9696 citations in Scopus. Excluding two highly cited papers [77,102] concerning KG climate classification, the average citation per article of the remaining 96 articles is about 35. The top 10 most cited articles from our list are shown in Table 5. The largest number of articles was published in Energy Additionally, Buildings, Building and Environment (previously Building Science), and Energy Conversion and Management (previously Energy Conversion) journals. The list of the most significant journals is shown in Figure 14. There are eight most-contributing authors (with four or more papers): Carpio [41,51,83,109,122]; Verichev [22,51,83,109,122]; Walsh [23–25,32,100]; Yang [35,52,71,84,87]; Zamorano [41,51,83,109]; Attia [7,49,80,106]; Cóstola [23–25,32]; and Labaki [23–25,32]. Figure 15 shows the top 20 most contributing authors (with two or more published papers). Articles were also analyzed by citation over time. A trend line and articles cited above the trend were determined (Figure 16). This indicates an increased interest in particular articles [6,7,73,89,94,95,125,190] from the scientific community.

Table 5. The most-cited publications in the review.

#	Authors	Title	Year	Journal	Cited by	Document Type	Reference
1	Peel M.C.	Updated world map of the Köppen–Geiger climate classification	2007	Hydrology and Earth System Sciences	5580	Article	[102]
2	Beck H.E.	Present and future Köppen–Geiger climate classification maps at 1 km resolution	2018	Scientific Data	1091	Article	[77]
3	Fovell R.G.	Climate zones of the conterminous United States defined using cluster analysis	1993	Journal of Climate	286	Article	[95]
4	Unal Y.	Redefining the climate zones of Turkey using cluster analysis	2003	International Journal of Climatology	235	Article	[94]
5	Wang H.	Impact of climate change heating and cooling energy use in buildings in the United States	2014	Energy and Buildings	193	Article	[89]
6	Attia S.	Overview and future challenges of nearly zero energy buildings (nZEB) design in Southern Europe	2017	Energy and Buildings	156	Article	[7]
7	Sarricolea P.	Climatic regionalisation of continental Chile	2017	Journal of Maps	121	Article	[73]
8	Asimakopoulos D.A.	Modeling the energy demand projection of the building sector in Greece in the 21st century	2012	Energy and Buildings	115	Article	[6]
9	Nguyen A.T.	An investigation on climate responsive design strategies of vernacular housing in Vietnam	2011	Building and Environment	88	Article	[190]
10	Gangolells M.	Energy mapping of existing building stock in Spain	2016	Journal of Cleaner Production	79	Article	[125]

Since the early days of bibliometric research, the concept of visualizing bibliometric networks, often known as "science mapping" has gotten a lot of attention. Visualization is an effective method for analyzing a wide range of bibliometric networks [68]. In the next section, the results of performed bibliographic analysis will be provided.

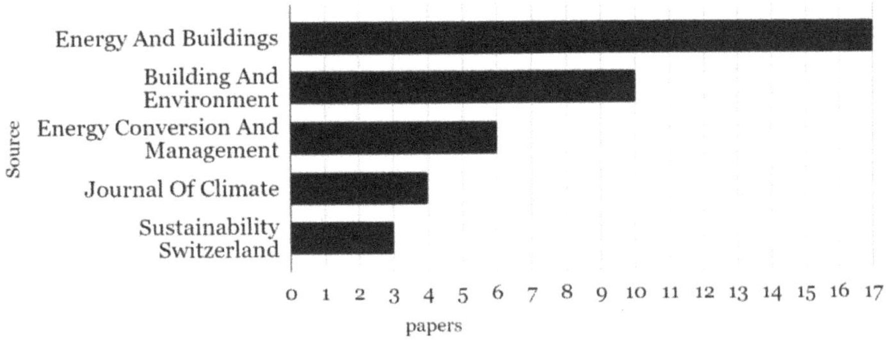

Figure 14. The bar chart of the journals that published CZB papers.

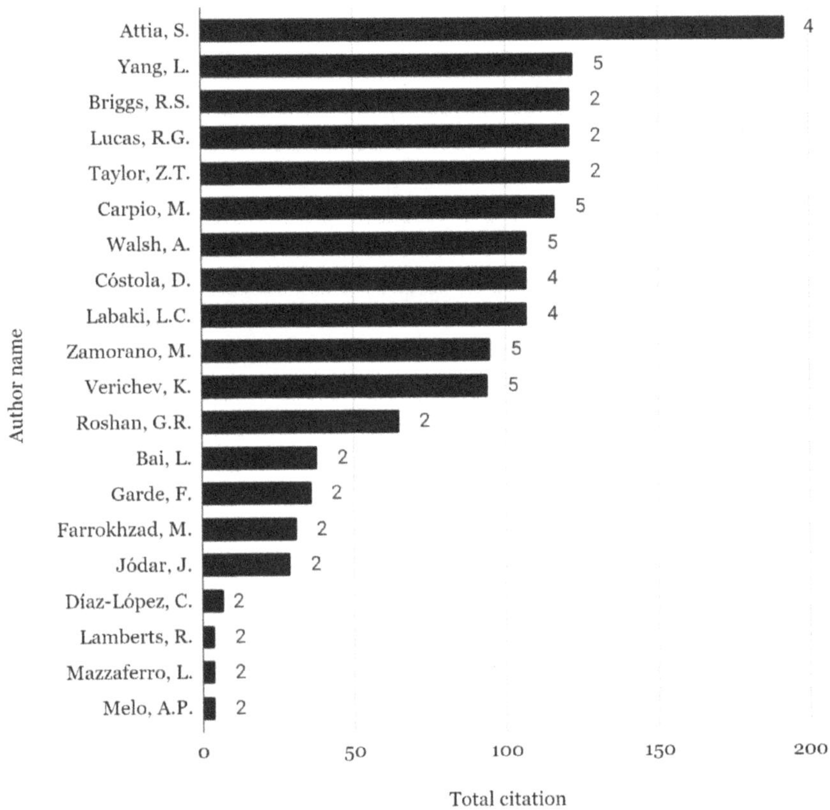

Figure 15. The bar chart of the most-contributing authors in the CZB area. The number of published CZB papers is shown at the outside end of each bar.

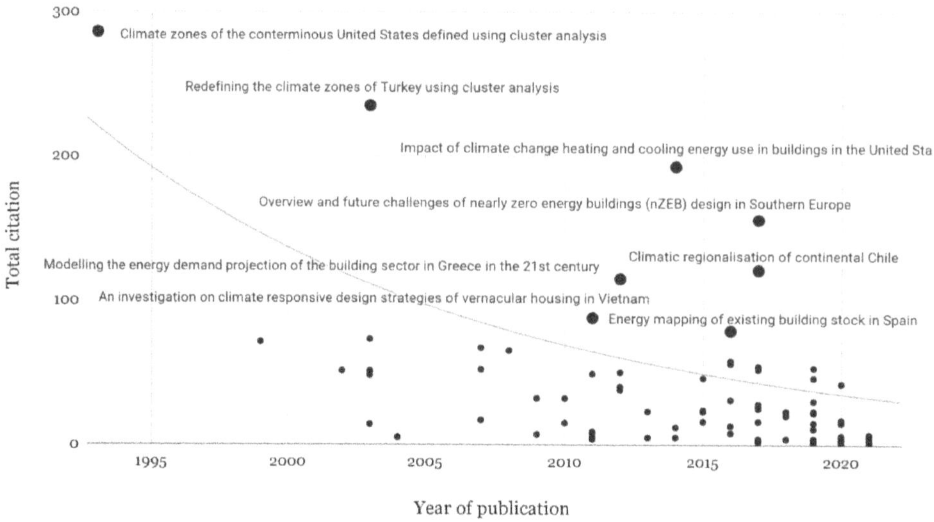

Figure 16. The ratio of citations to the year of publication and papers cited above trend.

10. Bibliographic Analysis

Co-citation analysis in this review is represented by the co-citation networks of researchers (Figure 17). To better perceive the visualization, the number of its nodes has been reduced, and the top 36 authors out of 5036 who passed the threshold level with the minimum number of citations of 15 were used for visualization. Each circle in the visualization represents an author. The size of a circle reflects the number of citations an author has received. Authors that are located close to each other in the visualization tend to be more strongly related, based on co-citations rather than authors/journals that are located far away from each other. Three clusters of most-cited authors are revealed.

1. Koppen, W.; Rubel, F.; and Kottek, M. are authors whose research topic is Köppen–Geiger climate classification in the upper left area [78,191–194];
2. Santamouris, M.; Attia, S.; Givoni, B.; and Carlucci, S. are authors that publish articles on the energy-efficiency of buildings, bioclimatic studies, and urban or local climate zoning research in the upper right area (green cluster) [7,177,178,195–200];
3. Costola, D.; Labaki, L.C.; Carpio, M.; Verichev, K.; and Yang, L. are directly focused on climate classification methods for buildings [16,23–25,32,41,51,83,84,90,122].

Additionally, the bibliographic coupling network of the top 100 authors (Figure 18) was created by analogy with the bibliographic coupling network of journals. Here, the closer two researchers are located to each other in the visualization, the more strongly they are related to each other based on bibliographic coupling. In other words, researchers that are located close to each other tend to cite the same publications. The network shows three groups of closely related authors:

1. Yang, L. and Walsh, A;
2. Almeida, M., Attia, S., and Roshan, G.;
3. Carpio, M., Verichev, K., and Diaz Lopez, C.

A direct citation network (Figure 19) based on articles that have at least one citation (90 out of 98) was constructed. We used normalized citations to correct the fact that older documents have had more time to receive citations than more recent documents. In VOSviewer the number of citations in a document is normalized by dividing it by the average number of citations in all publications issued in the same year and used for the VOSviewer files [70]. Using overlay visualization means that not only can a direct indication of the relatedness of publications be recognized, but the papers that have a high direct citation in this group of documents can be recognized, which indicates high interest from researchers working at the moment.

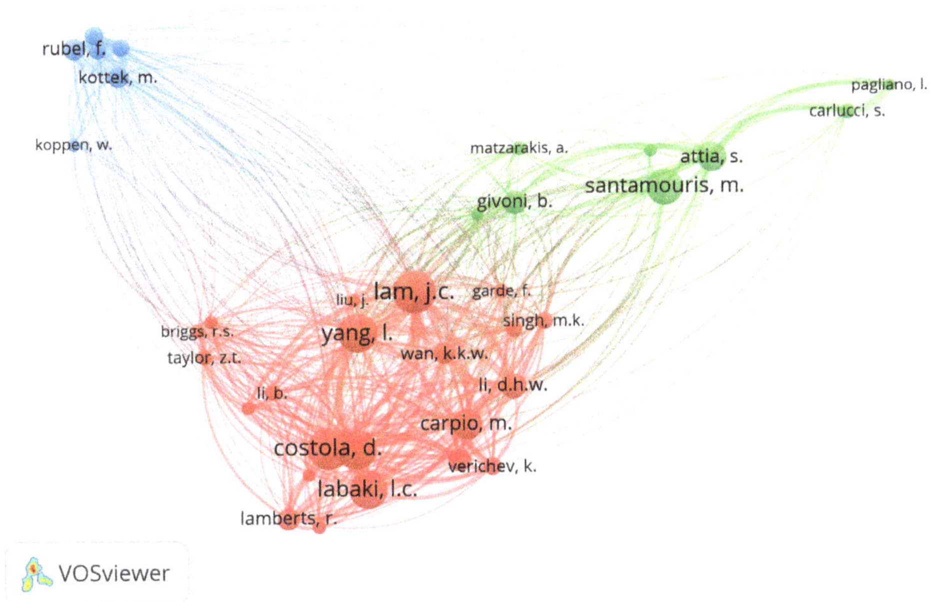

Figure 17. Co-citation network of researchers.

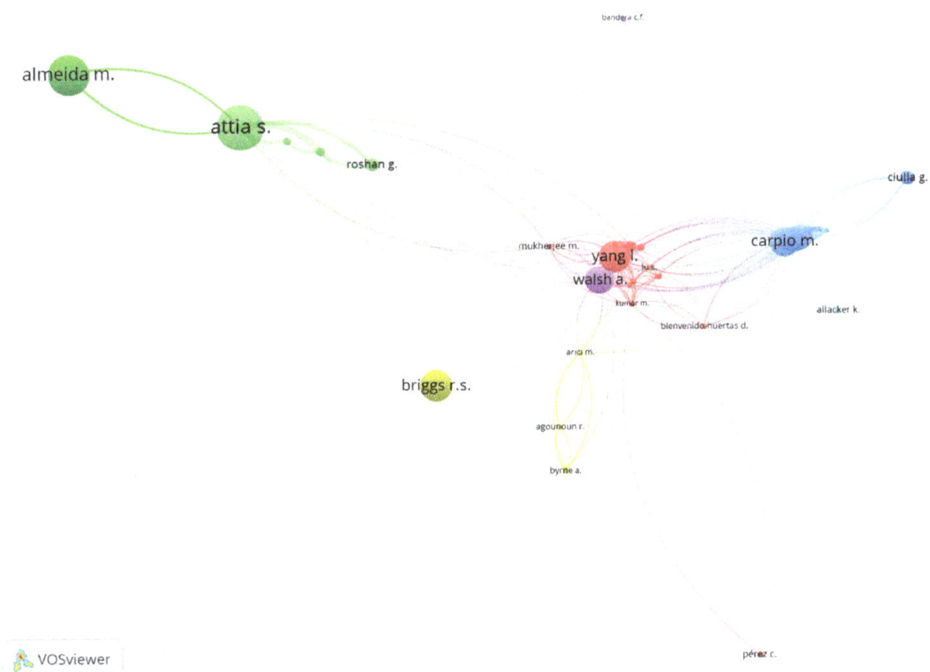

Figure 18. Bibliographic coupling network of authors.

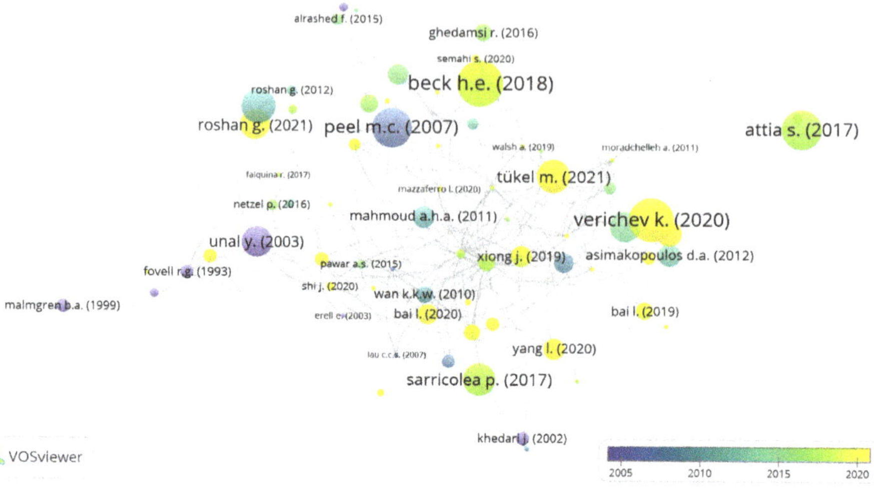

Figure 19. Direct citation network of articles.

The network of co-occurrences of keywords has also been analyzed (Figure 20). The main objective here was to obtain a sense of the terms that researchers used and to see whether there were any new subtopics. Of 2433 keywords, 58 keywords that occurred more than 10 times met the threshold and were used in the analysis. Keywords were extracted from the title and abstract of publications. Here, the larger the circle, the more often a keyword appears in the title, abstract, or keyword list of publications, and the closer keywords are to each other in the network, the greater the frequency of co-occurrences of the two terms in one publication. Using overlay visualization, the average publication year of each keyword can be recognized with a differentiation of colors from purple for the average publication year around 2012 to yellow for 2018. The top three keywords were: "buildings", "energy efficiency", and "climate change" with 31, 31, and 21 occurrences, respectively.

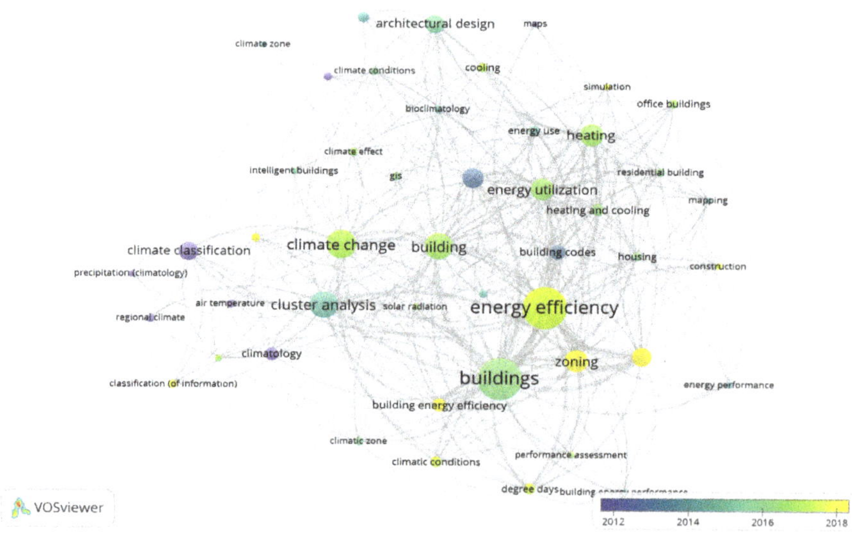

Figure 20. Co-occurrence network of keywords.

Keyword analysis shows that in the area of energy-efficient buildings, the issue of climate change is becoming quite important. To design energy-efficient and climate-resilient buildings, it is critical to gain insights into the energy demand across the building's service life from the early design stage onwards. Thirteen articles connected with climate change were found in the studied literature. However, in this article, we limit ourselves to specific information. As it becomes trendy to incorporate climate change into CZB studies, a more detailed review of this field may become the topic of our future publications. Here, we only give information about the number of publications in this review concerning future climate scenarios (Figure 21) and other details such as predicted period, climate model, and scenario type (Table 6).

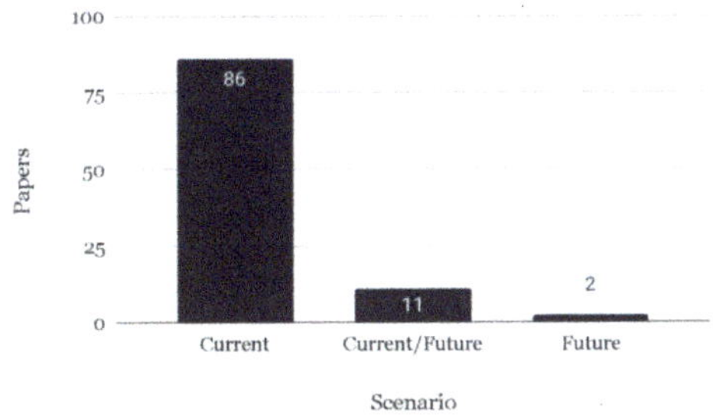

Figure 21. Histogram of publications in this review concerning future climate scenarios.

In addition to the above-discussed map of affiliations or public organizations, the bibliometric coupling network of countries is shown in Figure 22. However, unlike the above-discussed map, which is based on the number of publications, Figure 22 is based on bibliographic coupling information, which is, in general, an overlap in the reference lists of publications. Here, the larger the number of references two countries have in common, the stronger the bibliographic coupling relation between the countries. All 46 countries were mapped. By analogy with the bibliographic coupling network of journals, here, each circle represents a country. Large circles represent countries that have high normalized citation weight, and small circles are countries with a low number of citations. Generally, the closer two countries are placed in the visualization, the more closely they are connected based on bibliographic coupling. A group of Spain, China, the US, Australia, and Chile has close relations, with the strongest link between Spain and Chile.

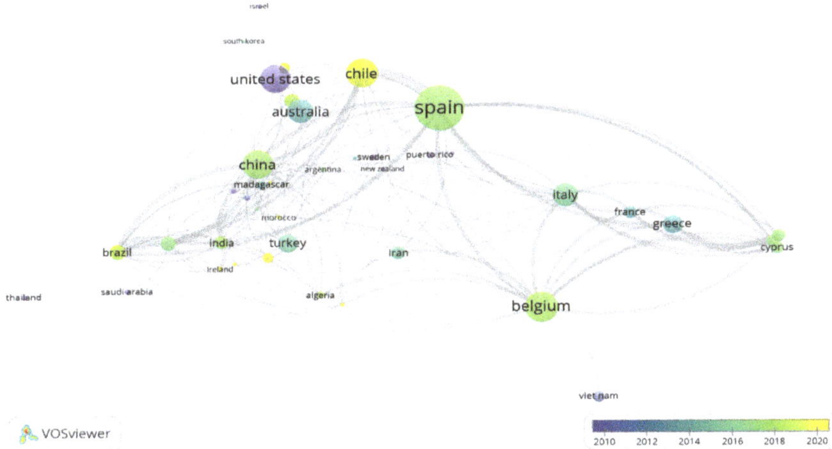

Figure 22. Bibliographic coupling of countries.

Table 6. Summary of papers for future climate scenarios.

Reference	Country/Region		Scenario	Predicted Period	Climate Model Name	Scenario Type
[9]	Entire territory of	Belgium	Current/Future	2070–2098	EC-Earth driven CPM for the Belgian domain extended with land-surface scheme TERRA_URB	RCP 8.5
[10]	Entire territory of	Iran	Current/Future	2025, 2050, and 2075	MAGICC and SCENGEN	
[97]		World	Current/Future	2081–2100	CCSM CSIRO-Mk3.6.0 MPI-ESM-MR GFDL-CM3 GISS-E2-R IPSL-CM5A-LR MRI-CGCM3 HadGEM2-ES	RCP8.5
[77]		World	Current/Future	2071–2100	CMIP5 32 climate model projections	RCP8.5
[103]		World	Current/Future	2015-2100	"SSP5-8.5" climate change scenario, created by the Institute Pierre Simon Laplace (IPSL)	SSP5-8.5
[109]	Entire territory of	Spain	Current/Future	2055 and 2085		RCP 4.5 and RCP 8.5
[110]	Entire territory of	Algeria	Current/Future	2040		
[111]	Entire territory of	Belarus	Current/Future	2021–2040, 2041–2060, 2061–2080, and 2081–2099	92 regional circulation model (RCM) from EURO-CORDEX RCM ensemble	RCP2.6, RCP4.5, and RCP8.5
[22]	Entire territory of	Spain	Current/Future	2055 and 2085	IPCC AR5	RCP2.6 and RCP8.5
[6]	Entire territory of	Greece	Current/Future	2041–2050 and 2091–2100	IPCC AR5 12 different regional climate models of the European ENSEMBLES project	RCP2.6, RCP4.5, and RCP8.5
[40]	Part of	Spain (Andalusia)	Current/Future	2050 and 2010	18 climate models included in the 2007 IPCC report	A2 scenario of the Special Report on Emissions Scenarios (SRES)
[89]	Entire territory of	United States	Future	2020, 2050, and 2080	HadCM3	RCP2.6, RCP4.5, and RCP8.5
[51]	Part of	Chile	Future	2050–2065	IPCC AR5	RCP2.6 and RCP8.5

11. Discussion

In this literature review, extensive detailed data were extracted from the available documents regarding CZB. Publication years and their type, authors, country or region of study, climate variables used, climate zoning methods used and their combinations, period of climate observations, number of climate zones, and other pertinent details served as the foundation for the subsequent analysis. Climate data sources and the period for climate observations were also highlighted.

For the categorization of CZ variables and CZ methods, detailed criteria were established. Ten major variables and the number of variables employed simultaneously for CZB were found. As anticipated, AT was the most prevalent variable used alone or in combination with other variables (63.6%). AT and RH were the most often occurring combination of variables across the articles reviewed. CZB employed two climatic variables simultaneously in 29.3% of cases. The use of a single variable or a combination of three variables was marginally less common, accounting for 28.3% and 22.2% of all cases, respectively. The use of more than three variables simultaneously for climate categorization was uncommon. In addition, when only one variable was used for zoning, DDs and TMY were the most common options. AT and RH were the most prevalent pairing of two variables in CZB. Typically, national meteorological services were the source of climate data. Less often used, but still popular, were web databases and software applications such as EnergyPlus, Autodesk Green Building Studio (GBS), and Meteonorm. Climate models were used in 6.7% of cases, basically for future climate scenarios. It is preferable to use contemporary climate data (last 20–30 years) for appropriate climatic zoning due to climate change.

CZB is characterized by a wide variety of applied methods and not all of them are directly related to building energy consumption. The twelve most commonly used methods were identified. MLM, DDM, and BES were the three most popular approaches. Often, only one method was used for CZ. One approach was used in 65% of cases, a combination of two methods was used in 28%, and the simultaneous use of three or more approaches was identified in 7% of cases. In addition, when only one method was used for CZB, the most prevalent choices were MLM, DDM, BES, and BCM.

Given its primary advantages, MLM provides significant potential to go deeper into the CZB and acquire more reliable, previously unavailable results. However, often, MLM is supplemented with other approaches, such as BES, to generate more accurate results or to evaluate them.

Due to its well-documented connection to building energy, DDM can offer high-quality CZ for a variety of applications. However, DDM primarily uses the outside AT and ignores other significant climatic variables that have an impact on a building's energy use. It should be underlined that the mean daily degree-hours technique is favored over all others for more accurate DDs calculation if more comprehensive meteorological information encompassing hourly outdoor AT of the chosen location is available.

BES is currently regarded as the most accurate method for predicting thermal building performance and has demonstrated significant promise as a policy tool when applied to CZB. Detailed climatic data and BES, according to several sources, could aid in the construction of a more robust climatic categorization [23,46,83,174].

Multiple methods can be used simultaneously as a strategy to improve CZB or validate its results. Any combinations based on DDM, BES, and MLM techniques tend to be the most powerful, efficient, and promising, providing the most reliable results. The main quality criteria of CZB assume that the energy consumption of one building archetype within one climatic zone should be nearly identical.

The dynamics of climate change are forcing scientists to predict future climate conditions and adapt buildings accordingly. To ensure the growth of a building stock that is sustainable and resilient, it is crucial to design and construct buildings today that can take on the dynamics of the environment during their entire life cycle. Several studies regarding future climate scenarios in CZB were revealed [37,51,54,97,98,109,176]. In this review, 12% of publications dealt with future climate zoning for buildings. BES can be easily used to gain insight into the future energy consumption of buildings and, based on that, analyze the possible future climate zones.

A bibliometric analysis was performed to support the main part of this review, which made it possible to evaluate and analyze the performance of research activities in the CZB field, including evaluating the scientific progress, recognizing the most authoritative journals, and identifying significant scientific performers. This evaluation and analysis were performed using bibliometric networks of co-citation, bibliographic coupling, keyword co-occurrence, and co-authorship techniques.

12. Conclusions

The energy consumption of buildings is affected by environmental or climatic conditions and varies with climate variability. Proper climate zoning is essential for most building energy-efficiency policies, with great importance in meeting growth targets for energy security and reducing GHG emissions. Thus, consideration should be given to the use of climatic zoning as a tool in the formulation of construction guidelines that address the energy-efficiency of buildings. Most sectors are implementing stricter energy-efficiency standards due to concerns over climate change and the diminishing availability of natural resources; here, CZB should be an essential factor to take into account. The ability to correctly categorize climates is fundamental to sustainable design, which can reduce the need for heating and cooling by a significant amount. Defining zones not only makes it possible to identify and mitigate the adverse impacts of climate on buildings by specifying basic zonal construction criteria, but also makes it possible to support the effective utilization of climate resources. Understanding the relationship between energy conservation and climate conditions in buildings can be beneficial to the design of housing that is climate-appropriate for a variety of geographical locations.

With this review, we set the goal of understanding at what level of scientific progress the CZB is now, and whether there have been positive cardinal changes in the study of climate zoning for buildings. We analyzed scientific publications in the field of CZB which have already been put into practice or are just being developed. In the last 10 years, there has been a significant increase in the number of publications on the CZB topic.

Now, climatic zoning approaches vary, and there is no "standard" strategy for CZB, though several are generally recognized and used. At the current stage, two global climate zoning systems are used for the needs of buildings and construction. The KG map is popular among researchers but can be hardly used for characterizing the performance of energy-efficiency measures for buildings. KG did not allow for the accumulation of accurate data needed to address the issue of CZB. Multiple sources compared the precision of a KG classification to that of an MLM and BES, revealing that ML and BES exceed traditional KG classification quality [72,74,201]. The degree-day-based global map of ASHRAE Standard, first presented in 2014, remains essentially the only solution providing data on climate zones for buildings globally. It is worth noting that, based only on degree-days, the ASHRAE map inherits its shortcomings, such as the use of outside AT and eliminating other environmental variables that affect a building's energy consumption. Additionally, the usage of only seven main zones with rather wide intervals of degree-days can lead to simplification of zoning. The use of more advanced zoning methods such as MLM or BES is still limited to the territories of countries and continents [7,50,81,105]. The official government standards for climate zoning, especially in developing countries, are very often criticized by the scientific community for not being able to be used with high accuracy and reliability [2,27,43,47,50,53,104,142]. The introduction of a standard global climate zoning approach or map for energy-efficient buildings based on methods recognized as the most effective among scientists could positively influence the problem of energy use in buildings at the global level. This action currently appears to be reasonably realistic given the sufficiency and quality of the available climatic data on a global scale.

It was revealed that BES and MLM have shown great potential when applied to climate zoning. It is reasonable to claim that there is now a solid scientific basis for applying BES and MLM for climate zoning needs. In addition, the significance of using BES to validate climatic zoning for buildings was proved by several publications. Overall, BES and MLM methods are simple to implement and have shown to be reliable in defining climate zones by transitioning from a climate-based to a performance-based approach. Additionally, a combination of approaches yields much better and more robust zoning classification results. It can also be assumed that over the next decade, we might see the gradual introduction of MLM and BES into the official standards of different countries, which could certainly have a positive impact on the energy-efficiency of buildings in particular and may also have a positive impact on the global climate change problem.

It is important to note that uncommon methods such as EBS and FDV can be used to create new CZ or to amend and enhance existing CZ. The real energy consumption of buildings in the EBS approach is very useful information that can help to understand how certain aspects (including climate) affect a building's final energy consumption. However, there are no publications to date that compare these methods to others to assess their quality.

Although the problem of studying climate and its classification is location-oriented, none of the found sources use the principles of spatial analysis or did not mention it. Spatial analysis has become a standard in many research areas (such as epidemiology, sociology, ecology, and tourism), but this has yet to be applied in the field of CZB. The role of the spatial aspect in CZB research is underestimated, and the understanding of the working and representation of space, spatial patterns, and processes is limited. One of the strongest future improvements in the field is the recognition of the spatial component in CZB, which has the potential to be favorable and is predicted to produce results that are more accurate and robust.

Author Contributions: Conceptualization, A.R. and S.A.M.; methodology, A.R. and S.A.M.; software, A.R.; validation, A.R. and S.A.M.; formal analysis, A.R.; investigation, A.R.; resources, S.A.M. and J.R.K.; data curation, S.A.M.; writing—original draft preparation, A.R.; writing—review and editing, A.R. and S.A.M.; visualization, A.R.; supervision, S.A.M.; project administration, S.A.M.; funding acquisition, S.A.M. All authors have read and agreed to the published version of the manuscript.

Funding: This research was funded by Nazarbayev University, Kazakhstan Faculty development competitive research grants numbers 021220FD0651.

Data Availability Statement: All data collected and analyzed in this review are presented in Table 1 of this article.

Conflicts of Interest: The authors declare no conflict of interest.

Abbreviations

Al	Altitude
AP	Atmospheric Pressure
ASHRAE	The American Society of Heating, Refrigerating, and Air-Conditioning Engineers
AT	Air Temperature
BCM	Bioclimatic Charts Method
BES	Building Energy Simulation
CA	Cluster Analysis
CSIM	Climate Severity Index Method
CDD	Cooling Degree-Day
CDH	Cooling Degree-Hour
CZ	Climatic Zoning
CZB	Climatic Zoning For Buildings
DBT	Dry-Bulb Temperature
DDs	Degree-Days
DHs	Degree-Hours
DDM	Degree-Days Methods
EBSM	Existing Building Stock Performance Method
FDV	A Frequency Distribution Of Climate Variables
GHG	Greenhouse Gas
GHI	Global Horizontal Irradiation
HCI	Heating Or Cooling Index
HC	Hierarchical Clustering
HDD	Heating Degree-Day
HDH	Heating Degree-Hour
HVAC	Heating, Ventilation, and Air-Conditioning Systems
IJM	Interval Judgment Method
IPCC	Intergovernmental Panel on Climate Change
KG	Köppen–Geiger
KGM	Köppen–Geiger Method
LCZ	Local Climate Zoning
ML	Machine Learning

MLM	Machine Learning Methods
MM	Mahoney Method
PCA	Principal Component Analysis
PMA	Percentage Misclassified Areas
Pr	Precipitation
PW	The Pressure of Water Vapor
RCP	Representative Concentration Pathway
RH	Relative Humidity
SR	Solar Radiation
TCCM	Thornthwaite Climate Classification Method
TMY	Typical Meteorological Year
W	Wind
WBT	Wet-Bulb Temperature

References

1. Li, D.H.W.; Yang, L.; Lam, J.C. Impact of climate change on energy use in the built environment in different climate zones—A review. *Energy* **2012**, *42*, 103–112. [CrossRef]
2. van Ruijven, B.J.; De Cian, E.; Wing, I.S. Amplification of future energy demand growth due to climate change. *Nat. Commun.* **2019**, *10*, 2762. [CrossRef]
3. UN Environment Programme. Global Status Report for Buildings and Construction: Towards a Zero-Emission, Efficient and Resilient Buildings and Construction Sector, Global Status Report. Available online: https://globalabc.org/sites/default/files/2022-11/FULL%20REPORT_2022%20Buildings-GSR_1.pdf (accessed on 1 May 2022).
4. IEA. Perspectives for the Clean Energy Transition. The Critical Role of Buildings. 2019. Available online: https://www.iea.org/reports/the-critical-role-of-buildings (accessed on 1 May 2022).
5. Tootkaboni, M.P.; Ballarini, I.; Corrado, V. Analysing the future energy performance of residential buildings in the most populated Italian climatic zone: A study of climate change impacts. *Energy Rep.* **2021**, *7*, 8548–8560. [CrossRef]
6. Asimakopoulos, D.A.; Santamouris, M.; Farrou, I.; Laskari, M.; Saliari, M.; Zanis, G.; Giannakidis, G.; Tigas, K.; Kapsomenakis, J.; Douvis, C.; et al. Modelling the energy demand projection of the building sector in Greece in the 21st century. *Energy Build.* **2012**, *49*, 488–498. [CrossRef]
7. Attia, S.; Eleftheriou, P.; Xeni, F.; Morlot, R.; Ménézo, C.; Kostopoulos, V.; Betsi, M.; Kalaitzoglou, I.; Pagliano, L.; Cellura, M.; et al. Overview and future challenges of nearly zero energy buildings (nZEB) design in Southern Europe. *Energy Build.* **2017**, *155*, 439–458. [CrossRef]
8. Bell, N.O.; Bilbao, J.I.; Kay, M.; Sproul, A.B. Future climate scenarios and their impact on heating, ventilation and air-conditioning system design and performance for commercial buildings for 2050. *Renew. Sustain. Energy Rev.* **2022**, *162*, 112363. [CrossRef]
9. Ramon, D.; Allacker, K.; De Troyer, F.; Wouters, H.; van Lipzig, N.P.M. Future heating and cooling degree days for Belgium under a high-end climate change scenario. *Energy Build.* **2020**, *216*, 109935. [CrossRef]
10. Roshan, G.; Orosa, J.A.; Nasrabadi, T. Simulation of climate change impact on energy consumption in buildings, case study of Iran. *Energy Policy* **2012**, *49*, 731–739. [CrossRef]
11. Li, L.; Sun, W.; Hu, W.; Sun, Y. Impact of natural and social environmental factors on building energy consumption: Based on bibliometrics. *J. Build. Eng.* **2021**, *37*, 102136. [CrossRef]
12. Santamouris, M. Green commercial buildings save energy. *Nat. Sustain.* **2018**, *1*, 613–614. [CrossRef]
13. WHO. Housing and Health Guidelines. 2018. Available online: https://apps.who.int/iris/bitstream/handle/10665/276001/9789241550376-eng.pdf (accessed on 22 June 2022).
14. IEA. Electricity Market Report. July 2022—Update. Available online: https://www.iea.org/reports/electricity-market-report-july-2022/executive-summary (accessed on 25 August 2022).
15. Bohne, R.A.; Huang, L.; Lohne, J. A global overview of residential building energy consumption in eight climate zones. *Int. J. Sustain. Build. Technol. Urban Dev.* **2016**, *7*, 38–51. [CrossRef]
16. Verichev, K.; Zamorano, M.; Salazar-Concha, C.; Carpio, M. Analysis of Climate-Oriented Researches in Building. *Appl. Sci.* **2021**, *11*, 3251. [CrossRef]
17. De Silva, M.N.; Sandanayake, Y.G. Building energy consumption factors: A literature review and future research agenda. In *World Construction Conference 2012—Global Challenges in Construction Industry*; University of Moratuwa: Colombo, Sri Lanka, 2012.
18. Inambao, A.I.F. Review of Factors Affecting Energy Efficiency in Commercial Buildings. *Int. J. Mech. Eng. Technol.* **2019**, *10*, 232–244.
19. Torres, M.G.; Pérez-Lombard, L.; Coronel, J.; Maestre, I.; Yan, D. A review on buildings energy information: Trends, end-uses, fuels and drivers. *Energy Rep.* **2022**, *8*, 626–637. [CrossRef]
20. Fathi, A.; El Bakkush, A.; Bondinuba, F.; Harris, D. The Effect of Outdoor Air Temperature on the Thermal Performance of a Residential Building. *J. Multidiscip. Eng. Sci. Technol.* **2015**, *2*, 3159–3240.
21. Li, M.; Shi, J.; Guo, J.; Cao, J.; Niu, J.; Xiong, M. Climate Impacts on Extreme Energy Consumption of Different Types of Buildings. *PLoS ONE* **2015**, *10*, e0124413. [CrossRef]

22. Díaz-López, C.; Verichev, K.; Holgado-Terriza, J.A.; Zamorano, M. Evolution of climate zones for building in Spain in the face of climate change. *Sustain. Cities Soc.* **2021**, *74*, 103223. [CrossRef]
23. Walsh, A.; Cóstola, D.; Labaki, L.C. Comparison of three climatic zoning methodologies for building energy efficiency applications. *Energy Build.* **2017**, *146*, 111–121. [CrossRef]
24. Walsh, A.; Cóstola, D.; Labaki, L.C. Review of methods for climatic zoning for building energy efficiency programs. *Build. Environ.* **2017**, *112*, 337–350. [CrossRef]
25. Walsh, A.; Cóstola, D.; Labaki, L.C. Performance-based validation of climatic zoning for building energy efficiency applications. *Appl. Energy* **2018**, *212*, 416–427. [CrossRef]
26. Butera, F.; Aste, N.; Adhikari, R. *Sustainable Building Design for Tropical Climates*; United Nations Human Settlements Programme (UN-Habitat): Nairobi, Kenya, 2015.
27. Xiong, J.; Yao, R.; Grimmond, S.; Zhang, Q.; Li, B. A hierarchical climatic zoning method for energy efficient building design applied in the region with diverse climate characteristics. *Energy Build.* **2019**, *186*, 355–367. [CrossRef]
28. Wang, R.; Lu, S.; Feng, W.; Zhai, X.; Li, X. Sustainable framework for buildings in cold regions of China considering life cycle cost and environmental impact as well as thermal comfort. *Energy Rep.* **2020**, *6*, 3036–3050. [CrossRef]
29. Rahman, M. Impact of Climatic Zones of Bangladesh on Office Building Energy Performance. *J. Build. Sustain.* **2018**, *1*, 55–63.
30. Albogami, S.; Boukhanouf, R. Residential building energy performance evaluation for different climate zones. *IOP Conf. Ser. Earth Environ. Sci.* **2019**, *329*, 012026. [CrossRef]
31. Chen, X.; Yang, H. Integrated energy performance optimization of a passively designed high-rise residential building in different climatic zones of China. *Appl. Energy* **2018**, *215*, 145–158. [CrossRef]
32. Walsh, A.; Cóstola, D.; Labaki, L.C. Validation of the climatic zoning defined by ASHRAE standard 169-2013. *Energy Policy* **2019**, *135*, 111016. [CrossRef]
33. Khambadkone, N.K.; Jain, R. A bioclimatic analysis tool for investigation of the potential of passive cooling and heating strategies in a composite Indian climate. *Build. Environ.* **2017**, *123*, 469–493. [CrossRef]
34. Wang, R.; Lu, S. A novel method of building climate subdivision oriented by reducing building energy demand. *Energy Build.* **2020**, *216*, 109999. [CrossRef]
35. Bai, L.; Yang, L.; Song, B.; Liu, N. A new approach to develop a climate classification for building energy efficiency addressing Chinese climate characteristics. *Energy* **2020**, *195*, 116982. [CrossRef]
36. López-Ochoa, L.M.; Las-Heras-Casas, J.; López-González, L.M.; Olasolo-Alonso, P. Environmental and energy impact of the EPBD in residential buildings in hot and temperate Mediterranean zones: The case of Spain. *Energy* **2018**, *161*, 618–634. [CrossRef]
37. Yang, Y.; Javanroodi, K.; Nik, V.M. Climate change and energy performance of European residential building stocks—A comprehensive impact assessment using climate big data from the coordinated regional climate downscaling experiment. *Appl. Energy* **2021**, *298*, 117246. [CrossRef]
38. William, M.A.; Suárez-López, M.J.; Soutullo, S.; Fouad, M.M.; Hanafy, A.A. Enviro-economic assessment of buildings decarbonization scenarios in hot climates: Mindset toward energy-efficiency. *Energy Rep.* **2022**, *8*, 172–181. [CrossRef]
39. Benevides, M.N.; Teixeira, D.B.D.S.; Carlo, J.C. Climatic zoning for energy efficiency applications in buildings based on multivariate statistics: The case of the Brazilian semiarid region. *Front. Archit. Res.* **2022**, *11*, 161–177. [CrossRef]
40. Bienvenido-Huertas, D.; Marín-García, D.; Carretero-Ayuso, M.J.; Rodríguez-Jiménez, C.E. Climate classification for new and restored buildings in Andalusia: Analysing the current regulation and a new approach based on k-means. *J. Build. Eng.* **2021**, *43*, 102829. [CrossRef]
41. Carpio, M.; Jódar, J.; Rodíguez, M.L.; Zamorano, M. A proposed method based on approximation and interpolation for determining climatic zones and its effect on energy demand and CO_2 emissions from buildings. *Energy Build.* **2015**, *87*, 253–264. [CrossRef]
42. Cory, S.; Lenoir, A.; Donn, M.; Garde, F. Formulating a building climate classification method. In Proceedings of the 12th Conference of International Building Performance Simulation Association Building Simulation 2011, BS 2011, Sydney, NSW, Australia, 14–16 November 2011; pp. 1662–1669.
43. de la Flor, F.J.S.; Domínguez, S.A.; Félix, J.L.M.; Falcón, R.G. Climatic zoning and its application to Spanish building energy performance regulations. *Energy Build.* **2008**, *40*, 1984–1990. [CrossRef]
44. Deng, X.; Tan, Z.; Tan, M.; Chen, W. A clustering-based climatic zoning method for office buildings in China. *J. Build. Eng.* **2021**, *42*, 102778. [CrossRef]
45. Jain, K.; Gupta, G.; Verma, K.K.; Agarwal, A. Climatic Classification of India for Building Design Using Data Analytics. *Natl. Acad. Sci. Lett.* **2022**, *45*, 235–239. [CrossRef]
46. Mazzaferro, L.; Machado, R.M.S.; Melo, A.P.; Lamberts, R. Do we need building performance data to propose a climatic zoning for building energy efficiency regulations? *Energy Build.* **2020**, *225*, 110303. [CrossRef]
47. Moral, F.J.; Pulido, E.; Ruíz, A.; López, F. Climatic zoning for the calculation of the thermal demand of buildings in Extremadura (Spain). *Theor. Appl. Climatol.* **2017**, *129*, 881–889. [CrossRef]
48. Kishore, K.N.; Rekha, J. A bioclimatic approach to develop spatial zoning maps for comfort, passive heating and cooling strategies within a composite zone of India. *Build. Environ.* **2018**, *128*, 190–215. [CrossRef]
49. Roshan, G.; Farrokhzad, M.; Attia, S. Climatic clustering analysis for novel atlas mapping and bioclimatic design recommendations. *Indoor Built Environ.* **2021**, *30*, 313–333. [CrossRef]

50. Tükel, M.; Tunçbilek, E.; Komerska, A.; Keskin, G.A.; Arıcı, M. Reclassification of climatic zones for building thermal regulations based on thermoeconomic analysis: A case study of Turkey. *Energy Build.* **2021**, *246*, 111121. [CrossRef]
51. Verichev, K.; Zamorano, M.; Carpio, M. Effects of climate change on variations in climatic zones and heating energy consumption of residential buildings in the southern Chile. *Energy Build.* **2020**, *215*, 109874. [CrossRef]
52. Yang, L.; Lyu, K.; Li, H.; Liu, Y. Building climate zoning in China using supervised classification-based machine learning. *Build. Environ.* **2020**, *171*, 106663. [CrossRef]
53. Zeleke, B.; Kumar, M.; Rajasekar, E. A Novel Building Performance Based Climate Zoning for Ethiopia. *Front. Sustain. Cities* **2022**, *4*, 3. [CrossRef]
54. Andric, I.; Al-Ghamdi, S.G. Climate change implications for environmental performance of residential building energy use: The case of Qatar. *Energy Rep.* **2020**, *6*, 587–592. [CrossRef]
55. Awadh, O. Sustainability and green building rating systems: LEED, BREEAM, GSAS and Estidama critical analysis. *J. Build. Eng.* **2017**, *11*, 25–29. [CrossRef]
56. Rezaallah, A.; Bolognesi, C.; Khoraskani, R.A. LEED and BREEAM; Comparison between policies, assessment criteria and calculation methods. In Proceedings of the 1st International Conference on Building Sustainability Assessment (BSA 2012), Porto, Portugal, 23–25 May 2012.
57. Sanderson, M. The Classification of Climates from Pythagoras to Koeppen. *Bull. Am. Meteorol. Soc.* **1999**, *80*, 669–673. [CrossRef]
58. Oliver, J. The history, status and future of climatic classification. *Phys. Geogr.* **2013**, *12*, 231–251. [CrossRef]
59. Robinson, A.H.; Wallis, H.M. Humboldt's Map of Isothermal Lines: A Milestone in Thematic Cartography. *Cartogr. J.* **1967**, *4*, 119–123. [CrossRef]
60. Construction Committee of the Russian Soviet Federative Socialist Republic. *Rules and Regulations for the Development of Populated Areas, Design and Construction of Buildings and Structures*; State Technical Publishing House: Moscow, Russia, 1930.
61. *National Building Code: 1941*; National Research Council of Canada: Ottawa, ON, Canada, 1941.
62. American Society of Heating, Engineers. *ASHRAE Standard 90-75: Energy Conservation in New Building Design*; American Society of Heating, Refrigerating, and Air-Conditioning Engineers, Incorporated. 1975. Available online: https://www.ojp.gov/ncjrs/virtual-library/abstracts/energy-conservation-new-building-design-impact-assessment-ashrae (accessed on 20 July 2022).
63. Laustsen, J. *Energy Efficiency Requirements in Building Codes: Policies for New Buildings*; International Energy Agency (IEA): Paris, France, 2008.
64. Moral-Munoz, J.A.; López-Herrera, A.G.; Herrera-Viedma, E.; Cobo, M.J. Science Mapping Analysis Software Tools: A Review. In *Springer Handbook of Science and Technology Indicators*; Glänzel, W., Moed, H.F., Schmoch, U., Thelwall, M., Eds.; Springer International Publishing: Cham, Switzerland, 2019; pp. 159–185.
65. Small, H.G. Co-citation in the scientific literature: A new measure of the relationship between two documents. *J. Am. Soc. Inf. Sci.* **1973**, *24*, 265–269. [CrossRef]
66. Kessler, M.M. Bibliographic coupling between scientific papers. *Am. Doc.* **1963**, *14*, 10–25. [CrossRef]
67. Glänzel, W. National characteristics in international scientific co-authorship relations. *Scientometrics* **2001**, *51*, 69–115. [CrossRef]
68. van Eck, N.J.; Waltman, L. Visualizing Bibliometric Networks. In *Measuring Scholarly Impact: Methods and Practice*; Ding, Y., Rousseau, R., Wolfram, D., Eds.; Springer International Publishing: Cham, Switzerland, 2014; pp. 285–320.
69. Selek, B.; Tuncok, I.K.; Selek, Z. Changes in climate zones across Turkey. *J. Water Clim. Chang.* **2018**, *9*, 178–195. [CrossRef]
70. van Eck, N.J.; Waltman, L.R. *VOS Viewer: A Computer Program for Bibliometric Mapping*; Research Paper, 01/01; Erasmus Research Institute of Management (ERIM), ERIM Is the Joint Research Institute of the Rotterdam School of Management; Erasmus University and the Erasmus School of Economics (ESE) at Erasmus Uni: Rotterdam, The Netherlands, 2009.
71. Shi, J.; Yang, L. A climate classification of China through k-nearest-neighbor and sparse subspace representation. *J. Clim.* **2020**, *33*, 243–262. [CrossRef]
72. Zscheischler, J.; Mahecha, M.D.; Harmeling, S. Climate classifications: The value of unsupervised clustering. In Proceedings of the 12th Annual International Conference on Computational Science, ICCS 2012, Omaha, NB, USA, 4–6 June 2012; pp. 897–906.
73. Sarricolea, P.; Herrera-Ossandon, M.; Meseguer-Ruiz, Ó. Climatic regionalisation of continental Chile. *J. Maps* **2017**, *13*, 66–73. [CrossRef]
74. Praene, J.P.; Malet-Damour, B.; Radanielina, M.H.; Fontaine, L.; Rivière, G. GIS-based approach to identify climatic zoning: A hierarchical clustering on principal component analysis. *Build. Environ.* **2019**, *164*, 106330. [CrossRef]
75. Netzel, P.; Stepinski, T. On using a clustering approach for global climate classification. *J. Clim.* **2016**, *29*, 3387–3401. [CrossRef]
76. Mahmoud, A.H.A. An analysis of bioclimatic zones and implications for design of outdoor built environments in Egypt. *Build. Environ.* **2011**, *46*, 605–620. [CrossRef]
77. Beck, H.E.; Zimmermann, N.E.; McVicar, T.R.; Vergopolan, N.; Berg, A.; Wood, E.F. Present and future köppen-geiger climate classification maps at 1-km resolution. *Sci. Data* **2018**, *5*, 180214. [CrossRef]
78. Kottek, M.; Grieser, J.; Beck, C.; Rudolf, B.; Rubel, F. World Map of the Köppen-Geiger Climate Classification Updated. *Meteorol. Z.* **2006**, *15*, 259–263. [CrossRef]
79. D'Amico, A.; Ciulla, G.; Panno, D.; Ferrari, S. Building energy demand assessment through heating degree days: The importance of a climatic dataset. *Appl. Energy* **2019**, *242*, 1285–1306. [CrossRef]
80. Roshan, G.R.; Farrokhzad, M.; Attia, S. Defining thermal comfort boundaries for heating and cooling demand estimation in Iran's urban settlements. *Build. Environ.* **2017**, *121*, 168–189. [CrossRef]

81. Tsikaloudaki, K.; Laskos, K.; Bikas, D. On the establishment of climatic zones in Europe with regard to the energy performance of buildings. *Energies* **2012**, *5*, 32–44. [CrossRef]
82. Rakoto-Joseph, O.; Garde, F.; David, M.; Adelard, L.; Randriamanantany, Z.A. Development of climatic zones and passive solar design in Madagascar. *Energy Convers. Manag.* **2009**, *50*, 1004–1010. [CrossRef]
83. Verichev, K.; Zamorano, M.; Carpio, M. Assessing the applicability of various climatic zoning methods for building construction: Case study from the extreme southern part of Chile. *Build. Environ.* **2019**, *160*, 106165. [CrossRef]
84. Wan, K.K.W.; Li, D.H.W.; Yang, L.; Lama, J.C. Climate classifications and building energy use implications in China. *Energy Build.* **2010**, *42*, 1463–1471. [CrossRef]
85. Pusat, S.; Ekmekci, I. A study on degree-day regions of Turkey. *Energy Effic.* **2016**, *9*, 525–532. [CrossRef]
86. Bawaneh, K.; Overcash, M.; Twomey, J. Climate zones and the influence on industrial nonprocess energy consumption. *J. Renew. Sustain. Energy* **2011**, *3*, 063113. [CrossRef]
87. Lau, C.C.S.; Lam, J.C.; Yang, L. Climate classification and passive solar design implications in China. *Energy Convers. Manag.* **2007**, *48*, 2006–2015. [CrossRef]
88. Khedari, J.; Sangprajak, A.; Hirunlabh, J. Thailand climatic zones. *Renew. Energy* **2002**, *25*, 267–280. [CrossRef]
89. Wang, H.; Chen, Q. Impact of climate change heating and cooling energy use in buildings in the United States. *Energy Build.* **2014**, *82*, 428–436. [CrossRef]
90. Verichev, K.; Salimova, A.; Carpio, M. Thermal and climatic zoning for construction in the southern part of Chile. *Adv. Sci. Res.* **2018**, *15*, 63–69. [CrossRef]
91. Singh, M.K.; Mahapatra, S.; Atreya, S.K. Development of bio-climatic zones in north-east India. *Energy Build.* **2007**, *39*, 1250–1257. [CrossRef]
92. Lamberts, R.; Roriz, M.; Ghisi, E. Bioclimatic zoning of Brazil: A proposal based on the givoni and mohoney methods. In Proceedings of the Sustaining the Future: Energy, Ecology, Architecture: Plea '99—The 16th International Conference Plea (Passive & Low Energy Architecture), Brisbane, Australia, 22–24 September 1999.
93. Erell, E.; Portnov, B.; Etzion, Y. Mapping the potential for climate-conscious design of buildings. *Build. Environ.* **2003**, *38*, 271–281. [CrossRef]
94. Unal, Y.; Kindap, T.; Karaca, M. Redefining the climate zones of Turkey using cluster analysis. *Int. J. Climatol.* **2003**, *23*, 1045–1055. [CrossRef]
95. Fovell, R.G.; Fovell, M.Y.C. Climate zones of the conterminous United States defined using cluster analysis. *J. Clim.* **1993**, *6*, 2103–2135. [CrossRef]
96. Alrashed, F.; Asif, M. Climatic Classifications of Saudi Arabia for Building Energy Modelling. In Proceedings of the 7th International Conference on Applied Energy, ICAE 2015, Abu Dhabi, United Arab Emirates, 28–31 March 2015; pp. 1425–1430.
97. Falquina, R.; Gallardo, C. Development and application of a technique for projecting novel and disappearing climates using cluster analysis. *Atmos. Res.* **2017**, *197*, 224–231. [CrossRef]
98. Chen, Y.; Li, M.; Cao, J.; Cheng, S.; Zhang, R. Effect of climate zone change on energy consumption of office and residential buildings in China. *Theor. Appl. Climatol.* **2021**, *144*, 353–361. [CrossRef]
99. Pernigotto, G.; Gasparella, A. Classification of European Climates for Building Energy Simulation Analyses. In Proceedings of the International Conference on High Performance, Tokyo, Japan, 28–31 January 2018.
100. Pernigotto, G.; Walsh, A.; Gasparella, A.; Hensen, J.L.M. Clustering of European climates and representative climate identification for building energy simulation analyses. In Proceedings of the 16th International Conference of the International Building Performance Simulation Association, Building Simulation, Rome, Italy, 2–4 September 2019; Volume 2019, pp. 4833–4840.
101. Aparecido, L.E.O.; Rolim, G.S.; Richetti, J.; de Souza, P.S.; Johann, J.A. Köppen, Thornthwaite and Camargo climate classifications for climatic zoning in the State of Paraná, Brazil. *Cienc. Agrotecnol.* **2016**, *40*, 405–417. [CrossRef]
102. Peel, M.C.; Finlayson, B.L.; McMahon, T.A. Updated world map of the Köppen-Geiger climate classification. *Hydrol. Earth Syst. Sci.* **2007**, *11*, 1633–1644. [CrossRef]
103. Ascencio-Vásquez, J.; Brecl, K.; Topič, M. Methodology of Köppen-Geiger-Photovoltaic climate classification and implications to worldwide mapping of PV system performance. *Sol. Energy* **2019**, *191*, 672–685. [CrossRef]
104. Bai, L.; Wang, S. Definition of new thermal climate zones for building energy efficiency response to the climate change during the past decades in China. *Energy* **2019**, *170*, 709–719. [CrossRef]
105. van Schijndel, A.W.M.; Schellen, H.L. The simulation and mapping of building performance indicators based on European weather stations. *Front. Archit. Res.* **2013**, *2*, 121–133. [CrossRef]
106. Semahi, S.; Benbouras, M.A.; Mahar, W.A.; Zemmouri, N.; Attia, S. Development of spatial distribution maps for energy demand and thermal comfort estimation in Algeria. *Sustainability* **2020**, *12*, 6066. [CrossRef]
107. Mazzaferro, L.; Machado, R.M.S.; Melo, A.P.; Lamberts, R. Climatic zoning methodology based on data-driven approach. In Proceedings of the 16th International Conference of the International Building Performance Simulation Association, Building Simulation, Rome, Italy, 2–4 September 2019; Volume 2019, pp. 3955–3962.
108. De Rosa, M.; Bianco, V.; Scarpa, F.; Tagliafico, L.A. Historical trends and current state of heating and cooling degree days in Italy. *Energy Convers. Manag.* **2015**, *90*, 323–335. [CrossRef]
109. Díaz-López, C.; Jódar, J.; Verichev, K.; Rodríguez, M.L.; Carpio, M.; Zamorano, M. Dynamics of changes in climate zones and building energy demand. A case study in Spain. *Appl. Sci.* **2021**, *11*, 4261. [CrossRef]

110. Ghedamsi, R.; Settou, N.; Gouareh, A.; Khamouli, A.; Saifi, N.; Recioui, B.; Dokkar, B. Modeling and forecasting energy consumption for residential buildings in Algeria using bottom-up approach. *Energy Build.* **2016**, *121*, 309–317. [CrossRef]
111. Danilovich, I.; Geyer, B. Estimates of current and future climate change in Belarus based on meteorological station data and the EURO-CORDEX-11 dataset. *Meteorol. Hydrol. Water Manag.* **2021**, *9*. [CrossRef]
112. Vondráková, A.; Vávra, A.; Voženílek, V. Climatic regions of the Czech Republic. *J. Maps* **2013**, *9*, 425–430. [CrossRef]
113. da Casa Martín, F.; Valiente, E.E.; Celis, D.F. Climate zoning for its application to bioclimatic design. Application in Galicia (Spain). *Inf. Constr.* **2017**, *69*. [CrossRef]
114. Muddu, R.D.; Gowda, D.M.; Robinson, A.J.; Byrne, A. Optimisation of retrofit wall insulation: An Irish case study. *Energy Build.* **2021**, *235*, 110720. [CrossRef]
115. Noh, B.; Choi, J.; Seo, D. A Study on the Classification Criteria of Climatic Zones in Korean Building Code Based on Heating Degree-Days. *Korean J. Air-Cond. Refrig. Eng.* **2015**, *27*, 574–580.
116. Ogunsote, O.; Prucnal-Ogunsote, B. Defining climatic zones for architectural design in Nigeria: A systematic delineation. *J. Environ. Technol.* **2002**, *1*, 1–14.
117. Mobolade, T.D.; Pourvahidi, P. Bioclimatic approach for climate classification of Nigeria. *Sustainability* **2020**, *12*, 4192. [CrossRef]
118. Hobaica, M.; Allard, F.; Belarbi, R. Passive Cooling Systems for Buildings. Potential of Use within Venezuela's Climatic Zones. Available online: https://hal.science/hal-00312480/ (accessed on 5 May 2022).
119. Malmgren, B.A.; Winter, A. Climate zonation in Puerto Rico based on principal components analysis and an artificial neural network. *J. Clim.* **1999**, *12*, 977–985. [CrossRef]
120. Izzo, M.; Rosskopf, C.M.; Aucelli, P.P.; Maratea, A.; Méndez, R.; Pérez, C.; Segura, H. A new climatic map of the dominican republic based on the thornthwaite classification. *Phys. Geogr.* **2010**, *31*, 455–472. [CrossRef]
121. Felix, L.D.P.J.; Izquierdo, R. Análisis comparativo de las diferentes zonas climáticas de la república dominicana. In Proceedings of the 1st Iberic Conference on Theoretical and Experimental Mechanics and Materials/11th National Congress on Experimental Mechanics, Porto, Portugal, 4–7 November 2018; pp. 865–876.
122. Verichev, K.; Carpio, M. Climatic zoning for building construction in a temperate climate of Chile. *Sustain. Cities Soc.* **2018**, *40*, 352–364. [CrossRef]
123. Hjortling, C.; Björk, F.; Berg, M.; Klintberg, T.A. Energy mapping of existing building stock in Sweden—Analysis of data from Energy Performance Certificates. *Energy Build.* **2017**, *153*, 341–355. [CrossRef]
124. Kunchornrat, A.; Namprakai, P.; Pont, P.T. The impacts of climate zones on the energy performance of existing Thai buildings. *Resour. Conserv. Recycl.* **2009**, *53*, 545–551. [CrossRef]
125. Gangolells, M.; Casals, M.; Forcada, N.; MacArulla, M.; Cuerva, E. Energy mapping of existing building stock in Spain. *J. Clean. Prod.* **2016**, *112*, 3895–3904. [CrossRef]
126. Dena, A.J.G.; Pascual, M.Á.; Bandera, C.F. Building energy model for mexican energy standard verification using physics-based open studio sgsave software simulation. *Sustainability* **2021**, *13*, 1521. [CrossRef]
127. Elmzughi, M.; Alghoul, S.; Mashena, M. Optimizing thermal insulation of external building walls in different climate zones in Libya. *J. Build. Phys.* **2021**, *45*, 368–390. [CrossRef]
128. Coskun, C.; Ertürk, M.; Oktay, Z.; Hepbasli, A. A new approach to determine the outdoor temperature distributions for building energy calculations. *Energy Convers. Manag.* **2014**, *78*, 165–172. [CrossRef]
129. Bodach, S. Developing bioclimatic zones and passive solar design strategies for Nepal. In Proceedings of the 30th International PLEA Conference: Sustainable Habitat for Developing Societies: Choosing the Way Forward—Proceedings, Ahmedabad, India, 16–18 December 2014; pp. 167–174.
130. Kim, Y.; Jang, H.K.; Yu, K.H. Study on extension of standard meteorological data for cities in South Korea Using ISO 15927-4. *Atmosphere* **2017**, *8*, 220. [CrossRef]
131. Moradchelleh, A. Construction design zoning of the territory of Iran and climatic modeling of civil buildings space. *J. King Saud Univ.—Sci.* **2011**, *23*, 355–369. [CrossRef]
132. Aliaga, V.S.; Ferrelli, F.; Piccolo, M.C. Regionalization of climate over the Argentine Pampas. *Int. J. Climatol.* **2017**, *37*, 1237–1247. [CrossRef]
133. Pineda-Martínez, L.F.; Carbajal, N.; Medina-Roldán, E. Regionalization and classification of bioclimatic zones in the central-northeastern region of México using principal component analysis (PCA). *Atmosfera* **2007**, *20*, 133–145.
134. Bhatnagar, M.; Mathur, J.; Garg, V. Climate zone classification of India using new base temperature. In Proceedings of the 16th International Conference of the International Building Performance Simulation Association, Building Simulation 2019, Rome, Italy, 2–4 September 2019; pp. 4841–4845.
135. Pawar, A.S.; Mukherjee, M.; Shankar, R. Thermal comfort design zone delineation for India using GIS. *Build. Environ.* **2015**, *87*, 193–206. [CrossRef]
136. Nair, R.J.; Brembilla, E.; Hopfe, C.; Mardaljevic, J. Weather data for building simulation: Grid resolution for climate zone delineation. In Proceedings of the 16th International Conference of the International Building Performance Simulation Association, Building Simulation 2019, Rome, Italy, 2–4 September 2019; pp. 3932–3939.
137. Jittawikul, A.; Saito, I.; Ishihara, O. Climatic Maps for Passive Cooling Methods Utilization in Thailand. *J. Asian Archit. Build. Eng.* **2004**, *3*, 109–114. [CrossRef]

138. Anas, H.; El Mghouchi, Y.; Halima, Y.; Nawal, A.; Mohamed, C. Novel climate classification based on the information of solar radiation intensity: An application to the climatic zoning of Morocco. *Energy Convers. Manag.* **2021**, *247*, 114770. [CrossRef]
139. Briggs, R.S.; Lucas, R.G.; Taylor, Z.T. Climate classification for building energy codes and standards: Part 1—Development process. In *Technical and Symposium Papers Presented At the 2003 Winter Meeting of The ASHRAE*; ASHRAE Transactions: Chicago, IL, USA, 2003; pp. 109–121.
140. Briggs, R.S.; Lucas, R.G.; Taylor, Z.T. Climate classification for building energy codes and standards: Part 2—Zone definitions, maps, and comparisons. In *Technical and Symposium Papers Presented At the 2003 Winter Meeting of The ASHRAE*; ASHRAE Transactions: Chicago, IL, USA, 2003; pp. 122–130.
141. Abebe, S.; Assefa, T. Development of climatic zoning and energy demand prediction for Ethiopian cities in degree days. *Energy Build.* **2022**, *260*, 111935. [CrossRef]
142. Lam, J.C.; Yang, L.; Liu, J. Development of passive design zones in China using bioclimatic approach. *Energy Convers. Manag.* **2006**, *47*, 746–762. [CrossRef]
143. Butera, F.M. *Sustainable Building Design for Tropical Climates. Principles and Applications for Eastern Africa*; Nairobi United Nations Human Settlements Programme (UN-Habitat): Nairobi, Kenya, 2014.
144. Ferstl, K. *Climate and Human Settlements—Integrating Climate into Urban Planning and Building Design in Africa*; UNEP: Nairobi, Kenya, 1991.
145. Poulsen, M.; Lauring, M.; Brunsgaard, C. A review of climate change adaptive measures in architecture within temperate climate zones. *J. Green Build.* **2020**, *15*, 113–130. [CrossRef]
146. Navarro, F.C. Arquitetura e Clima na Bolívia: Uma Proposta de Zoneamento Bioclimático. Ph.D. Thesis, Universidade Estadual de Campinas Faculdade de Engenharia Civil, Arquitetura e Urbanismo, Campinas, Spain, 2007. [CrossRef]
147. Asfaw, S.A. *Developing Parametric Solar Envelope for Ethiopian Cities*; School of Graduate Studies of Addis Ababa University in Partial Fulfillment of the Requirements of the Masters of Science in Urban Design and Development, Ethiopia Institute of Architecture Building Construction and City Development: Addis Ababa, Ethiopia, 2020.
148. Arens, E.A.; Williams, P.B. The effect of wind on energy consumption in buildings. *Energy Build.* **1977**, *1*, 77–84. [CrossRef]
149. Diaz, C.A.; Osmond, P. Influence of Rainfall on the Thermal and Energy Performance of a Low Rise Building in Diverse Locations of the Hot Humid Tropics. *Procedia Eng.* **2017**, *180*, 393–402. [CrossRef]
150. Jovanović, S.; Savić, S.; Bojić, M.; Djordjević, Z.; Nikolić, D. The impact of the mean daily air temperature change on electricity consumption. *Energy* **2015**, *88*, 604–609. [CrossRef]
151. Ministry of Environment and Sustainable Development Vice Ministry of Environment and Sustainable Development Directorate of Environmental. *Environmental Criteria for the Design and Construction of Urban Housing*; Ministerio de Ambiente y Desarrollo Sostenible: Bogotá, Colombia, 2012.
152. Corporal-Lodangco, I.L.; Leslie, L.M. Defining Philippine Climate Zones Using Surface and High-Resolution Satellite Data. *Procedia Comput. Sci.* **2017**, *114*, 324–332. [CrossRef]
153. Meng, F.; Li, M.; Cao, J.; Li, J.; Xiong, M.; Feng, X.; Ren, G. The effects of climate change on heating energy consumption of office buildings in different climate zones in China. *Theor. Appl. Climatol.* **2017**, *133*, 521–530. [CrossRef]
154. Bodach, S.; Lang, W.; Auer, T. Design guidelines for energy-efficient hotels in Nepal. *Int. J. Sustain. Built Environ.* **2016**, *5*, 411–434. [CrossRef]
155. Dowd, P.; Wang, H.; Pardo-Iguzquiza, E.; Yongguo, Y. Constrained Spatial Clustering of Climate Variables for Geostatistical Reconstruction of Optimal Time Series and Spatial Fields. In *Geostatistics Valencia 2016*; Springer: Berlin/Heidelberg, Germany, 2017; pp. 879–891.
156. Daly, C. Guidelines for assessing the stability of spatial climate data sets. *Int. J. Climatol.* **2006**, *26*, 707–721. [CrossRef]
157. Raymundo, C.E.; Oliveira, M.C.; Eleuterio, T.D.A.; André, S.R.; da Silva, M.G.; Queiroz, E.R.D.S.; Medronho, R.D.A. Spatial analysis of COVID-19 incidence and the sociodemographic context in Brazil. *PLoS ONE* **2021**, *16*, e0247794. [CrossRef] [PubMed]
158. Hall, A.; Jones, G.V. Spatial analysis of climate in winegrape-growing regions in Australia. *Aust. J. Grape Wine Res.* **2010**, *16*, 389–404. [CrossRef]
159. Hammer, R.B.; Stewart, S.I.; Winkler, R.L.; Radeloff, V.C.; Voss, P.R. Characterizing dynamic spatial and temporal residential density patterns from 1940–1990 across the North Central United States. *Landsc. Urban Plan.* **2004**, *69*, 183–199. [CrossRef]
160. Tobler, W. On the first law of geography: A reply. *Ann. Assoc. Am. Geogr.* **2004**, *94*, 304–310. [CrossRef]
161. Anselin, L. Local Indicators of Spatial Association—LISA. *Ann. Assoc. Am. Geogr.* **1995**, *27*, 93–115. [CrossRef]
162. Day, T. TM 41 Degree-days: Theory and application. In *TM 41 Degree-Days: Theory and Application*; Chartered Institution of Building Services Engineers (CIBSE): London, UK, 2006.
163. *ANSI/ASHRAE Standard 169-2021*; Climatic Data for Building Design Standards. American Society of Heating, Refrigerating and Air-Conditioning Engineers: Atlanta, GA, USA, 2021.
164. Guttman, N.B.; Lehman, R.L. Estimation of Daily Degree-hours. *J. Appl. Meteorol.* **1992**, *31*, 797. [CrossRef]
165. Hitchin, E.R. Estimating monthly degree-days. *Build. Serv. Eng. Res. Technol.* **1983**, *4*, 159–162. [CrossRef]
166. Schoenau, G.; Kehrig, R.A. Method for calculating degree-days to any base temperature. *Energy Build.* **1990**, *14*, 299–302. [CrossRef]
167. Omarov, B.; Memon, S.A.; Kim, J. A novel approach to develop climate classification based on degree days and building energy performance. *Energy* **2023**, *267*, 126514. [CrossRef]

168. Quayle, R.G.; Diaz, H.F. Heating Degree Day Data Applied to Residential Heating Energy Consumption. *J. Appl. Meteorol.* **1980**, *19*, 241–246. [CrossRef]
169. Le Comte, D.M.; Warren, H.E. Modeling the Impact of Summer Temperatures on National Electricity Consumption. *J. Appl. Meteorol.* **1981**, *20*, 1415–1419. [CrossRef]
170. Lehman, R.L.; Warren, H.E. Residential Natural Gas Consumption: Evidence That Conservation Efforts to Date Have Failed. *Science* **1978**, *199*, 879–882. [CrossRef] [PubMed]
171. Warren, H.E.; LeDuc, S.K. Impact of Climate on Energy Sector in Economic Analysis. *J. Appl. Meteorol.* **1981**, *20*, 1431–1439. [CrossRef]
172. Strahler, A.N. *Physical Geography*; John Willey: New York, NY, USA, 1969.
173. Bhatnagar, M.; Mathur, J.; Garg, V. Determining base temperature for heating and cooling degree-days for India. *J. Build. Eng.* **2018**, *18*, 270–280. [CrossRef]
174. Crawley, D. Building Performance Simulation: A Tool for Policymaking. Ph.D. Thesis, University of Strathclyde, Glasgow, UK, 2008.
175. Dias, J.B.; da Graça, G.C.; Soares, P.M.M. Comparison of methodologies for generation of future weather data for building thermal energy simulation. *Energy Build.* **2020**, *206*, 109556. [CrossRef]
176. Tootkaboni, M.P.; Ballarini, I.; Zinzi, M.; Corrado, V. A Comparative Analysis of Different Future Weather Data for Building Energy Performance Simulation. *Climate* **2021**, *9*, 37. [CrossRef]
177. Givoni, B. *Man, Climate, and Architecture*; Elsevier Publishing Company Limited: New York, NY, USA, 1969.
178. Givoni, B. Comfort, climate analysis and building design guidelines. *Energy Build.* **1992**, *18*, 11–23. [CrossRef]
179. Milne, B.G.M. Architectural design based on climate. *Energy Conserv. Through Build. Des.* **1979**, 96–113.
180. Olgyay, V. *Design with Climate: A Bioclimatic Approach to Architectural Regionalism*; Van Nostrand Reinhold: New York, NY, USA, 1992.
181. Pajek, L.; Košir, M. Implications of present and upcoming changes in bioclimatic potential for energy performance of residential buildings. *Build. Environ.* **2018**, *127*, 157–172. [CrossRef]
182. Walker, C.L.; Hasanzadeh, S.; Esmaeili, B.; Anderson, M.R.; Dao, B. Developing a winter severity index: A critical review. *Cold Reg. Sci. Technol.* **2019**, *160*, 139–149. [CrossRef]
183. Boustead, B.M.; Hilberg, S.; Shulski, M.; Hubbard, K. The Accumulated Winter Season Severity Index (AWSSI). *J. Appl. Meteorol. Climatol.* **2015**, *54*, 150326122910004. [CrossRef]
184. Federici, A.; Iatauro, D.; Romeo, C.; Signoretti, P.; Terrinoni, L. Climatic Severity Index: Definition of Summer Climatic Zones in Italy through the Assessment of Air Conditioning Energy Need in Buildings. In Proceedings of the Clima 2013 RHEVA Word Congress & International Conference on IAQVECAt, Prague, Czech Republic, 16–19 June 2013.
185. Koppen, W. *Das geographische System der Klimate*; Gebrüder Borntraeger: Berlin, Germany, 1936.
186. Humphreys, M. Outdoor temperatures and comfort indoors. *Batim. Int. Build. Res. Pract.* **1978**, *6*, 92. [CrossRef]
187. Huang, J.; Ritschard, R.L.; Bull, J.C.; Chang, L. Climatic indicators for estimating residential heating and cooling loads. In *ASHRAE Transactions*; Lawrence Berkeley Lab.: Berkeley, CA, USA, 1987; pp. 72–111.
188. Eto, J.H. On using degree-days to account for the effects of weather on annual energy use in office buildings. *Energy Build.* **1988**, *12*, 113–127. [CrossRef]
189. Kershaw, T.; Eames, M.; Coley, D. Comparison of multi-year and reference year building simulations. *Build. Serv. Eng. Res. Technol.* **2010**, *31*, 357–369. [CrossRef]
190. Nguyen, A.T.; Tran, Q.B.; Tran, D.Q.; Reiter, S. An investigation on climate responsive design strategies of vernacular housing in Vietnam. *Build. Environ.* **2011**, *46*, 2088–2106. [CrossRef]
191. Rubel, F.; Brugger, K.; Haslinger, K.; Auer, I. The climate of the European Alps: Shift of very high resolution Köppen-Geiger climate zones 1800–2100. *Meteorol. Z.* **2017**, *26*, 115–125. [CrossRef]
192. Rubel, F.; Kottek, M. Comments on: "The thermal zones of the Earth by Wladimir Köppen (1884). *Meteorol. Z.* **2011**, *20*, 361–365. [CrossRef]
193. Rubel, F.; Kottek, M. Observed and projected climate shifts 1901–2100 depicted by world maps of the Köppen-Geiger climate classification. *Meteorol. Z.* **2010**, *19*, 135–141. [CrossRef]
194. Köppen, W.; Volken, E.; Brönnimann, S. The thermal zones of the Earth according to the duration of hot, moderate and cold periods and to the impact of heat on the organic world. *Meteorol. Z.* **2011**, *20*, 351–360. [CrossRef] [PubMed]
195. Santamouris, M. *Energy and Climate in the Urban Built Environment*; Routledge: Oxfordshire, UK, 2013.
196. Santamouris, M.; Papanikolaou, N.; Livada, I.; Koronakis, I.; Georgakis, C.; Argiriou, A.; Assimakopoulos, D.N. On the impact of urban climate on the energy consumption of building. *Solar Energy* **2001**, *70*, 201–216. [CrossRef]
197. Attia, S. *Net Zero Energy Buildings (NZEB): Concepts, Frameworks and Roadmap for Project Analysis and Implementation*; Butterworth-Heinemann: Oxford, UK, 2018.
198. Attia, S.; Gratia, E.; De Herde, A.; Hensen, J.L.M. Simulation-based decision support tool for early stages of zero-energy building design. *Energy Build.* **2012**, *49*, 2–15. [CrossRef]
199. Attia, S.; Hamdy, M.; O'Brien, W.; Carlucci, S. Assessing gaps and needs for integrating building performance optimization tools in net zero energy buildings design. *Energy Build.* **2013**, *60*, 110–124. [CrossRef]

200. Givoni, B. Indoor temperature reduction by passive cooling systems. *Sol. Energy* **2011**, *85*, 1692–1726. [CrossRef]
201. Attia, S.; Lacombe, T.; Rakotondramiarana, H.T.; Garde, F.; Roshan, G. Analysis tool for bioclimatic design strategies in hot humid climates. *Sustain. Cities Soc.* **2019**, *45*, 8–24. [CrossRef]

MDPI
St. Alban-Anlage 66
4052 Basel
Switzerland
www.mdpi.com

Buildings Editorial Office
E-mail: buildings@mdpi.com
www.mdpi.com/journal/buildings

www.ingramcontent.com/pod-product-compliance
Lightning Source LLC
Chambersburg PA
CBHW041945260926
51850CB00018B/890